Rotation Transforms for Computer Graphics

John Vince

Rotation Transforms for Computer Graphics

Springer

Professor Emeritus, John Vince, MTech, PhD,
DSc, CEng, FBCS
Bournemouth University, Bournemouth, UK
url: www.johnvince.co.uk

ISBN 978-0-85729-153-0 e-ISBN 978-0-85729-154-7
DOI 10.1007/978-0-85729-154-7
Springer London Dordrecht Heidelberg New York

British Library Cataloguing in Publication Data
A catalogue record for this book is available from the British Library

Cover design: deblik

Printed on acid-free paper

Springer is part of Springer Science+Business Media (www.springer.com)

This book is dedicated to my grandchildren,
Megan, Mia and Lucie

Preface

Every time I complete a manuscript my attention turns quickly to the title of my next book. And after completing the latest version of *Mathematics for Computer Graphics*, I began to think of what should follow. It didn't take too long to identify the subject of this book: rotation transforms, which have always interested me throughout my career in computer graphics.

I knew that I was not alone in finding some of the ideas difficult, as every time I searched the Internet using search keys such as '*quaternions*', '*Euler angles*', '*rotors*', etc., I would come across websites where groups were discussing the meaning of gimbal lock, the matrix representation of a quaternion, eigenvectors, etc., and I knew straight away that I had to do my bit to clarify the subject.

One of the main problems why there is so much confusion arises through the different forms of vector and matrix notation. Some authors work with matrices that involve row vectors, rather than column vectors, which leads to a transposed matrix. In some cases, the direction of rotation is clockwise, rather than the normal positive, anti-clockwise direction. Quaternions are treated as a four-dimensional object where one has to visualise a hyper-sphere before they can be mastered. Some of the algorithms for extracting eigenvectors and their associated eigenvalue can be very sensitive to the type of matrix in use. This is all rather disconcerting.

The experienced mathematician will take all of this in his or her stride, but to a cg programmer trying to implement the best rotation algorithm and design some stable code, this is not good news. So about a year ago, I started to collect my thoughts on how to approach this subject. After a few false starts and chapter rewrites, I decided to write an introductory book that would take the reader through the foundations of rotation transforms from complex numbers to Clifford algebra rotors, touching on vectors, matrices and quaternions on the journey.

Illustrations are vital to understanding rotation transforms, especially the difference between rotated points and rotated frames. I came across many websites, technical literature and books where the illustrations confused rather than clarified what was going on, and I explored various approaches before settling for a unit cube with numbered vertices. This book contains over a hundred illustrations, which, I hope will help the reader understand the underlying mathematics.

In order to create some sort of structure, I have separated transforms for rotating points in a fixed frame, from transforms that rotate frames with fixed points. I have also separated transforms in the plane from transforms in 3D space. In all, there are thirteen chapters, including an introduction and summary chapters.

Chapter 2 provides a quick introduction to complex numbers and the rotational qualities of imaginary i. The reader should be comfortable with these objects, as we find imaginary quantities in quaternions and multivectors.

Chapter 3 covers vectors and their products, whilst Chap. 4 describes matrices and their associated algebra. It also explores other relevant topics such as matrix inversion, symmetric and antisymmetric matrices, eigenvectors and eigenvalues.

Chapter 5 covers quaternions and their various forms, but I leave their rotational abilities for Chap. 11. I play down their four-dimensional attributes as I don't believe that this characteristic is too important within this introductory book.

Chapter 6 introduces multivector rotors that are part of Clifford's geometric algebra, and again, their rotational qualities are delayed until Chap. 12.

Chapter 7 covers rotation transforms in the plane and establishes strategies used for transforming points in space, whilst Chap. 8 addresses rotating frames of reference in the plane.

Chapter 9 is an important chapter as it introduces the classic techniques for handling 3D rotations, composite rotations, gimbal lock, and provides a stable technique for extracting eigenvectors and eigenvalues from a matrix.

Chapter 10 develops the ideas of Chap. 9 to explain how coordinates are computed in rotating frames of reference.

Chapter 11 takes quaternions from Chap. 5 and shows how they provide a powerful tool for rotating points and frames about an arbitrary axis.

Chapter 12 takes the multivectors from Chap. 6 and shows how they provide a unified system for handling rotors. Finally, Chap. 13 draws the book to a conclusion.

I would like to take this opportunity to acknowledge the authors of books, technical papers and websites who have influenced my writing over recent years. From these dedicated people I have discovered new writing techniques, how to format equations, and how to communicate complex ideas in an easy manner. Without them this book would not have been possible. However, there is one author that I must acknowledge: Michael J. Crowe. His book *A History of Vector Analysis* [1] is an amazing description of how vectors and quaternions evolved, and is highly recommended.

In particular, I would like to thank Dr Tony Crilly, Reader Emeritus at Middlesex University, who read a draft manuscript and made many important recommendations. Tony read the book through the eyes of a novice and questioned my writing style when clarity started to sink below the surface. Forty years ago, when I was struggling with gimbal lock and Euler transforms, Tony brought to my attention the rotation transform developed by Olinde Rodrigues, who had invented quaternions before Hamilton, but that's another story. I included this transform in my animation software system PICASO, running on a mainframe computer with a 24 KB store! I was very nervous about using it as sines and cosines were evaluated at a software level and extremely slow.

I would also like to thank Prof. Patrick Riley for providing me with a *harmonogram* that has formed the book's cover design, and for his feedback on early drafts of the manuscript when I needed to know whether I was managing to communicate my ideas effectively.

Once again, I am indebted to Beverley Ford, General Manager, Springer UK, and Helen Desmond, Assistant Editor for Computer Science, for their support and reminding me of the importance of deadlines. I would also like to thank Springer's technical support team for their help with $\LaTeX\,2_{\varepsilon}$.

Ringwood John Vince

Contents

Chapter 1
Introduction

1.1 Rotation Transforms

In computer graphics the position of an object is expressed by two transforms: translation and rotation. It is relatively easy to visualise a translation and express it mathematically, however rotations do present problems. Furthermore, it is not just objects that require rotating and translating – frames of reference have to be positioned within the world coordinate system in order to secure different views of the virtual world. In order to do this, it is necessary to combine rotation and translation transforms.

When rotating and translating objects, the angles and translation offsets are often relative to a fixed frame of reference. However, when rotating and translating frames of reference, the angles and offsets are relative to a changing frame of reference, which requires careful handling. Primarily, this book is about rotation transforms, and how they are used for moving objects and frames of reference in the plane and in 3D space. But in order to do this within a real computer graphics context, it is necessary to include the translation transform, which introduces some realism to the final solution.

The world of mathematics offers a wide variety of rotation techniques to choose from such as direction cosines, Euler angles, quaternions and multivectors. Each has strengths and weaknesses, advocates and critics, therefore no attempt will be made to identify a 'best' technique. However, I will attempt to draw your attention to their qualities in order that you can draw your own conclusions.

1.2 Mathematical Techniques

Six branches of mathematics play an important role in rotations: trigonometry, complex numbers, vectors, matrices, quaternions and multivectors, which are described in the following chapters and ensure that this book is self contained. We only require to consider certain aspects of trigonometry which will become foundations for the

J. Vince, *Rotation Transforms for Computer Graphics*,
DOI 10.1007/978-0-85729-154-7_1, © Springer-Verlag London Limited 2011

other topics. Complex numbers are extremely useful from two perspectives: the first is that they pave the way to the idea of a rotational operator, and second, they play an intrinsic part in quaternions and multivectors. Vectors provide a mechanism for representing oriented lines, and together with complex numbers form the basis for quaternions, which provide a mechanism for rotating points about an arbitrary axis. Lastly, multivectors introduce the concept of oriented areas and volumes, and provide an algebra for undertaking a wide range of geometric operations, especially rotations.

1.3 The Reader

This is an introductory book and is aimed at readers studying or working in computer graphics who require an overview of the mathematics behind rotation transforms. They are probably the same people I have encountered asking questions on Internet forums about Euler angles, quaternions, gimbal lock and how to extract a direction vector from a rotation matrix.

Many years ago, when writing a computer animation software, I encountered gimbal lock and had to find a way around the problem. Today, students and programmers are still discovering gimbal lock for the first time, and that certain mathematical techniques are not completely stable, and that special cases require detection if their software is to remain operational.

1.4 Aims and Objectives of This Book

The aim of this book is to take the reader through the important ideas and mathematical techniques associated with rotation transforms, without becoming too pedantic about mathematical terminology. My objective is to make the reader confident and comfortable with the algebra associated with complex numbers, vectors, matrices, quaternions and rotors, which seems like a daunting task. However, I believe that this is achievable, and is why I have included a large number of worked examples, and shown what happens when we ignore important rules.

1.5 Assumptions Made in This Book

I only expect the reader to be competent in handling algebraic expansions, and to have a reasonable understanding of trigonometry and geometry. They will probably be familiar with vectors but not necessarily with matrices, which is why I have included chapters on these topics.

1.6 How to Use the Book

The book has a linear narrative and readers with different backgrounds can jump in at any convenient point. Apart from explaining the mathematical techniques, I have tried to introduce the reader to the mathematicians behind the techniques. Mathematicians such as Hamilton, Cayley, Gibbs, Clifford, Euler, Laplace, Sarrus and Grassmann have all played a part in rotation transforms and associated mathematics, and are responsible for the techniques we use today. Hopefully, you will find this background material relevant and interesting.

Chapter 2
Complex Numbers

2.1 Introduction

Complex numbers have been described as the 'king' of numbers, probably because they resolve all sorts of mathematical problems where ordinary real numbers fail. For example, the rather innocent looking equation

$$1 + x^2 = 0$$

has no real solution, which seems amazing when one considers the equation's simplicity. But one does not need a long equation to show that the algebra of real numbers is unable to cope with objects such as

$$x = \sqrt{-1}.$$

However, this did not prevent mathematicians from finding a way around such an inconvenience, and fortuitously the solution turned out to be an incredible idea that is used everywhere from electrical engineering to cosmology. The simple idea of declaring the existence of a quantity i, such that $i^2 = -1$, permits us to express the solution to the above equation as

$$x = \pm i.$$

All very well, you might say, but what is i? What is mathematics? One could also ask, and spend an eternity searching for an answer! i is simply a mathematical object whose square is -1. Let us continue with this strange object and see how it leads us into the world of rotations.

2.2 Complex Numbers

A complex number has two parts: a *real* part and an *imaginary* part. The real part is just an ordinary number that may be zero, positive or negative, and the imaginary part is another real number multiplied by i. For example, $2 + 3i$ is a complex number

J. Vince, *Rotation Transforms for Computer Graphics*,
DOI 10.1007/978-0-85729-154-7_2,

where 2 is the real part and $3i$ is the imaginary part. The following are all complex numbers:

$$2, \quad 2+2i, \quad 1-3i, \quad -4i, \quad 17i.$$

Note the convention to place the real part first, followed by i. However, if i is associated with a trigonometric function such as *sin* or *cos*, it is usual to place i in front of the function: $i\sin\theta$ or $i\cos\theta$, to avoid any confusion that it is part of the function's angle.

All that we have to remember is that whenever we manipulate complex numbers, the occurrence of i^2 is replaced by -1.

2.2.1 *Axioms*

The axioms defining the behaviour of complex numbers are identical to those associated with real numbers. For example, given two complex numbers z_1 and z_2 they obey the following rules:
Addition:

$$\begin{aligned} \text{Commutative} \quad & z_1 + z_2 = z_2 + z_1 \\ \text{Associative} \quad & (z_1 + z_2) + z_3 = z_1 + (z_2 + z_3). \end{aligned}$$

Multiplication:

$$\begin{aligned} \text{Commutative} \quad & z_1 z_2 = z_2 z_1 \\ \text{Associative} \quad & (z_1 z_2) z_3 = z_1 (z_2 z_3) \\ \text{Distributive} \quad & z_1 (z_2 + z_3) = z_1 z_2 + z_1 z_3 \\ & (z_1 + z_2) z_3 = z_1 z_3 + z_2 z_3. \end{aligned}$$

2.3 The Modulus

The *modulus* of a complex number $a+bi$ is defined as $\sqrt{a^2+b^2}$. For example, the modulus of $3+4i$ is 5. In general, the modulus of a complex number z is written $|z|$:

$$\begin{aligned} z &= a + bi \\ |z| &= \sqrt{a^2 + b^2}. \end{aligned}$$

We'll see why this is so when we cover the polar representation of a complex number.

2.4 Addition and Subtraction

Given two complex numbers:

$$
\begin{aligned}
z_1 &= a + bi \\
z_2 &= c + di \\
z_1 \pm z_2 &= (a \pm c) + (b \pm d)\, i
\end{aligned}
$$

where the real and imaginary parts are added or subtracted, respectively. For example:

$$
\begin{aligned}
z_1 &= 5 + 3i \\
z_2 &= 3 + 2i \\
z_1 + z_2 &= 8 + 5i \\
z_1 - z_2 &= 2 + i.
\end{aligned}
$$

2.5 Multiplication by a Scalar

A scalar is just an ordinary number, and may be used to multiply a complex number using normal algebraic rules. For example, the complex number $a + bi$ is multiplied by the scalar λ as follows:

$$\lambda\,(a + bi) = \lambda a + \lambda bi$$

and a specific example:

$$2\,(3 + 5i) = 6 + 10i.$$

2.6 Product of Two Complex Numbers

The product of two complex numbers is evaluated by creating all the terms algebraically, and collecting up the real and imaginary terms:

$$
\begin{aligned}
z_1 &= a + bi \\
z_2 &= c + di \\
z_1 z_2 &= (a + bi)\,(c + di) \\
&= ac + adi + bci + bdi^2 \\
&= (ac - bd) + (ad + bc)\, i
\end{aligned}
$$

which is another complex number. For example:

$$\begin{aligned}
z_1 &= 3+4i \\
z_2 &= 5-2i \\
z_1 z_2 &= (3+4i)(5-2i) \\
&= 15-6i+20i-8i^2 \\
&= 15+14i+8 \\
&= 23+14i.
\end{aligned}$$

Remember that the addition, subtraction and multiplication of complex numbers obey the normal axioms of algebra. Also, the multiplication of two complex numbers, and their addition always results in a complex number, that is, the two operations are closed.

2.7 The Complex Conjugate

A special case exists when we multiply two complex numbers together where the only difference between them is the sign of the imaginary part:

$$\begin{aligned}
(a+bi)(a-bi) &= a^2-abi+abi-b^2i^2 \\
&= a^2+b^2.
\end{aligned}$$

As this real value is such an interesting result, $a-bi$ is called the *complex conjugate* of $a+bi$. In general, the complex conjugate of

$$z = a+bi$$

is written either with a bar $\bar{z}$ symbol or an asterisk z^* as

$$z^* = a-bi$$

and implies that

$$zz^* = a^2+b^2 = |z|^2.$$

2.8 Division of Two Complex Numbers

The complex conjugate provides us with a mechanism to divide one complex number by another. For instance, consider the quotient

$$\frac{a+bi}{c+di}.$$

This can be resolved by multiplying the numerator and denominator by the complex conjugate $c-di$ to create a real denominator:

$$\begin{aligned}\frac{a+bi}{c+di} &= \frac{(a+bi)(c-di)}{(c+di)(c-di)}\\ &= \frac{ac-adi+bci-bdi^2}{c^2+d^2}\\ &= \left(\frac{ac+bd}{c^2+d^2}\right)+\left(\frac{bc-ad}{c^2+d^2}\right)i.\end{aligned}$$

Another special case is when $a=1$ and $b=0$:

$$\frac{1}{c+di} = (c+di)^{-1} = \left(\frac{c}{c^2+d^2}\right)-\left(\frac{d}{c^2+d^2}\right)i$$

which is the *inverse* of a complex number.

Let's evaluate the quotient:

$$\frac{4+3i}{3+4i}.$$

Multiplying top and bottom by the complex conjugate $3-4i$ we have

$$\begin{aligned}\frac{4+3i}{3+4i} &= \frac{(4+3i)(3-4i)}{(3+4i)(3-4i)}\\ &= \frac{12-16i+9i-12i^2}{25} = \frac{24}{25}-\frac{7}{25}i.\end{aligned}$$

2.9 The Inverse

Although we have already discovered the inverse of a complex number, let's employ another strategy by declaring

$$z_1 = \frac{1}{z}$$

where z is a complex number.

Next, we divide both sides by the complex conjugate of z to create

$$\frac{z_1}{z^*} = \frac{1}{zz^*}.$$

But we have previously shown that $zz^* = |z|^2$, therefore,

$$\frac{z_1}{z^*} = \frac{1}{|z|^2}$$

and rearranging, we have

$$z_1 = \frac{z^*}{|z|^2}.$$

In general

$$\frac{1}{z} = z^{-1} = \frac{z^*}{|z|^2}.$$

As an illustration let's find the inverse of $3+4i$

$$\begin{aligned}\frac{1}{3+4i} &= (3+4i)^{-1}\\ &= \frac{3-4i}{25}\\ &= \frac{3}{25}-\frac{4}{25}i.\end{aligned}$$

Let's test this result by multiplying z by its inverse:

$$(3+4i)\left(\frac{3}{25}-\frac{4}{25}i\right)=\frac{9}{25}-\frac{12}{25}i+\frac{12}{25}i+\frac{16}{25}=1$$

which confirms the correctness of the inverse.

2.10 The Complex Plane

Leonhard Euler (1707–1783) (whose name rhymes with *boiler*) played a significant role in putting complex numbers on the map. His ideas on rotations are also used in computer graphics to locate objects and virtual cameras in space, as we shall see later on.

Consider the scenario depicted in Fig. 2.1. Any number on the number line is related to the same number with the opposite sign via a rotation of 180°. For example, when 2 is rotated 180° about zero, it becomes -2, and when -3 is rotated 180° about zero it becomes 3.

But as we know that $i^2=-1$ we can write

$$-n=i^2n.$$

If we now regard i^2 as a rotation through 180°, then i could be a rotation through 90°!

Figure 2.2 shows how complex numbers can be interpreted as 2D coordinates using the *complex plane* where the real part is the horizontal coordinate and the

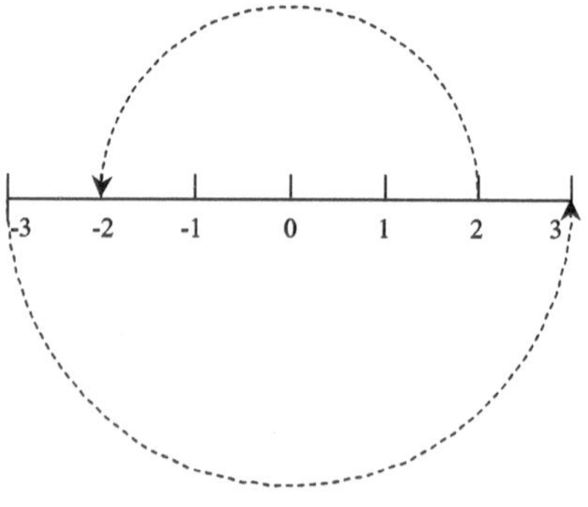

Fig. 2.1 Rotating numbers through 180° reverses their sign

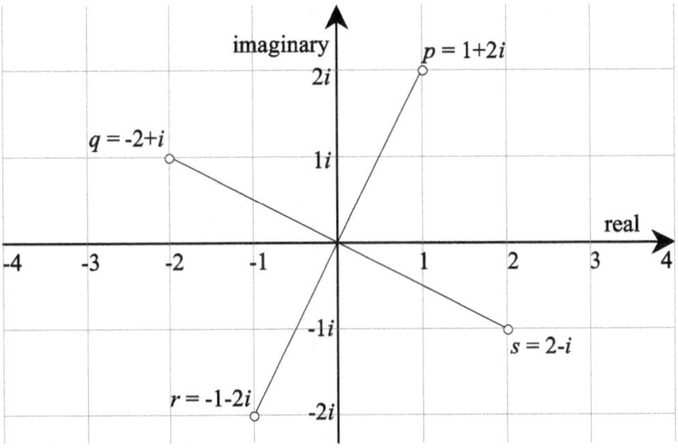

Fig. 2.2 The graphical representation of complex numbers

imaginary part is the vertical coordinate. The figure also shows four complex numbers:

$$p = 1 + 2i, \quad q = -2 + i, \quad r = -1 - 2i, \quad s = 2 - i$$

which happen to be 90° apart. For example, the complex number p in Fig. 2.2 is rotated 90° to q by multiplying it by i:

$$\begin{aligned} i\,(1 + 2i) &= i + 2i^2 \\ &= -2 + i. \end{aligned}$$

The point q is rotated another 90° to r by multiplying it by i:

$$\begin{aligned} i\,(-2 + i) &= -2i + i^2 \\ &= -1 - 2i. \end{aligned}$$

The point r is rotated another 90° to s by multiplying it by i:

$$\begin{aligned} i\,(-1 - 2i) &= -i - 2i^2 \\ &= 2 - i. \end{aligned}$$

Finally, the point s is rotated 90° back to p by multiplying it by i:

$$\begin{aligned} i\,(2 - i) &= 2i - i^2 \\ &= 1 + 2i. \end{aligned}$$

2.11 Polar Representation

The complex plane provides a simple mechanism to represent complex numbers graphically. This in turn makes it possible to use a *polar representation* as shown

Fig. 2.3 Polar representation of a complex number

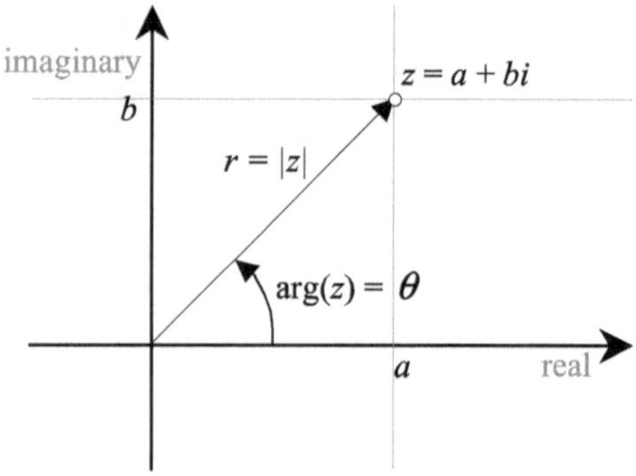

in Fig. 2.3 where we see the complex number $z = a + bi$ representing the oriented line r. The length of r is obviously $\sqrt{a^2 + b^2}$, which is why the modulus of a complex number has the same definition. We can see from Fig. 2.3 that the horizontal component of z is $r\cos\theta$ and the vertical component is $r\sin\theta$, which permits us to write

$$\begin{aligned} z &= a + bi \\ &= r\cos\theta + ri\sin\theta \\ &= r(\cos\theta + i\sin\theta). \end{aligned}$$

Note that i has been placed in front of the sin function.

The angle θ between r and the real axis is called the *argument* and written $\arg(z)$, and in this case

$$\arg(z) = \theta.$$

One of Euler's discoveries concerns the relationship between the series for exponential e, sin and cos:

$$e^{i\theta} = \cos\theta + i\sin\theta$$

which enables us to write

$$z = re^{i\theta}.$$

We are now in a position to revisit the product and quotient of two complex numbers using polar representation. For example:

$$\begin{aligned} z &= r(\cos\theta + i\sin\theta) \\ w &= s(\cos\phi + i\sin\phi) \\ zw &= rs(\cos\theta + i\sin\theta)(\cos\phi + i\sin\phi) \\ &= rs\big(\cos\theta\cos\phi + i\cos\theta\sin\phi + i\sin\theta\cos\phi + i^2\sin\theta\sin\phi\big) \\ &= rs\big((\cos\theta\cos\phi - \sin\theta\sin\phi) + i(\sin\theta\cos\phi + \cos\theta\sin\phi)\big) \end{aligned}$$

and as

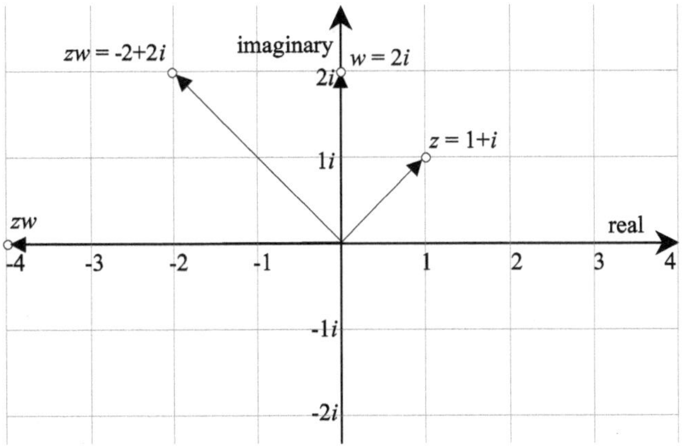

Fig. 2.4 The product of two complex numbers

$$\cos(\theta+\phi) = \cos\theta\cos\phi - \sin\theta\sin\phi$$
$$\sin(\theta+\phi) = \sin\theta\cos\phi + \cos\theta\sin\phi$$
$$zw = rs\big(\cos(\theta+\phi) + i\sin(\theta+\phi)\big).$$

So the product of two complex numbers creates a third one with modulus

$$|zw| = rs$$

and argument

$$\arg(zw) = \arg(z) + \arg(w) = \theta + \phi.$$

Let's illustrate this with an example. Figure 2.4 shows two complex numbers

$$z = 1 + i, \quad w = 2i$$

therefore,

$$|z| = \sqrt{2}, \quad \arg(z) = 45^\circ$$
$$|w| = 2, \quad \arg(w) = 90^\circ$$
$$|zw| = 2\sqrt{2}$$
$$\arg(zw) = 135^\circ$$

which is another complex number $-2 + 2i$.

2.12 Rotors

The above observations imply that multiplying a complex number by another, whose modulus is unity, causes no scaling. For example, multiplying $3 + 4i$ by $1 + 0i$ creates the same complex number, unscaled and unrotated. However, multiplying $3 + 4i$ by $0 + i$ rotates it by 90° without any scaling.

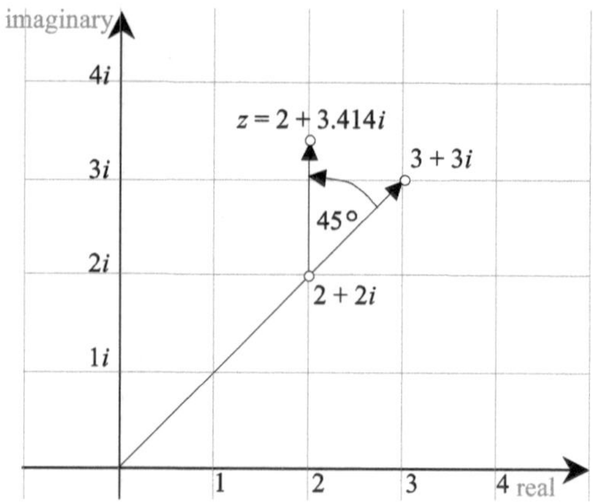

Fig. 2.5 Rotating a complex number about another complex number

So to rotate $2+2i$ by 45° we must multiply it by

$$\cos 45^\circ + i\sin 45^\circ = \frac{\sqrt{2}}{2} + \frac{\sqrt{2}}{2}i$$

$$\begin{aligned}\left(\frac{\sqrt{2}}{2} + \frac{\sqrt{2}}{2}i\right)(2+2i) &= \sqrt{2} + \sqrt{2}i + \sqrt{2}i + \sqrt{2}i^2\\ &= 2\sqrt{2}i.\end{aligned}$$

So now we have a *rotor* to rotate a complex number through any angle. In general, the rotor to rotate a complex number $a+bi$ through an angle θ is

$$\mathbf{R}_\theta = \cos\theta + i\sin\theta.$$

Now let's consider the problem of rotating $3+3i$, 45° about $2+2i$ as shown in Fig. 2.5. From the figure, the result is $z \approx 2 + 3.414i$, but let's calculate it by subtracting $2+2i$ from $3+3i$ to shift the operation to the origin, then multiply the result by $\sqrt{2}/2 + \sqrt{2}/2i$, and then add back $2+2i$:

$$\begin{aligned}z &= \left(\frac{\sqrt{2}}{2} + \frac{\sqrt{2}}{2}i\right)\big((3+3i) - (2+2i)\big) + 2 + 2i\\ &= \left(\frac{\sqrt{2}}{2} + \frac{\sqrt{2}}{2}i\right)(1+i) + 2 + 2i\\ &= \frac{\sqrt{2}}{2} + \frac{\sqrt{2}}{2}i + \frac{\sqrt{2}}{2}i - \frac{\sqrt{2}}{2} + 2 + 2i\\ &= 2 + (2+\sqrt{2})i\\ &\approx 2 + 3.414i\end{aligned}$$

which is correct. Therefore, to rotate any point (x, y) through an angle θ we convert it into a complex number $x+yi$ and multiply by the rotor $\cos\theta + i\sin\theta$:

$$\begin{aligned}x' + y'i &= (\cos\theta + i\sin\theta)(x + yi)\\ &= (x\cos\theta - y\sin\theta) + (x\sin\theta + y\cos\theta)\,i\end{aligned}$$

where (x', y') is the rotated point.

But as we shall see in Chap. 4, this is the transform for rotating a point (x, y) about the origin:

$$\begin{bmatrix}x'\\ y'\end{bmatrix} = \begin{bmatrix}\cos\theta & -\sin\theta\\ \sin\theta & \cos\theta\end{bmatrix}\begin{bmatrix}x\\ y\end{bmatrix}.$$

Before moving on let's consider the effect the complex conjugate of a rotor has on rotational direction, and we can do this by multiplying $x + yi$ by the rotor $\cos\theta - i\sin\theta$:

$$\begin{aligned}x' + y'i &= (\cos\theta - i\sin\theta)(x + yi)\\ &= x\cos\theta + y\sin\theta - (x\sin\theta + y\cos\theta)\,i\end{aligned}$$

which in matrix form is

$$\begin{bmatrix}x'\\ y'\end{bmatrix} = \begin{bmatrix}\cos\theta & \sin\theta\\ -\sin\theta & \cos\theta\end{bmatrix}\begin{bmatrix}x\\ y\end{bmatrix}$$

which is a rotation of $-\theta$.

Therefore, we define a rotor $\mathbf{R}_\theta$ and its conjugate $\mathbf{R}_\theta^\dagger$ as

$$\begin{aligned}\mathbf{R}_\theta &= \cos\theta + i\sin\theta\\ \mathbf{R}_\theta^\dagger &= \cos\theta - i\sin\theta\end{aligned}$$

where $\mathbf{R}_\theta$ rotates $+\theta$, and $\mathbf{R}_\theta^\dagger$ rotates $-\theta$. The dagger symbol '†' is chosen as it is used for rotors in multivectors, which are covered later.

2.13 Summary

There is no doubt that complex numbers are amazing objects and arise simply by introducing the symbol i which squares to -1. It is unfortunate that the names 'complex' and 'imaginary' are used to describe them as they are neither complex nor imaginary, but very simple. We will come across them again in later chapters and see how they provide a way of rotating 3D points.

In this chapter we have seen that complex numbers can be added, subtracted, multiplied and divided, and they can even be raised to a power. We have also come across new terms such as: *complex conjugate*, *modulus* and *argument*. We have also discovered the *rotor* which permits us to rotate 2D points.

In the mid-19th century, mathematicians started to look for the 3D equivalent of complex numbers, and after many years of work, Sir William Rowan Hamilton invented *quaternions* which are the subject of a later chapter.

2.13.1 Summary of Complex Operations

Complex number

$$z = a + bi \quad \text{where } i^2 = -1.$$

Addition and subtraction

$$\begin{aligned} z_1 &= a + bi \\ z_2 &= c + di \\ z_1 \pm z_2 &= (a \pm c) + (b \pm d)\, i. \end{aligned}$$

Scalar product

$$\lambda z = \lambda a + \lambda b i.$$

Modulus

$$|z| = \sqrt{a^2 + b^2}.$$

Product

$$z_1 z_2 = (ac - bd) + (ad + bc)\, i.$$

Complex conjugate

$$z^* = a - bi.$$

Division

$$\frac{z_1}{z_2} = \left(\frac{ac + bd}{c^2 + d^2}\right) + \left(\frac{bc - ad}{c^2 + d^2}\right) i.$$

Inverse

$$z^{-1} = \frac{z^*}{|z|^2}.$$

Polar form

$$\begin{aligned} z &= r\,(\cos\theta + i\sin\theta) \\ r &= |z| \\ \theta &= \arg(z) \\ z &= re^{i\theta}. \end{aligned}$$

Rotors

$$\begin{aligned} \mathbf{R}_\theta &= \cos\theta + i\sin\theta \\ \mathbf{R}_\theta^\dagger &= \cos\theta - i\sin\theta. \end{aligned}$$

Chapter 3
Vectors

3.1 Introduction

Vectors can be used to represent all sorts of data from weather maps to magnetic fields, and in computer graphics they are used to represent oriented lines and locate points in space. In 1853 Sir William Rowan Hamilton (1805–1865) published his book *Lectures on Quaternions* [2] in which he described terms such as *vector*, *transvector* and *provector*. Hamilton had been looking for a 3D equivalent to complex numbers and discovered quaternions. However his work was not widely accepted until 1884, when the American mathematician Josiah Willard Gibbs (1839–1903) published his treatise *Elements of Vector Analysis*, [3] describing modern *vector analysis*.

3.2 Vector Notation

As a vector contains two or more numbers, its symbolic name is generally printed using a **bold** font to distinguish it from a scalar variable. Examples being **n**, **i** and **q**. When a vector is assigned its numeric values, the following notation is used

$$\mathbf{n} = \begin{bmatrix} 2 \\ 3 \end{bmatrix}.$$

The numbers 2 and 3 are the *components* of **n** and their position within the brackets is very important.

Two types of notation are in use today: *column vectors* and *row vectors*. In this book we use column vectors, although they can be transposed into a row vector using this notation: $\mathbf{n} = [2 \quad 3]^{\mathrm{T}}$. The superscript ^{T} reminds us of the column to row transposition.

J. Vince, *Rotation Transforms for Computer Graphics*,
DOI 10.1007/978-0-85729-154-7_3, © Springer-Verlag London Limited 2011

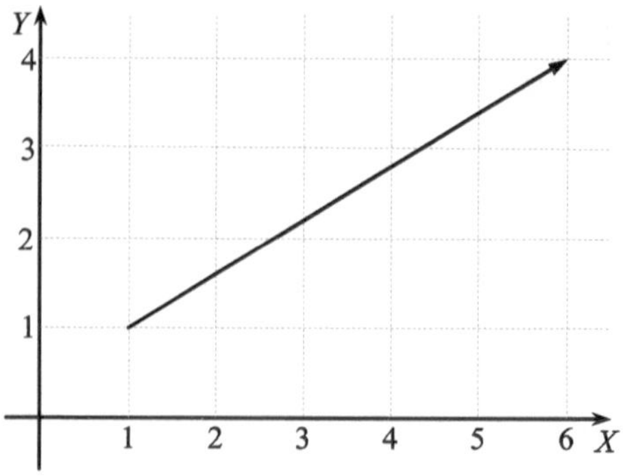

Fig. 3.1 A vector is represented by an oriented line segment

3.3 Graphical Representation of Vectors

Cartesian coordinates provide an excellent mechanism for representing vectors and allows them to be incorporated within the classical framework of mathematics. Figure 3.1 shows an oriented line segment used to represent a vector. The length of the line represents the vector's magnitude, and the line's orientation and arrow define its direction.

The line's direction is determined by the vector's head (x_h, y_h) and tail (x_t, y_t) from which we compute its x- and y-components Δ_x and Δ_y:

$$\Delta_x = x_h - x_t$$

$$\Delta_y = y_h - y_t.$$

For example, in Fig. 3.1 the vector's head is $(6, 4)$ and its tail is $(1, 1)$, which makes its components $\Delta_x = 5$ and $\Delta_y = 3$ or $[5 \quad 3]^{\mathrm{T}}$. If the vector is pointing in the opposite direction, its components become $\Delta_x = -5$ and $\Delta_y = -3$ or $[-5 \quad -3]^{\mathrm{T}}$.

One can readily see from this notation that a vector does not have a unique position in space. It does not matter where we place a vector, so long as we preserve its length and orientation its components will not alter.

3.4 Magnitude of a Vector

The length or *magnitude* of a vector **r** is written $|\mathbf{r}|$ and is computed by applying the theorem of Pythagoras to its components Δ_x and Δ_y:

$$|\mathbf{r}| = \sqrt{\Delta_x^2 + \Delta_y^2}.$$

For example, the magnitude of vector $[3 \quad 4]^{\mathrm{T}}$ is $\sqrt{3^2 + 4^2} = 5$. Figure 3.2 shows eight vectors, with their geometric properties listed in Table 3.1. The subscripts $_h$ and $_t$ stand for *head* and *tail* respectively.

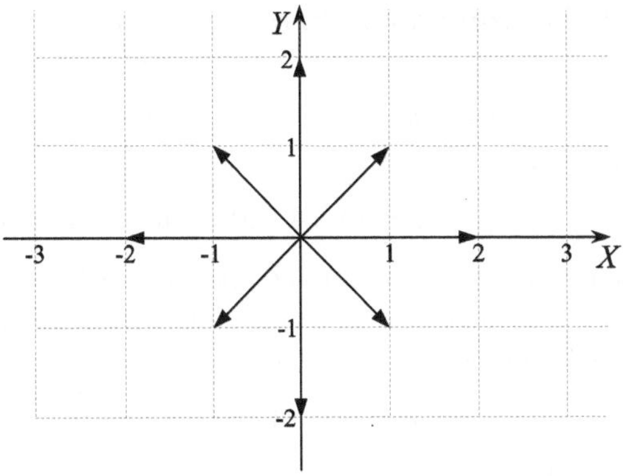

Fig. 3.2 Eight vectors whose coordinates are shown in Table 3.1

Table 3.1 Values associated with the eight vectors in Fig. 3.2

x_h	y_h	x_t	y_t	Δ_x	Δ_y	**|vector|**
2	0	0	0	2	0	2
0	2	0	0	0	2	2
−2	0	0	0	−2	0	2
0	−2	0	0	0	−2	2
1	1	0	0	1	1	$\sqrt{2}$
−1	1	0	0	−1	1	$\sqrt{2}$
−1	−1	0	0	−1	−1	$\sqrt{2}$
1	−1	0	0	1	−1	$\sqrt{2}$

3.5 3D Vectors

A 3D vector simply requires an extra component to represent its z-component Δ_z:

$$\mathbf{r} = \begin{bmatrix} \Delta_x & \Delta_y & \Delta_z \end{bmatrix}^{\mathrm{T}}$$

and its length is given by

$$|\mathbf{r}| = \sqrt{\Delta_x^2 + \Delta_y^2 + \Delta_z^2}.$$

3.6 Vector Manipulation

Vectors are very different to scalars, and rules have been developed to control how the two mathematical entities interact with one another. For instance, we need to consider vector addition, subtraction and multiplication, and how a vector is modified by a scalar. Let's begin with multiplying a vector by a scalar.

3.6.1 Multiplying a Vector by a Scalar

Given a vector **n**, 2**n** means that the vectors components are doubled. For example, given

$$\mathbf{n} = [3 \quad 4 \quad 5]^{\mathrm{T}} \quad \text{then} \quad 2\mathbf{n} = [6 \quad 8 \quad 10]^{\mathrm{T}}.$$

Similarly, dividing **n** by 2, its components are halved. Note that the vector's direction remains unchanged – only its magnitude changes. However, the vector's direction is reversed if the scalar is negative:

$$\lambda = -2 \quad \text{then} \quad \lambda\mathbf{n} = [-6 \quad -8 \quad -10]^{\mathrm{T}}.$$

In general, given

$$\mathbf{n} = \begin{bmatrix} n_1 \\ n_2 \\ n_3 \end{bmatrix} \quad \text{then} \quad \pm\lambda\mathbf{n} = \begin{bmatrix} \pm\lambda n_1 \\ \pm\lambda n_2 \\ \pm\lambda n_3 \end{bmatrix} \quad \text{where } \lambda \text{ is a scalar.}$$

3.6.2 Vector Addition and Subtraction

Given vectors **r** and **s**, $\mathbf{r} \pm \mathbf{s}$ is defined as

$$\mathbf{r} = \begin{bmatrix} x_r \\ y_r \\ z_r \end{bmatrix}, \quad \mathbf{s} = \begin{bmatrix} x_s \\ y_s \\ z_s \end{bmatrix} \quad \text{then} \quad \mathbf{r} \pm \mathbf{s} = \begin{bmatrix} x_r \pm x_s \\ y_r \pm y_s \\ z_r \pm z_s \end{bmatrix}.$$

Vector addition is commutative:

$$\mathbf{a} + \mathbf{b} = \mathbf{b} + \mathbf{a}$$

$$\text{e.g.} \quad \begin{bmatrix} 1 \\ 2 \\ 3 \end{bmatrix} + \begin{bmatrix} 4 \\ 5 \\ 6 \end{bmatrix} = \begin{bmatrix} 4 \\ 5 \\ 6 \end{bmatrix} + \begin{bmatrix} 1 \\ 2 \\ 3 \end{bmatrix}.$$

However, like scalar subtraction, vector subtraction is not commutative

$$\mathbf{a} - \mathbf{b} \neq \mathbf{b} - \mathbf{a}$$

$$\text{e.g.} \quad \begin{bmatrix} 4 \\ 5 \\ 6 \end{bmatrix} - \begin{bmatrix} 1 \\ 2 \\ 3 \end{bmatrix} \neq \begin{bmatrix} 1 \\ 2 \\ 3 \end{bmatrix} - \begin{bmatrix} 4 \\ 5 \\ 6 \end{bmatrix}.$$

3.7 Position Vectors

Given any point $P(x, y, z)$, a *position vector* **p** is created by assuming that P is the vector's head and the origin is its tail. Because the tail coordinates are $(0, 0, 0)$ the vector's components are x, y, z. Consequently, the vector's length $|\mathbf{p}|$ equals

$\sqrt{x^2+y^2+z^2}$. For example, the point $P(4,5,6)$ creates a position vector **p** relative to the origin:

$$\mathbf{p} = [4 \quad 5 \quad 6]^{\mathrm{T}} \quad \text{and} \quad |\mathbf{p}| = \sqrt{4^2+5^2+6^2} \approx 20.88.$$

3.8 Unit Vectors

By definition, a *unit vector* has a length of 1. A simple example is **i** where

$$\mathbf{i} = [1 \quad 0 \quad 0]^{\mathrm{T}} \quad \text{and} \quad |\mathbf{i}| = 1.$$

Converting a vector into a unit form is called *normalising* and is achieved by dividing the vector's components by its length. To formalise this process consider the vector $\mathbf{r} = [x \quad y \quad z]^{\mathrm{T}}$ with length $|\mathbf{r}| = \sqrt{x^2+y^2+z^2}$. The unit form of **r** is given by

$$\hat{\mathbf{r}} = \frac{1}{|\mathbf{r}|}[x \quad y \quad z]^{\mathrm{T}}.$$

This process is confirmed by showing that the length of $\hat{\mathbf{r}}$ is 1:

$$\begin{aligned}
|\hat{\mathbf{r}}| &= \sqrt{\left(\frac{x}{|\hat{\mathbf{r}}|}\right)^2 + \left(\frac{y}{|\hat{\mathbf{r}}|}\right)^2 + \left(\frac{z}{|\hat{\mathbf{r}}|}\right)^2} \\
&= \frac{1}{|\hat{\mathbf{r}}|}\sqrt{x^2+y^2+z^2} \\
|\hat{\mathbf{r}}| &= 1.
\end{aligned}$$

To put this into context, consider the conversion of $\mathbf{r} = [1 \quad 2 \quad 3]^{\mathrm{T}}$ into a unit form:

$$\begin{aligned}
|\mathbf{r}| &= \sqrt{1^2+2^2+3^2} = \sqrt{14} \\
\hat{\mathbf{r}} &= \frac{1}{\sqrt{14}}\begin{bmatrix}1\\2\\3\end{bmatrix} \approx \begin{bmatrix}0.267\\0.535\\0.802\end{bmatrix}.
\end{aligned}$$

3.9 Cartesian Vectors

We begin by defining three Cartesian unit vectors **i**, **j**, **k** aligned with the x-, y- and z-axes respectively:

$$\mathbf{i} = \begin{bmatrix}1\\0\\0\end{bmatrix}, \quad \mathbf{j} = \begin{bmatrix}0\\1\\0\end{bmatrix}, \quad \mathbf{k} = \begin{bmatrix}0\\0\\1\end{bmatrix}.$$

Any vector aligned with the x-, y- or z-axes can be defined by a scalar multiple of the unit vectors **i**, **j** and **k** respectively. For example, a vector 10 units long aligned with the x-axis is $10\mathbf{i}$, and a vector 20 units long aligned with the z-axis is $20\mathbf{k}$. By employing the rules of vector addition and subtraction we can compose a vector **r** by summing three *Cartesian unit vector* as follows:

$$\mathbf{r} = a\mathbf{i} + b\mathbf{j} + c\mathbf{k}$$

which is equivalent to writing **r** as

$$\mathbf{r} = \begin{bmatrix} a \\ b \\ c \end{bmatrix}$$

and means that the length of **r** is computed as

$$|\mathbf{r}| = \sqrt{a^2 + b^2 + c^2}.$$

Any pair of Cartesian vectors such as **r** and **s** are combined as follows

$$\begin{aligned} \mathbf{r} &= a\mathbf{i} + b\mathbf{j} + c\mathbf{k} \\ \mathbf{s} &= d\mathbf{i} + e\mathbf{j} + f\mathbf{k} \\ \mathbf{r} \pm \mathbf{s} &= (a \pm d)\,\mathbf{i} + (b \pm e)\,\mathbf{j} + (c \pm f)\,\mathbf{k}. \end{aligned}$$

For example:

$$\begin{aligned} \mathbf{r} &= 2\mathbf{i} + 3\mathbf{j} + 4\mathbf{k} \\ \mathbf{s} &= 5\mathbf{i} + 6\mathbf{j} + 7\mathbf{k} \\ \mathbf{r} + \mathbf{s} &= 7\mathbf{i} + 9\mathbf{j} + 11\mathbf{k}. \end{aligned}$$

3.10 Scalar Product

The mathematicians who defined the structure of vector analysis provided two ways to multiply vectors together: one gives rise to a scalar result and the other a vector result. For example, we could multiply two vectors **r** and **s** by using the product of their magnitudes: $|\mathbf{r}||\mathbf{s}|$. Although this is a valid operation it ignores the orientation of the vectors, which is one of their important features. The idea, however, is readily developed into a useful operation by including the angle between the vectors.

Figure 3.3 shows two vectors **r** and **s** that have been drawn, for convenience, such that their tails touch. Taking **s** as the reference vector – which is an arbitrary choice – we compute the projection of **r** on **s**, which takes into account their relative orientation. The length of **r** on **s** is $|\mathbf{r}|\cos\beta$. We can now multiply the magnitude of **s** by the projected length of **r**: $|\mathbf{s}||\mathbf{r}|\cos\beta$.

This scalar product is written

$$\mathbf{r} \cdot \mathbf{s} = |\mathbf{r}||\mathbf{s}|\cos\beta. \tag{3.1}$$

The dot symbol '·' is used to denote a scalar multiplication, which is why the product is often referred to as the *dot product*. We now need to discover how to compute it.

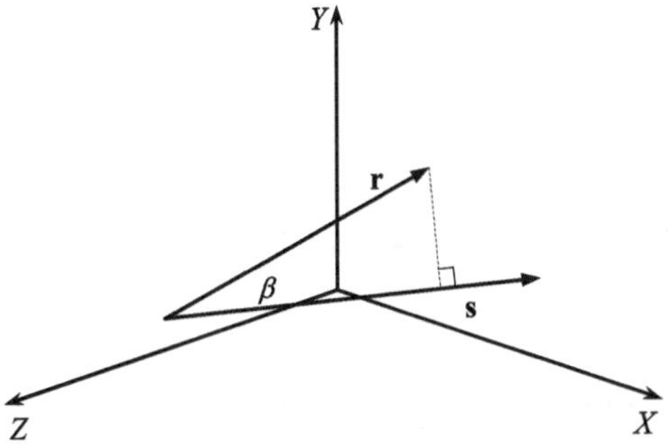

Fig. 3.3 Visualising the scalar product

To begin, we define two Cartesian vectors **r** and **s**, and proceed to multiply them together using the dot product definition:

$$\mathbf{r} = a\mathbf{i} + b\mathbf{j} + c\mathbf{k}$$
$$\mathbf{s} = d\mathbf{i} + e\mathbf{j} + f\mathbf{k}$$

therefore,

$$\begin{aligned}\mathbf{r}\cdot\mathbf{s} &= (a\mathbf{i} + b\mathbf{j} + c\mathbf{k})\cdot(d\mathbf{i} + e\mathbf{j} + f\mathbf{k})\\ &= a\mathbf{i}\cdot(d\mathbf{i} + e\mathbf{j} + f\mathbf{k}) + b\mathbf{j}\cdot(d\mathbf{i} + e\mathbf{j} + f\mathbf{k}) + c\mathbf{k}\cdot(d\mathbf{i} + e\mathbf{j} + f\mathbf{k})\\ &= ad\mathbf{i}\cdot\mathbf{i} + ae\mathbf{i}\cdot\mathbf{j} + af\mathbf{i}\cdot\mathbf{k} + bd\mathbf{j}\cdot\mathbf{i} + be\mathbf{j}\cdot\mathbf{j} + bf\mathbf{j}\cdot\mathbf{k}\\ &\quad + cd\mathbf{k}\cdot\mathbf{i} + ce\mathbf{k}\cdot\mathbf{j} + cf\mathbf{k}\cdot\mathbf{k}.\end{aligned}$$

Before we proceed any further, we can see that we have created various dot product terms such as $\mathbf{i}\cdot\mathbf{i}$, $\mathbf{i}\cdot\mathbf{j}$, $\mathbf{i}\cdot\mathbf{k}$, etc. These terms can be divided into two groups: those that reference the same unit vector, and those that reference different unit vectors.

Using the definition of the dot product (3.1), terms such as $\mathbf{i}\cdot\mathbf{i}$, $\mathbf{j}\cdot\mathbf{j}$ and $\mathbf{k}\cdot\mathbf{k} = 1$, because the angle between **i** and **i**, **j** and **j**, or **k** and **k**, is 0°, and $\cos 0° = 1$. But because the other vector combinations are separated by 90°, and $\cos 90° = 0$, all remaining terms collapse to zero. Bearing in mind that the magnitude of a unit vector is 1, we can write

$$\mathbf{r}\cdot\mathbf{s} = |\mathbf{r}||\mathbf{s}|\cos\beta = ad + be + cf.$$

This result confirms that the dot product is indeed a scalar quantity.

Let's use the scalar product to find the angle between two vectors **r** and **s**:

$$\begin{aligned}\mathbf{r} &= [2 \quad 0 \quad 4]^{\mathrm{T}}\\ \mathbf{s} &= [5 \quad 6 \quad 10]^{\mathrm{T}}\\ |\mathbf{r}| &= \sqrt{2^2 + 0^2 + 4^2} \approx 4.472\\ |\mathbf{s}| &= \sqrt{5^2 + 6^2 + 10^2} \approx 12.689\\ |\mathbf{r}||\mathbf{s}|\cos\beta &= 2\times 5 + 0\times 6 + 4\times 10 = 50\\ 12.689 \times 4.472 \times \cos\beta &= 50\end{aligned}$$

$$\cos\beta = \frac{50}{12.689 \times 4.472} \approx 0.8811$$
$$\beta = \cos^{-1} 0.8811 \approx 28.22^\circ.$$

The angle between the two vectors is approximately 28.22°, and β is always the smallest angle associated with the geometry.

3.11 The Vector Product

The second way to multiply vectors is by using the *vector product*, which is also called the *cross product* due to the '×' symbol used in its notation. It is based on the observation that two vectors **r** and **s** can be multiplied together to produce a third vector **t**:

$$\mathbf{r} \times \mathbf{s} = \mathbf{t}$$

where

$$|\mathbf{t}| = |\mathbf{r}||\mathbf{s}| \sin\beta \tag{3.2}$$

and β is the angle between **r** and **s**.

The vector **t** is normal (90°) to the plane containing the vectors **r** and **s**, which makes it an ideal way of computing surface normals in computer graphics applications. Once again, let's define two vectors and proceed to multiply them together using the '×' operator:

$$\begin{aligned}
\mathbf{r} &= a\mathbf{i} + b\mathbf{j} + c\mathbf{k} \\
\mathbf{s} &= d\mathbf{i} + e\mathbf{j} + f\mathbf{k} \\
\mathbf{r} \times \mathbf{s} &= (a\mathbf{i} + b\mathbf{j} + c\mathbf{k}) \times (d\mathbf{i} + e\mathbf{j} + f\mathbf{k}) \\
&= a\mathbf{i} \times (d\mathbf{i} + e\mathbf{j} + f\mathbf{k}) + b\mathbf{j} \times (d\mathbf{i} + e\mathbf{j} + f\mathbf{k}) + c\mathbf{k} \times (d\mathbf{i} + e\mathbf{j} + f\mathbf{k}) \\
&= ad\mathbf{i} \times \mathbf{i} + ae\mathbf{i} \times \mathbf{j} + af\mathbf{i} \times \mathbf{k} + bd\mathbf{j} \times \mathbf{i} + be\mathbf{j} \times \mathbf{j} + bf\mathbf{j} \times \mathbf{k} \\
&\quad + cd\mathbf{k} \times \mathbf{i} + ce\mathbf{k} \times \mathbf{j} + cf\mathbf{k} \times \mathbf{k}.
\end{aligned}$$

As we found with the dot product, there are two groups of vector terms: those that reference the same unit vector, and those that reference different unit vectors.

Using the definition for the cross product (3.2), operations such as $\mathbf{i} \times \mathbf{i}$, $\mathbf{j} \times \mathbf{j}$ and $\mathbf{k} \times \mathbf{k}$ result in a vector whose magnitude is 0. This is because the angle between the vectors is 0°, and $\sin 0^\circ = 0$. Consequently these terms vanish and we are left with

$$\mathbf{r} \times \mathbf{s} = ae\mathbf{i} \times \mathbf{j} + af\mathbf{i} \times \mathbf{k} + bd\mathbf{j} \times \mathbf{i} + bf\mathbf{j} \times \mathbf{k} + cd\mathbf{k} \times \mathbf{i} + ce\mathbf{k} \times \mathbf{j}. \tag{3.3}$$

The mathematician Sir William Rowan Hamilton struggled for many years to generalise complex numbers – and in so doing created a means of describing 3D rotations. At the time, he was not using vectors – as they had yet to be defined – but the imaginary terms i, j and k. Hamilton's problem was to resolve the products ij, jk, ki and their opposites ji, kj and ik.

One day in 1843, when he was out walking, thinking about this problem, he thought the impossible: $ij = k$, but $ji = -k$, $jk = i$, but $kj = -i$, and $ki = j$, but $ik = -j$. To his surprise, this worked, but it contradicted the commutative multiplication law of scalars. Although Hamilton had discovered "3D complex numbers", to which he gave the name *quaternion*, they were not popular with everyone. And as mentioned above, Josiah Gibbs saw that converting the imaginary i, j and k terms into the unit vectors **i**, **j** and **k** created a non-complex algebra for manipulating vectors, and for over a century we have been using Gibbs' vector notation.

Let's continue with Hamilton's rules and reduce the cross product terms of (3.3) to

$$\mathbf{r}\times\mathbf{s} = ae\mathbf{k} - af\mathbf{j} - bd\mathbf{k} + bf\mathbf{i} + cd\mathbf{j} - ce\mathbf{i}. \tag{3.4}$$

Equation (3.4) can be tidied up to bring like terms together:

$$\mathbf{r}\times\mathbf{s} = (bf - ce)\mathbf{i} + (cd - af)\mathbf{j} + (ae - bd)\mathbf{k}. \tag{3.5}$$

Now let's repeat the original vector equations to see how (3.5) is computed:

$$\begin{aligned}\mathbf{r} &= a\mathbf{i} + b\mathbf{j} + c\mathbf{k}\\ \mathbf{s} &= d\mathbf{i} + e\mathbf{j} + f\mathbf{k}\\ \mathbf{r}\times\mathbf{s} &= (bf - ce)\mathbf{i} + (cd - af)\mathbf{j} + (ae - bd)\mathbf{k}.\end{aligned} \tag{3.6}$$

To compute **i**'s scalar we consider the scalars associated with the other two unit vectors, i.e. b, c, e, and f, and cross-multiply and subtract them to form $(bf - ce)$.

To compute **j**'s scalar we consider the scalars associated with the other two unit vectors, i.e. a, c, d, and f, and cross-multiply and subtract them to form $(cd - af)$.

To compute **k**'s scalar we consider the scalars associated with the other two unit vectors, i.e. a, b, d, and e, and cross-multiply and subtract them to form $(ae - bd)$.

The middle operation seems out of step with the other two, but in fact it preserves a cyclic symmetry often found in mathematics. Nevertheless, some authors reverse the sign of the **j** scalar term and cross-multiply and subtract the terms to produce $-(af - cd)$ which maintains a visual pattern for remembering the cross-multiplication. Equation (3.6) now becomes

$$\mathbf{r}\times\mathbf{s} = (bf - ce)\mathbf{i} - (af - cd)\mathbf{j} + (ae - bd)\mathbf{k}. \tag{3.7}$$

Although we have not yet covered *determinants*, their notation allows us to write (3.7) as

$$\mathbf{r}\times\mathbf{s} = \begin{vmatrix} b & c \\ e & f \end{vmatrix}\mathbf{i} - \begin{vmatrix} a & c \\ d & f \end{vmatrix}\mathbf{j} + \begin{vmatrix} a & b \\ d & e \end{vmatrix}\mathbf{k}.$$

A 2×2 determinant is the difference between the product of the diagonal terms.

Therefore, to derive the cross product of two vectors we first write the vectors in the correct sequence. Remembering that $\mathbf{r}\times\mathbf{s}$ does not equal $\mathbf{s}\times\mathbf{r}$. Second, we compute the three scalar terms and form the resultant vector, which is perpendicular to the plane containing the original vectors.

Let's illustrate the vector product with two examples. First, we will confirm that the vector product works with the unit vectors **i**, **j** and **k**. We start with

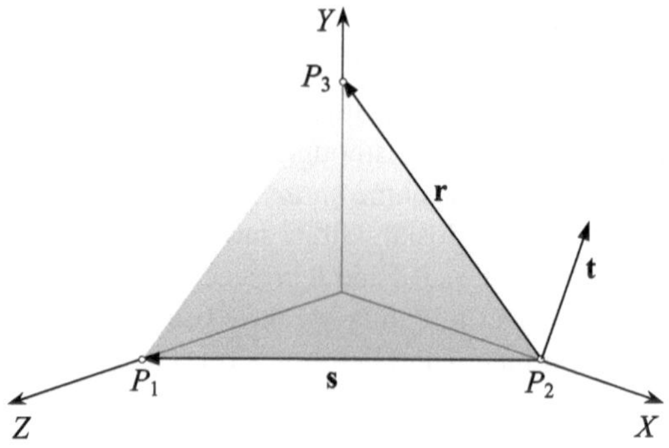

Fig. 3.4 The vector **t** is normal to the vectors **r** and **s**

$$\mathbf{r} = 1\mathbf{i} + 0\mathbf{j} + 0\mathbf{k}$$
$$\mathbf{s} = 0\mathbf{i} + 1\mathbf{j} + 0\mathbf{k}$$

and then compute (3.7)

$$\mathbf{r} \times \mathbf{s} = (0 \times 0 - 0 \times 1)\mathbf{i} - (1 \times 0 - 0 \times 0)\mathbf{j} + (1 \times 1 - 0 \times 0)\mathbf{k}.$$

The **i** scalar and **j** scalar terms are both zero, but the **k** scalar term is 1, which makes $\mathbf{i} \times \mathbf{j} = \mathbf{k}$.

Now let's show what happens when we reverse the vectors:

$$\mathbf{s} \times \mathbf{r} = (1 \times 0 - 0 \times 0)\mathbf{i} - (1 \times 0 - 0 \times 0)\mathbf{j} + (0 \times 0 - 1 \times 1)\mathbf{k}.$$

The **i** scalar and **j** scalar terms are both zero, but the **k** scalar term is -1, which makes $\mathbf{j} \times \mathbf{i} = -\mathbf{k}$. So we see that the vector product is *antisymmetric*, i.e. there is a sign reversal when the vectors are reversed. Similarly, it can be shown that

$$\mathbf{j} \times \mathbf{k} = \mathbf{i}$$
$$\mathbf{k} \times \mathbf{i} = \mathbf{j}$$
$$\mathbf{k} \times \mathbf{j} = -\mathbf{i}$$
$$\mathbf{i} \times \mathbf{k} = -\mathbf{j}.$$

Now let's consider two vectors **r** and **s** and compute the normal vector **t**. The vectors are chosen so that we can anticipate approximately the answer. For the sake of clarity, the vector equations include the scalar multipliers 0 and 1. Normally, these would be omitted. Figure 3.4 shows the vectors **r** and **s** and the normal vector **t**, and Table 3.2 contains the coordinates of the vertices forming the two vectors.

$$\mathbf{r} = \begin{bmatrix} x_3 - x_2 \\ y_3 - y_2 \\ z_3 - z_2 \end{bmatrix}, \quad \mathbf{s} = \begin{bmatrix} x_1 - x_2 \\ y_1 - y_2 \\ z_1 - z_2 \end{bmatrix}$$

then

$$P_1 = (0, 0, 1), \quad P_2 = (1, 0, 0), \quad P_3 = (0, 1, 0)$$
$$\mathbf{r} = -1\mathbf{i} + 1\mathbf{j} + 0\mathbf{k}$$

Table 3.2 Coordinates of the vertices used in Fig. 3.4

Vertex	x	y	z
P_1	0	0	1
P_2	1	0	0
P_3	0	1	0

$$\begin{aligned}\mathbf{s} &= -1\mathbf{i} + 0\mathbf{j} + 1\mathbf{k}\\ \mathbf{r}\times\mathbf{s} &= (1\times 1 - 0\times 0)\mathbf{i} - (-1\times 1 - (-1)\times 0)\mathbf{j} + (-1\times 0 - (-1)\times 1)\mathbf{k}\\ \mathbf{t} &= \mathbf{i}+\mathbf{j}+\mathbf{k}.\end{aligned}$$

This confirms what we expected from Fig. 3.4. Now let's reverse the vectors to illustrate the importance of vector sequence:

$$\begin{aligned}\mathbf{s}\times\mathbf{r} &= (0\times 0 - 1\times 1)\mathbf{i} - (-1\times 0 - (-1)\times 1)\mathbf{j} + (-1\times 1 - (-1)\times 0)\mathbf{k}\\ \mathbf{t} &= -\mathbf{i}-\mathbf{j}-\mathbf{k}\end{aligned}$$

which is in the opposite direction to $\mathbf{r}\times\mathbf{s}$ and confirms that the vector product is non-commutative.

3.12 The Right-Hand Rule

When we cover multivectors we will see that lines, planes and volumes are all oriented and can be described mathematically. In particular, 3D space is described as being left- or right-handed, and in this book we choose to work with a right-handed space. Consequently, the *right-hand rule* is an *aide mémoire* for working out the orientation of the cross product vector. Given the operation $\mathbf{r}\times\mathbf{s}$, if the right-hand thumb is aligned with $\mathbf{r}$, the first finger with $\mathbf{s}$, and the middle finger points in the direction of $\mathbf{t}$.

3.13 Deriving a Unit Normal Vector

Figure 3.5 shows a triangle with vertices defined in an anti-clockwise sequence from its visible side. This is the side from which we want the surface normal to point. Using the following information we will compute the surface normal using the cross product and then convert it to a unit normal vector.

Create vector $\mathbf{r}$ between P_1 and P_3, and vector $\mathbf{s}$ between P_2 and P_3:

$$\begin{aligned}\mathbf{r} &= -1\mathbf{i} + 1\mathbf{j} + 0\mathbf{k}\\ \mathbf{s} &= -1\mathbf{i} + 0\mathbf{j} + 2\mathbf{k}\\ \mathbf{r}\times\mathbf{s} &= (1\times 2 - 0\times 0)\mathbf{i} - (-1\times 2 - 0\times -1)\mathbf{j} + (-1\times 0 - 1\times -1)\mathbf{k}\\ \mathbf{t} &= 2\mathbf{i} + 2\mathbf{j} + \mathbf{k}\end{aligned}$$

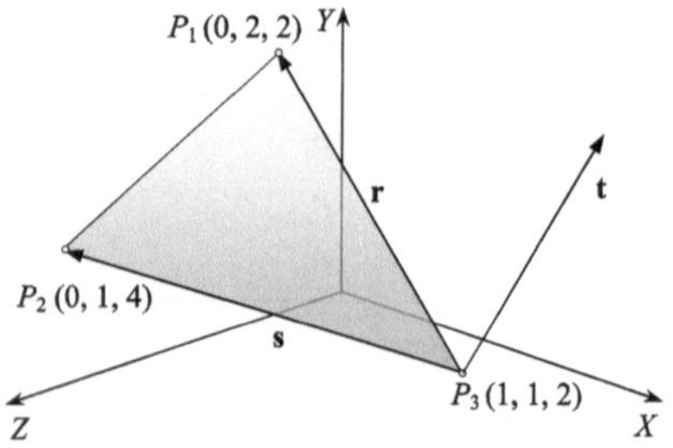

Fig. 3.5 The normal vector **t** is derived from the cross product $\mathbf{r} \times \mathbf{s}$

$$|\mathbf{t}| = \sqrt{2^2 + 2^2 + 1^2} = 3$$
$$\hat{\mathbf{t}}_u = \frac{2}{3}\mathbf{i} + \frac{2}{3}\mathbf{j} + \frac{1}{3}\mathbf{k}.$$

3.14 Interpolating Vectors

In computer animation we need to vary quantities such as height, width, depth, light intensity, radius, etc., such that they change over a sequence of animation frames. The change may be linear or non-linear, and a variety of techniques exist for changing one numeric value into another. This process is called *interpolating*.

To interpolate between two values v_1 and v_2 we often use the linear interpolant:

$$v = (1 - t)v_1 + tv_2$$

where the parameter t varies between 0 and 1. For example, given $v_1 = 2$ and $v_2 = 10$ we can compute a half-way point by making $t = 0.5$:

$$v = 0.5 \times 2 + 0.5 \times 10 = 6$$

where t is linked to the animation frame number.

However, this technique cannot be used for changing quantities such as a light source direction, dust-cloud particle velocity, or the direction and intensity of a flame. This is because these quantities possess both magnitude and direction – they are vector quantities.

For example, if we interpolated the x- and y-components of the vectors $[2 \quad 3]^T$ and $[4 \quad 7]^T$, the in-between vectors would carry the change of orientation but ignore the change in magnitude. To preserve both, we must design a spherical interpolant that is sensitive to a vector's length *and* orientation.

Figure 3.6 shows two unit vectors $\mathbf{v}_1$ and $\mathbf{v}_2$ separated by an angle θ. The interpolated vector $\mathbf{v}$ can be defined as a portion of $\mathbf{v}_1$ and a portion of $\mathbf{v}_2$:

$$\mathbf{v} = a\mathbf{v}_1 + b\mathbf{v}_2.$$

Let's define the values of a and b such that they are a function of the separating angle θ. Vector $\mathbf{v}$ is $t\theta$ from $\mathbf{v}_1$ and $(1 - t)\theta$ from $\mathbf{v}_2$, and it is evident from Fig. 3.6 that using the sine rule

$$\frac{a}{\sin(1 - t)\theta} = \frac{b}{\sin t\theta}. \tag{3.8}$$

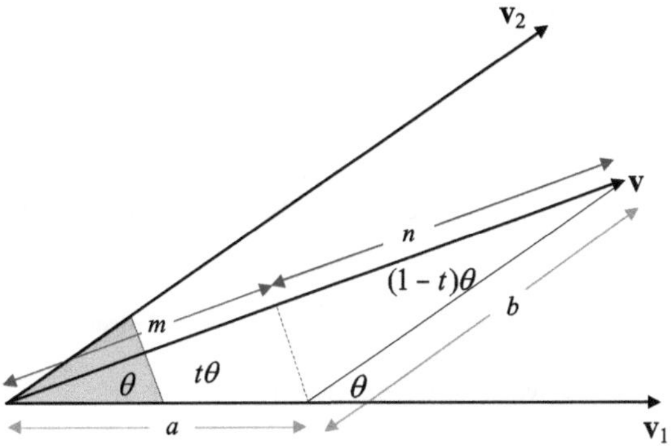

Fig. 3.6 Vector **v** is derived from a part of $\mathbf{v}_1$ and b part of $\mathbf{v}_2$

Furthermore,

$$m = a\cos t\theta$$
$$n = b\cos(1-t)\theta$$

where

$$m + n = 1. \tag{3.9}$$

From (3.8)

$$b = \frac{a\sin t\theta}{\sin(1-t)\theta}$$

and from (3.9) we get

$$a\cos t\theta + \frac{a\sin t\theta\cos(1-t)\theta}{\sin(1-t)\theta} = 1.$$

Solving for a we find that

$$a = \frac{\sin(1-t)\theta}{\sin\theta}$$
$$b = \frac{\sin t\theta}{\sin\theta}.$$

Therefore, the final spherical interpolant is

$$\mathbf{v} = \frac{\sin(1-t)\theta}{\sin\theta}\mathbf{v}_1 + \frac{\sin t\theta}{\sin\theta}\mathbf{v}_2. \tag{3.10}$$

To see how this operates, let's consider a simple exercise of interpolating between two unit vectors $[1 \quad 0]^{\mathrm{T}}$ and $[-1/\sqrt{2} \quad 1/\sqrt{2}]^{\mathrm{T}}$. The angle θ between the vectors is 135°. Equation (3.10) is used to interpolate individually the x- and the y-components individually:

$$v_x = \frac{\sin(1-t)135^\circ}{\sin 135^\circ} \times (1) + \frac{\sin t135^\circ}{\sin 135^\circ} \times \left(-\frac{1}{\sqrt{2}}\right)$$
$$v_y = \frac{\sin(1-t)135^\circ}{\sin 135^\circ} \times (0) + \frac{\sin t135^\circ}{\sin 135^\circ} \times \left(\frac{1}{\sqrt{2}}\right).$$

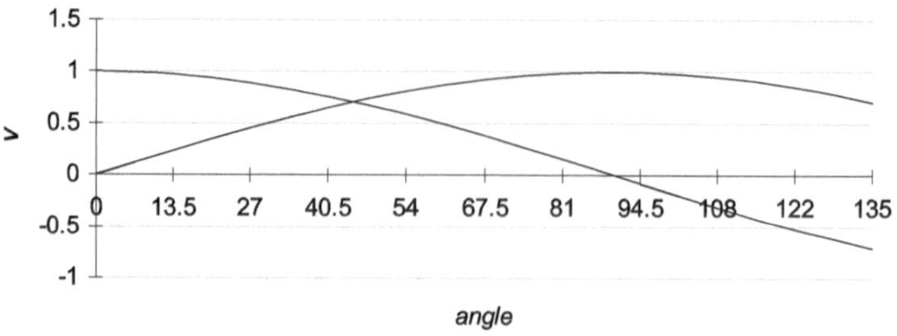

Fig. 3.7 Curves of the interpolated angles

Fig. 3.8 A trace of the interpolated vectors between $[1 \quad 0]^{\mathrm{T}}$ and $[-\frac{1}{\sqrt{2}} \quad \frac{1}{\sqrt{2}}]^{\mathrm{T}}$

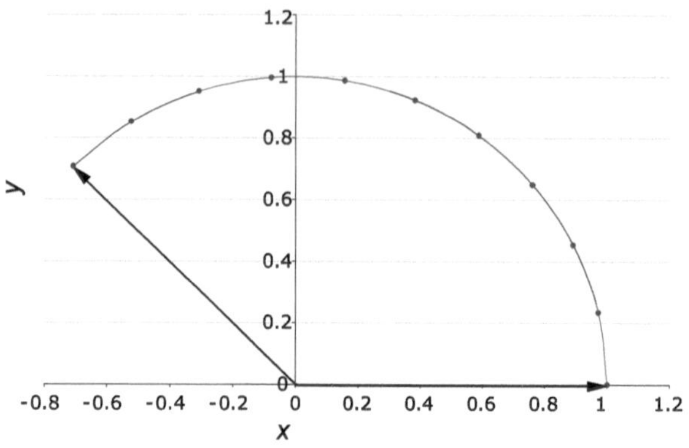

Figure 3.7 shows the interpolating curves and Fig. 3.8 shows the positions of the interpolated vectors, and a trace of the interpolated vectors.

Two observations to note about (3.10):

- First, the angle θ is the angle between the two vectors, which, if not known, can be computed using the dot product.
- Second, the range of θ is given by $0 < \theta < 180°$, for when $\theta = 0°, 180°$ the denominator collapses to zero. To illustrate this we will repeat (3.10) for $\theta = 179°$.

The result is shown in Fig. 3.9, which reveals clearly that the interpolant works normally over this range. One more degree, however, and it fails! Nevertheless, one could still leave the range equal to 180° and test for the conditions $t = 0$ then $\mathbf{v} = \mathbf{v}_1$ and when $t = 180°$ then $\mathbf{v} = \mathbf{v}_2$.

So far, we have only considered unit vectors. Now let's see how the interpolant responds to vectors of different magnitudes. As a test, we can input the following vectors to (3.10):

$$\mathbf{v}_1 = \begin{bmatrix} 2 \\ 0 \end{bmatrix} \quad \text{and} \quad \mathbf{v}_2 = \begin{bmatrix} 0 \\ 1 \end{bmatrix}.$$

The separating angle $\theta = 90°$, and the result is shown in Fig. 3.10. Note how the initial length of $\mathbf{v}_1$ reduces from 2 to 1 over 90°. It is left to the reader to examine other combinations of vectors.

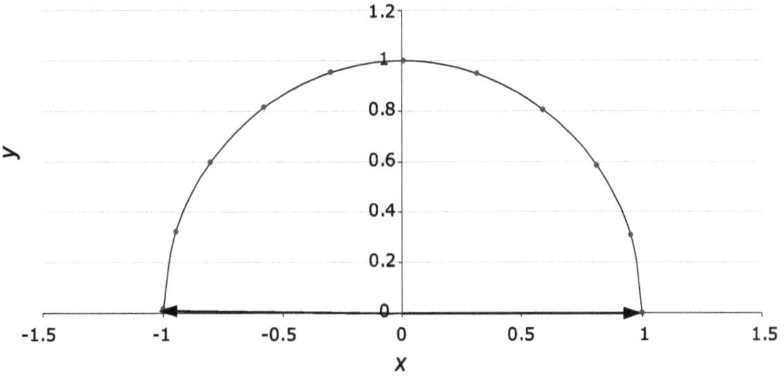

Fig. 3.9 Interpolating between two unit vectors 179° apart

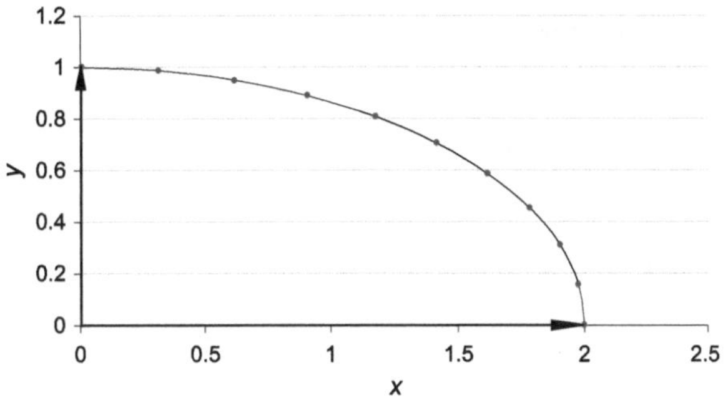

Fig. 3.10 Interpolating between the vectors $[2 \quad 0]^{\mathrm{T}}$ and $[0 \quad 1]^{\mathrm{T}}$

3.15 Summary

This chapter has covered the important features of vectors relevant to rotations. Basically, we need to know how to create a position vector, normalise a vector, and multiply two vectors using the scalar and vector product. In Chap. 6, we explore the ideas of multivectors, which build upon the contents of this chapter.

3.15.1 Summary of Vector Operations

Vector

$$\mathbf{v} = [x \quad y \quad z]^{\mathrm{T}}$$
$$\mathbf{v} = x\mathbf{i} + y\mathbf{j} + z\mathbf{k}.$$

Addition and subtraction

$$\mathbf{v}_1 = x_1\mathbf{i} + y_1\mathbf{j} + z_1\mathbf{k}$$
$$\mathbf{v}_2 = x_2\mathbf{i} + y_2\mathbf{j} + z_2\mathbf{k}$$
$$\mathbf{v}_1 \pm \mathbf{v}_2 = (x_1 \pm x_2)\mathbf{i} + (y_1 \pm y_2)\mathbf{j} + (z_1 \pm z_2)\mathbf{k}.$$

Scalar product

$$\lambda \mathbf{v} = \lambda x \mathbf{i} + \lambda y \mathbf{j} + \lambda z \mathbf{k}.$$

Magnitude

$$|\mathbf{v}| = \sqrt{x^2 + y^2 + z^2}.$$

Unit vector

$$|\mathbf{v}| = 1.$$

Scalar product

$$\mathbf{v}_1 \cdot \mathbf{v}_2 = |\mathbf{v}_1||\mathbf{v}_2| \cos \beta$$
$$\mathbf{v}_1 \cdot \mathbf{v}_2 = x_1 x_2 + y_1 y_2 + z_1 z_2.$$

Vector product

$$\mathbf{v}_1 \times \mathbf{v}_2 = \mathbf{t}$$
$$|\mathbf{t}| = |\mathbf{v}_1||\mathbf{v}_2| \sin \beta$$
$$\mathbf{v}_1 \times \mathbf{v}_2 = (y_1 z_2 - z_1 y_2)\,\mathbf{i} + (z_1 x_2 - x_1 z_2)\,\mathbf{j} + (x_1 y_2 - y_1 x_2)\,\mathbf{k}.$$

Interpolating vectors

$$\mathbf{v} = \frac{\sin(1-t)\theta}{\sin \theta}\mathbf{v}_1 + \frac{\sin t\theta}{\sin \theta}\mathbf{v}_2 \quad [0 < t < 1].$$

Chapter 4
Matrices

4.1 Introduction

Matrix notation was investigated by the British mathematician, Arthur Cayley (1821–1895), in 1858, fifteen years after the invention of quaternions. Cayley and others had realised that it was possible to express a collection of equations by separating constants and variables. For example, the following simultaneous equations

$$2x + 3y = 18 \tag{4.1}$$

$$4x - y = 8 \tag{4.2}$$

have a solution $x = 3$ and $y = 4$, which can be discovered by eliminating one variable, such as x, and computing y, which in turn can be substituted into one of the equations to reveal the value of x. However, matrix notation allows us to express the equations as follows

$$\begin{bmatrix} 2 & 3 \\ 4 & -1 \end{bmatrix} \begin{bmatrix} x \\ y \end{bmatrix} = \begin{bmatrix} 18 \\ 8 \end{bmatrix} \tag{4.3}$$

where the array of four numbers is a *matrix* and the other two columns are *vectors*. When multiplying the matrix and the vector $[x \quad y]^{\mathrm{T}}$ together we must multiply the respective terms of the top row of the matrix with the column vector to equal 18 and create (4.1). Similarly, we must multiply the respective terms of the bottom row of the matrix with the column vector to equal 8 and create (4.2).

Matrix notation also allows us to express these equations as

$$\mathbf{Av} = \mathbf{c} \tag{4.4}$$

where

$$\mathbf{A} = \begin{bmatrix} 2 & 3 \\ 4 & -1 \end{bmatrix}, \quad \mathbf{v} = \begin{bmatrix} x \\ y \end{bmatrix}, \quad \mathbf{c} = \begin{bmatrix} 18 \\ 8 \end{bmatrix}.$$

There happens to be a special matrix such that when it multiplies a vector it results in no change – this matrix is called an *identity matrix* and has the form

$$\mathbf{I} = \begin{bmatrix} 1 & 0 \\ 0 & 1 \end{bmatrix}.$$

J. Vince, *Rotation Transforms for Computer Graphics*,
DOI 10.1007/978-0-85729-154-7_4, © Springer-Verlag London Limited 2011

It is also possible to compute another matrix $\mathbf{A}^{-1}$, called the *inverse* of $\mathbf{A}$, such that $\mathbf{A}^{-1}\mathbf{A} = \mathbf{I}$. Therefore, if we assume that $\mathbf{A}$ is still

$$\mathbf{A} = \begin{bmatrix} 2 & 3 \\ 4 & -1 \end{bmatrix}$$

and is invertible, we can compute $\mathbf{A}^{-1}$ and use it to multiply both sides of (4.4), we have

$$\begin{aligned} \mathbf{A}^{-1}\mathbf{A}\mathbf{v} &= \mathbf{A}^{-1}\mathbf{c} \\ \mathbf{I}\mathbf{v} &= \mathbf{A}^{-1}\mathbf{c} \\ \mathbf{v} &= \mathbf{A}^{-1}\mathbf{c} \end{aligned}$$

which reveals the components of the vector $\mathbf{v}$, and the solution to the equations. Without showing its derivation, the inverse of $\mathbf{A}$ is

$$\mathbf{A}^{-1} = \frac{1}{14}\begin{bmatrix} 1 & 3 \\ 4 & -2 \end{bmatrix} \tag{4.5}$$

and when we multiply $\mathbf{c}$ by $\mathbf{A}^{-1}$ we get

$$\mathbf{A}^{-1}\mathbf{c} = \frac{1}{14}\begin{bmatrix} 1 & 3 \\ 4 & -2 \end{bmatrix}\begin{bmatrix} 18 \\ 8 \end{bmatrix} = \begin{bmatrix} 3 \\ 4 \end{bmatrix}$$

which is the desired result.

Matrices can also be regarded as rectangular arrays of numbers, which may possess various properties. For instance, we can imagine a matrix where all the elements have the same value. We could also create a matrix where all the elements are zero, apart from the diagonal elements. There are many more such configurations, which are explored in this chapter.

If this is the first time you have met matrices, then the author's book *Mathematics for Computer Graphics* [4] will provide you with the necessary background. So for the moment, let's continue and discover more about matrices.

4.2 Matrices

Let's begin by defining a matrix as a rectangular array of numbers with *row* rows and *col* columns, where any element in the matrix is addressed by $a_{row,col}$. The matrix of numbers can be represented in shorthand as

$$\mathbf{A} = [a_{row,col}]$$

where *row* and *col* are natural numbers. However, matrices representing 2D and 3D rotations are all square, where the number of rows equals the number of columns. The following are all examples of square matrices:

$$\begin{bmatrix} 1 & 2 \\ 3 & 4 \end{bmatrix}, \quad \begin{bmatrix} 1 & 2 & 3 \\ 4 & 5 & 6 \\ 7 & 8 & 9 \end{bmatrix}, \quad \begin{bmatrix} 1 & 2 & 3 & 4 \\ 5 & 6 & 7 & 8 \\ 9 & 8 & 7 & 6 \\ 5 & 4 & 3 & 2 \end{bmatrix}.$$

We will discover in later chapters that a 4×4 matrix is the largest matrix we will require to represent a 3D rotation. Now let's look at some of the ways we manipulate matrices.

4.3 The Transpose of a Matrix

One useful matrix operation is the *transpose* where every element $a_{row,col}$ is exchanged with its transpose $a_{col,row}$, and is written

$$\mathbf{A}^{\mathrm{T}} = [a_{row,col}]^{\mathrm{T}} = [a_{col,row}].$$

For example, here is a matrix $\mathbf{A}$ and its transpose $\mathbf{A}^{\mathrm{T}}$

$$\mathbf{A} = \begin{bmatrix} 1 & 2 \\ 3 & 4 \end{bmatrix}, \quad \mathbf{A}^{\mathrm{T}} = \begin{bmatrix} 1 & 3 \\ 2 & 4 \end{bmatrix}.$$

It is possible that the elements of $\mathbf{A}$ are such that $\mathbf{A} = \mathbf{A}^{\mathrm{T}}$. Such a matrix is called a *symmetric* matrix, and we will examine this later.

4.4 The Identity Matrix

As mentioned above, the *identity matrix* $\mathbf{I}$ is a matrix such that

$$\mathbf{IA} = \mathbf{AI} = \mathbf{A}.$$

The three identity matrices we will encounter in later chapters are

$$\begin{bmatrix} 1 & 0 \\ 0 & 1 \end{bmatrix}, \quad \begin{bmatrix} 1 & 0 & 0 \\ 0 & 1 & 0 \\ 0 & 0 & 1 \end{bmatrix}, \quad \begin{bmatrix} 1 & 0 & 0 & 0 \\ 0 & 1 & 0 & 0 \\ 0 & 0 & 1 & 0 \\ 0 & 0 & 0 & 1 \end{bmatrix}$$

and it should be obvious that $\mathbf{I}^{\mathrm{T}} = \mathbf{I}$.

4.5 Adding and Subtracting Matrices

It is possible to add and subtract matrices so long as they have the same number of rows and columns. For example, in matrix notation

$$\mathbf{A} \pm \mathbf{B} = [a_{row,col} \pm b_{row,col}].$$

For example:

$$\mathbf{A} = \begin{bmatrix} 5 & 6 \\ 7 & 8 \end{bmatrix}, \quad \mathbf{B} = \begin{bmatrix} 1 & 2 \\ 3 & 4 \end{bmatrix}$$

then

$$\mathbf{A} + \mathbf{B} = \begin{bmatrix} 6 & 8 \\ 10 & 12 \end{bmatrix}, \quad \mathbf{A} - \mathbf{B} = \begin{bmatrix} 4 & 4 \\ 4 & 4 \end{bmatrix}.$$

4.6 Multiplying a Matrix by a Scalar

Multiplying a matrix by a scalar λ is the same as multiplying an equation by the same scalar. Therefore,

$$\pm\lambda\mathbf{A} = [\pm\lambda a_{row,col}].$$

For example, if $\lambda = 2$

$$\mathbf{A} = \begin{bmatrix} 1 & 2 \\ 3 & 4 \end{bmatrix}, \quad \lambda\mathbf{A} = \begin{bmatrix} 2 & 4 \\ 6 & 8 \end{bmatrix}.$$

4.7 Product of Two Matrices

As already mentioned, every element in a matrix has a unique address specified by its row and column: $a_{row,col}$ where a comma separates the values of *row* and *col*. However, these commas can make the notation very fussy and are not always employed. For example, a_{11} represents the element for $row = 1$ and $col = 1$, and a_{23} represents the element for $row = 2$ and $col = 3$. In this book, we never need to manipulate matrices with more that 4 rows or columns, therefore, there is no confusion.

Matrices have their origins in algebra, therefore matrix algebra must agree with its algebraic counterpart. Bearing this in mind, let's investigate the product of two matrices:

$$\mathbf{A} = \begin{bmatrix} a_{11} & a_{12} \\ a_{21} & a_{22} \end{bmatrix}, \quad \mathbf{B} = \begin{bmatrix} b_{11} & b_{12} \\ b_{21} & b_{22} \end{bmatrix}$$

then their product is given by

$$\mathbf{AB} = \begin{bmatrix} a_{11}b_{11} + a_{12}b_{21} & a_{11}b_{12} + a_{12}b_{22} \\ a_{21}b_{11} + a_{22}b_{21} & a_{21}b_{12} + a_{22}b_{22} \end{bmatrix}.$$

For example, given

$$\mathbf{A} = \begin{bmatrix} 5 & 6 \\ 7 & 8 \end{bmatrix}, \quad \mathbf{B} = \begin{bmatrix} 1 & 2 \\ 3 & 4 \end{bmatrix}$$

then

$$\begin{aligned} \mathbf{AB} &= \begin{bmatrix} 5\times 1 + 6\times 3 & 5\times 2 + 6\times 4 \\ 7\times 1 + 8\times 3 & 7\times 2 + 8\times 4 \end{bmatrix} \\ &= \begin{bmatrix} 5+18 & 10+24 \\ 7+24 & 14+32 \end{bmatrix} \\ &= \begin{bmatrix} 23 & 34 \\ 31 & 46 \end{bmatrix}. \end{aligned}$$

Now let's compute **BA**

$$\mathbf{BA} = \begin{bmatrix} 1\times5+2\times7 & 1\times6+2\times8 \\ 3\times5+4\times7 & 3\times6+4\times8 \end{bmatrix}$$
$$= \begin{bmatrix} 19 & 22 \\ 43 & 50 \end{bmatrix}$$

which confirms that, in general, matrix multiplication is non-commutative.

It can also be shown that

$$[\mathbf{AB}]^{\mathrm{T}} = \mathbf{B}^{\mathrm{T}}\mathbf{A}^{\mathrm{T}}$$

and

$$[\mathbf{A}+\mathbf{B}]^{\mathrm{T}} = \mathbf{A}^{\mathrm{T}} + \mathbf{B}^{\mathrm{T}}.$$

4.8 The Inverse Matrix

Most mathematics software systems include a matrix inversion package, but for completeness, let's demonstrate how we arrived at the inverse matrix $\mathbf{A}^{-1}$ (4.5):

$$\mathbf{A}^{-1} = \frac{1}{14}\begin{bmatrix} 1 & 3 \\ 4 & -2 \end{bmatrix}.$$

Initially, we started with the simultaneous equations

$$2x + 3y = 18 \tag{4.6}$$
$$4x - y = 8. \tag{4.7}$$

Solving these algebraically we multiply (4.7) by 3 and add it to (4.6) to eliminate y:

$$\begin{aligned} 2x + 3y &= 18 \\ 12x - 3y &= 24 \\ 14x &= 42 \\ x &= 3. \end{aligned}$$

Substituting $x = 3$ in (4.6) gives

$$6 + 3y = 18$$

which makes $y = 4$.

4.8.1 Calculation of Inverse

Now let's solve (4.6) and (4.7) using matrix notation. We begin with

$$\mathbf{Av} = \mathbf{c}$$

where

$$\mathbf{A} = \begin{bmatrix} 2 & 3 \\ 4 & -1 \end{bmatrix}, \quad \mathbf{v} = \begin{bmatrix} x \\ y \end{bmatrix}, \quad \mathbf{c} = \begin{bmatrix} 18 \\ 8 \end{bmatrix}.$$

Next we introduce an identity matrix, which does not disturb anything:

$$\mathbf{Av} = \mathbf{Ic} \tag{4.8}$$

$$\begin{bmatrix} 2 & 3 \\ 4 & -1 \end{bmatrix} \begin{bmatrix} x \\ y \end{bmatrix} = \begin{bmatrix} 1 & 0 \\ 0 & 1 \end{bmatrix} \begin{bmatrix} 18 \\ 8 \end{bmatrix}. \tag{4.9}$$

The objective is to multiply both sides of (4.8) by $\mathbf{A}^{-1}$ and turn the LHS matrix $\mathbf{A}$ into an identity matrix, and at the same time turn the RHS matrix $\mathbf{I}$ into $\mathbf{A}^{-1}$. But as we don't know $\mathbf{A}^{-1}$ we will have to do this in a number of steps. Like the above simultaneous equations, we can scale, add, subtract or divide matrix rows, so long as we manipulate the entire matrix equation.

We start by subtracting $2 \times row(1)$ from $row(2)$ in (4.9):

$$\begin{bmatrix} 2 & 3 \\ 0 & -7 \end{bmatrix} \begin{bmatrix} x \\ y \end{bmatrix} = \begin{bmatrix} 1 & 0 \\ -2 & 1 \end{bmatrix} \begin{bmatrix} 18 \\ 8 \end{bmatrix}. \tag{4.10}$$

Next, multiply $row(1) \times \frac{7}{3}$ in (4.10):

$$\begin{bmatrix} \frac{14}{3} & 7 \\ 0 & -7 \end{bmatrix} \begin{bmatrix} x \\ y \end{bmatrix} = \begin{bmatrix} \frac{7}{3} & 0 \\ -2 & 1 \end{bmatrix} \begin{bmatrix} 18 \\ 8 \end{bmatrix}. \tag{4.11}$$

Next, add $row(2)$ to $row(1)$ in (4.11):

$$\begin{bmatrix} \frac{14}{3} & 0 \\ 0 & -7 \end{bmatrix} \begin{bmatrix} x \\ y \end{bmatrix} = \begin{bmatrix} \frac{1}{3} & 1 \\ -2 & 1 \end{bmatrix} \begin{bmatrix} 18 \\ 8 \end{bmatrix}. \tag{4.12}$$

Next, multiply $row(1) \times \frac{3}{14}$ in (4.12):

$$\begin{bmatrix} 1 & 0 \\ 0 & -7 \end{bmatrix} \begin{bmatrix} x \\ y \end{bmatrix} = \begin{bmatrix} \frac{1}{14} & \frac{3}{14} \\ -2 & 1 \end{bmatrix} \begin{bmatrix} 18 \\ 8 \end{bmatrix}. \tag{4.13}$$

Finally, divide $row(2)$ by -7 in (4.13):

$$\begin{bmatrix} 1 & 0 \\ 0 & 1 \end{bmatrix} \begin{bmatrix} x \\ y \end{bmatrix} = \begin{bmatrix} \frac{1}{14} & \frac{3}{14} \\ \frac{2}{7} & -\frac{1}{7} \end{bmatrix} \begin{bmatrix} 18 \\ 8 \end{bmatrix}. \tag{4.14}$$

As the LHS matrix is an identity matrix, the RHS matrix in (4.14) must be $\mathbf{A}^{-1}$ and is tidied up to become

$$\mathbf{A}^{-1} = \frac{1}{14} \begin{bmatrix} 1 & 3 \\ 4 & -2 \end{bmatrix}.$$

Later on, we will explore another technique that does not involve any overt algebraic skills.

4.9 Determinant of a Matrix

When solving a pair of simultaneous equations such as

$$ax + by = r$$
$$cx + dy = s$$

the expression $ad - bc$ arises in the solution. For example, in the simultaneous equations (4.6) and (4.7) the corresponding expression has a value $2 \times (-1) - 3 \times 4 = -14$ whose magnitude appears in the solution of $\mathbf{A}^{-1}$. Because this expression is so useful, it is identified by the name *determinant* and is written

$$\det \mathbf{A} = |\mathbf{A}| = ad - bc$$

where

$$\mathbf{A} = \begin{bmatrix} a & b \\ c & d \end{bmatrix}.$$

Determinants are formed from square arrays, in that they have the same number of rows and columns, which permits us to classify them in terms of their *order*. Some texts classify a scalar quantity as a *first-order determinant* – for example a. A *second-order determinant* has two rows and columns – for example

$$\begin{vmatrix} a & b \\ c & d \end{vmatrix}.$$

When dealing with three simultaneous equations

$$ax + by + cz = r$$
$$dx + ey + fz = s$$
$$gx + hy + iz = t$$

the corresponding matrix is

$$\mathbf{A} = \begin{bmatrix} a & b & c \\ d & e & f \\ g & h & i \end{bmatrix}$$

and the equivalent determinant is

$$aei + bfg + cdh - ceg - afh - bdi.$$

4.9.1 Sarrus's Rule

The French mathematician, J.P. Sarrus (1789–1861), noted that a *third-order determinant* is easily computed by exploiting a pattern which is very obvious if the determinant's columns are extended as follows:

$$\begin{vmatrix} a & b & c & a & b \\ d & e & f & d & e \\ g & h & i & g & h \end{vmatrix}.$$

aei, bfg and cdh are strings of elements sloping downwards to the right, whereas, ceg, afh and bdi are strings of elements sloping downwards to the left. For example, the determinant

$$\det \mathbf{A} = \begin{vmatrix} 2 & 0 & 4 \\ 3 & 1 & 0 \\ 4 & 2 & 2 \end{vmatrix}$$

has a value of

$$\begin{aligned} |\mathbf{A}| &= (2 \times 1 \times 2) + (0 \times 0 \times 4) + (4 \times 3 \times 2) \\ &\quad - (4 \times 1 \times 4) - (2 \times 0 \times 2) - (0 \times 3 \times 2) \\ &= 4 + 0 + 24 - 16 - 0 - 0 \\ |\mathbf{A}| &= 12. \end{aligned}$$

In general

$$\det \mathbf{A} = |\mathbf{A}| = aei + bfg + cdh - ceg - afh - bdi \tag{4.15}$$

and the expansion is known as *Sarrus's rule.*

Equation (4.15) can be rearranged to reveal another pattern:

$$|\mathbf{A}| = a(ei - fh) + b(fg - di) + c(dh - eg)$$

where the expressions

$$(ei - fh), \quad (fg - di), \quad (dh - eg)$$

are regarded as *minor determinants*. Let's pause at this juncture and examine $|\mathbf{A}|$'s minor determinants:

$$\begin{vmatrix} a & b & c \\ d & e & f \\ g & h & i \end{vmatrix} = \begin{vmatrix} a & & \\ & e & f \\ & h & i \end{vmatrix} - \begin{vmatrix} & b & \\ d & & f \\ g & & i \end{vmatrix} + \begin{vmatrix} & & c \\ d & e & \\ g & h & \end{vmatrix}. \tag{4.16}$$

Equation (4.16) shows how a third-order determinant can be represented as the sum of three minor determinants:

$$\begin{vmatrix} a & b & c \\ d & e & f \\ g & h & i \end{vmatrix} = a\begin{vmatrix} e & f \\ h & i \end{vmatrix} - b\begin{vmatrix} d & f \\ g & i \end{vmatrix} + c\begin{vmatrix} d & e \\ g & h \end{vmatrix}.$$

Note that the middle minor determinant is prefixed with a negative sign. This ensures that the equation's value is the same as (4.15). The reason for this is that mathematics is not interested in forming convenient visual patterns on the page – its patterns are created from cycles of elements or subscripts. Consequently, if we prefer a visual pattern as an *aide-mémoire*, we must make the necessary algebraic adjustments to maintain the equation's integrity. For example, by repeating the first column of $|\mathbf{A}|$ as follows

$$\begin{vmatrix} a & b & c & a \\ d & e & f & d \\ g & h & i & g \end{vmatrix}$$

the pattern $(fg - di)$ is now very obvious. However, it is widely agreed that it is best to ignore this and embrace a simple visual pattern. The price to be paid for this is a negative sign as follows:

$$|\mathbf{A}| = a(ei - fh) - b(di - fg) + c(dh - eg).$$

4.9.2 The Laplace Expansion

The French mathematician, Pierre Simon de Laplace (1749–1827), developed a method of expanding a determinant in terms of its minors, which, with the associated change of sign, is called a *cofactor*. The cofactor $c_{row,col}$ of an element $a_{row,col}$ is the minor that remains after removing from the original determinant the *row* row and the *col* column.

For example, in (4.17) the minor of a_{11} is identified by removing the first row and the first column; the minor of a_{12} is identified by removing the first row and the second column; and the minor of a_{13} is identified by removing the first row and the third column:

$$\det\mathbf{A} = \begin{vmatrix} a_{11} & a_{12} & a_{13} \\ a_{21} & a_{22} & a_{23} \\ a_{31} & a_{32} & a_{33} \end{vmatrix}. \tag{4.17}$$

The three minor determinants for a_{11}, a_{12} and a_{13} are respectively:

$$A_{11} = \begin{vmatrix} a_{22} & a_{23} \\ a_{32} & a_{33} \end{vmatrix}, \qquad A_{12} = \begin{vmatrix} a_{21} & a_{23} \\ a_{31} & a_{33} \end{vmatrix}, \qquad A_{13} = \begin{vmatrix} a_{21} & a_{22} \\ a_{31} & a_{32} \end{vmatrix}$$

whereas, the three cofactors are

$$\begin{aligned} c_{11} &= +a_{11}A_{11} \\ c_{12} &= -a_{12}A_{12} \\ c_{13} &= +a_{13}A_{13}. \end{aligned}$$

In general, the minor of $a_{row,col}$ is denoted $A_{row,col}$.

Laplace proposed the following formulae for selecting the cofactor sign:

$$(-1)^{row+col}$$

which generates the pattern

$$\begin{vmatrix} + & - & + & \dots \\ - & + & - & \dots \\ + & - & + & \dots \\ \dots & \dots & \dots & \dots \end{vmatrix}.$$

Although we have chosen the first row to expand the above determinants, any row, or column may be used.

The above Laplace expansion can be used for any order determinant, and for the purposes of this book, the highest order we will encounter is a *fourth-order determinant*. So let's expand the following determinant

$$\det \mathbf{A} = \begin{vmatrix} 0 & 4 & -1 & 0 \\ 1 & 0 & 1 & 4 \\ 0 & 6 & 3 & 3 \\ 3 & 5 & 2 & 2 \end{vmatrix}.$$

The first column and the first row both contain two zeros, which helps our expansion, so let's expand $|\mathbf{A}|$ using the first row. The two relevant minor determinants are

$$A_{12} = \begin{vmatrix} 1 & 1 & 4 \\ 0 & 3 & 3 \\ 3 & 2 & 2 \end{vmatrix}, \qquad A_{13} = \begin{vmatrix} 1 & 0 & 4 \\ 0 & 6 & 3 \\ 3 & 5 & 2 \end{vmatrix}.$$

Next, we expand A_{12} and A_{13} with their cofactors as follows

$$\begin{aligned} c_{12} &= -4\left[1(-1)^{1+1}\begin{vmatrix} 3 & 3 \\ 2 & 2 \end{vmatrix} + 1(-1)^{1+2}\begin{vmatrix} 0 & 3 \\ 3 & 2 \end{vmatrix} + 4(-1)^{1+3}\begin{vmatrix} 0 & 3 \\ 3 & 2 \end{vmatrix}\right] \\ &= -4[0+9-36] \\ &= 108 \\ c_{13} &= -1\left[1(-1)^{1+1}\begin{vmatrix} 6 & 3 \\ 5 & 2 \end{vmatrix} + 4(-1)^{1+3}\begin{vmatrix} 0 & 6 \\ 3 & 5 \end{vmatrix}\right] \\ &= -1[-3-72] \\ &= 75. \end{aligned}$$

Therefore,

$$\det \mathbf{A} = c_{12} + c_{13} = 108 + 75 = 183.$$

4.10 Cofactors and Inverse of a Matrix

Although the idea of cofactors has been described in the context of determinants, they can also be applied to matrices. For example, let's start with the following matrix and its cofactor matrix

$$\mathbf{A} = \begin{bmatrix} 0 & 1 & 3 \\ 2 & 1 & 4 \\ 4 & 2 & 6 \end{bmatrix}$$

$$\text{cofactor matrix of } \mathbf{A} = \begin{bmatrix} A_{11} & A_{12} & A_{13} \\ A_{21} & A_{22} & A_{23} \\ A_{31} & A_{32} & A_{33} \end{bmatrix}$$

where

$$A_{11} = +\begin{vmatrix} a_{22} & a_{23} \\ a_{32} & a_{33} \end{vmatrix} = +\begin{vmatrix} 1 & 4 \\ 2 & 6 \end{vmatrix} = -2$$

$$A_{12} = -\begin{vmatrix} a_{21} & a_{23} \\ a_{31} & a_{33} \end{vmatrix} = -\begin{vmatrix} 2 & 4 \\ 4 & 6 \end{vmatrix} = 4$$

$$A_{13} = +\begin{vmatrix} a_{21} & a_{23} \\ a_{31} & a_{33} \end{vmatrix} = +\begin{vmatrix} 2 & 1 \\ 4 & 2 \end{vmatrix} = 0$$

$$A_{21} = -\begin{vmatrix} a_{22} & a_{23} \\ a_{32} & a_{33} \end{vmatrix} = -\begin{vmatrix} 1 & 3 \\ 2 & 6 \end{vmatrix} = 0$$

$$A_{22} = +\begin{vmatrix} a_{11} & a_{13} \\ a_{31} & a_{33} \end{vmatrix} = +\begin{vmatrix} 0 & 3 \\ 4 & 6 \end{vmatrix} = -12$$

$$A_{23} = -\begin{vmatrix} a_{11} & a_{12} \\ a_{31} & a_{32} \end{vmatrix} = -\begin{vmatrix} 0 & 1 \\ 4 & 2 \end{vmatrix} = 4$$

$$A_{31} = +\begin{vmatrix} a_{12} & a_{13} \\ a_{22} & a_{23} \end{vmatrix} = +\begin{vmatrix} 1 & 3 \\ 1 & 4 \end{vmatrix} = 1$$

$$A_{32} = -\begin{vmatrix} a_{11} & a_{13} \\ a_{21} & a_{23} \end{vmatrix} = -\begin{vmatrix} 0 & 3 \\ 2 & 4 \end{vmatrix} = 6$$

$$A_{33} = +\begin{vmatrix} a_{11} & a_{12} \\ a_{21} & a_{22} \end{vmatrix} = +\begin{vmatrix} 0 & 1 \\ 2 & 1 \end{vmatrix} = -2$$

therefore, the cofactor matrix of **A** is

$$\text{cofactor matrix of } \mathbf{A} = \begin{bmatrix} -2 & 4 & 0 \\ 0 & -12 & 4 \\ 1 & 6 & -2 \end{bmatrix}.$$

It can be shown that the product of a matrix with the transpose of its cofactor matrix has the following form:

$$\mathbf{A}\,(\text{cofactor matrix of } \mathbf{A})^{\mathrm{T}} = \begin{bmatrix} \det\mathbf{A} & 0 & \dots & 0 \\ 0 & \det\mathbf{A} & \dots & 0 \\ \dots & \dots & \dots & \dots \\ 0 & 0 & 0 & \det\mathbf{A} \end{bmatrix}$$

and multiplying throughout by $1/\det\mathbf{A}$ we have

$$(1/\det\mathbf{A})\,\mathbf{A}\,(\text{cofactor matrix of } \mathbf{A})^{\mathrm{T}} = \mathbf{I}$$

which implies that

$$\mathbf{A}^{-1} = \frac{(\text{cofactor matrix of } \mathbf{A})^{\mathrm{T}}}{\det\mathbf{A}}.$$

Naturally, this assumes that the inverse actually exists, and it will if $\det\mathbf{A} \neq 0$.

Let's find the inverse of the above matrix

$$\mathbf{A} = \begin{bmatrix} 0 & 1 & 3 \\ 2 & 1 & 4 \\ 4 & 2 & 6 \end{bmatrix}$$

$$(\text{cofactor matrix of } \mathbf{A}) = \begin{bmatrix} -2 & 4 & 0 \\ 0 & -12 & 4 \\ 1 & 6 & -2 \end{bmatrix}$$

$$(\text{cofactor matrix of } \mathbf{A})^{\mathrm{T}} = \begin{bmatrix} -2 & 0 & 1 \\ 4 & -12 & 6 \\ 0 & 4 & -2 \end{bmatrix}$$

$$\det \mathbf{A} = 1 \times 4 \times 4 + 3 \times 2 \times 2 - 1 \times 2 \times 6 - 3 \times 1 \times 4 = 4$$

$$\mathbf{A}^{-1} = \frac{1}{4} \begin{bmatrix} -2 & 0 & 1 \\ 4 & -12 & 6 \\ 0 & 4 & -2 \end{bmatrix}.$$

Let's check this result by multiplying $\mathbf{A}$ by $\mathbf{A}^{-1}$ which must equal $\mathbf{I}$:

$$\begin{aligned} \mathbf{A}\mathbf{A}^{-1} &= \begin{bmatrix} 0 & 1 & 3 \\ 2 & 1 & 4 \\ 4 & 2 & 6 \end{bmatrix} \frac{1}{4} \begin{bmatrix} -2 & 0 & 1 \\ 4 & -12 & 6 \\ 0 & 4 & -2 \end{bmatrix} \\ &= \frac{1}{4} \begin{bmatrix} 4 & 0 & 0 \\ 0 & 4 & 0 \\ 0 & 0 & 4 \end{bmatrix} \\ &= \begin{bmatrix} 1 & 0 & 0 \\ 0 & 1 & 0 \\ 0 & 0 & 1 \end{bmatrix}. \end{aligned}$$

Finally, let's compute the inverse matrix for (4.6) and (4.7) using cofactors:

$$\mathbf{A} = \begin{bmatrix} 2 & 3 \\ 4 & -1 \end{bmatrix}$$

$$(\text{cofactor matrix of } \mathbf{A}) = \begin{bmatrix} -1 & -4 \\ -3 & 2 \end{bmatrix}$$

$$(\text{cofactor matrix of } \mathbf{A})^{\mathrm{T}} = \begin{bmatrix} -1 & -3 \\ -4 & 2 \end{bmatrix}$$

$$\det \mathbf{A} = 2 \times (-1) - 3 \times 4 = -14$$

$$\mathbf{A}^{-1} = \frac{1}{14} \begin{bmatrix} 1 & 3 \\ 4 & -2 \end{bmatrix}$$

which confirms the original result.

In general, the inverse of a 2×2 matrix is given by

$$\mathbf{A} = \begin{bmatrix} a_{11} & a_{12} \\ a_{21} & a_{22} \end{bmatrix}$$

$$\mathbf{A}^{-1} = \frac{1}{a_{11}a_{22} - a_{12}a_{21}} \begin{bmatrix} a_{22} & -a_{12} \\ -a_{21} & a_{11} \end{bmatrix}$$

which, for the above matrix is

$$\mathbf{A}^{-1} = \frac{-1}{14} \begin{bmatrix} -1 & -3 \\ -4 & 2 \end{bmatrix} = \frac{1}{14} \begin{bmatrix} 1 & 3 \\ 4 & -2 \end{bmatrix}.$$

4.11 Orthogonal Matrix

Although many matrices have to be inverted using the transpose of their cofactor matrix, an *orthogonal matrix* implies that its transpose is also its inverse. For example:

$$\mathbf{A} = \begin{bmatrix} \cos\beta & -\sin\beta \\ \sin\beta & \cos\beta \end{bmatrix}$$

is orthogonal because

$$\mathbf{A}^{\mathrm{T}} = \begin{bmatrix} \cos\beta & \sin\beta \\ -\sin\beta & \cos\beta \end{bmatrix}$$

and

$$\mathbf{A}\mathbf{A}^{\mathrm{T}} = \begin{bmatrix} \cos\beta & -\sin\beta \\ \sin\beta & \cos\beta \end{bmatrix} \begin{bmatrix} \cos\beta & \sin\beta \\ -\sin\beta & \cos\beta \end{bmatrix} = \begin{bmatrix} 1 & 0 \\ 0 & 1 \end{bmatrix}.$$

Orthogonal matrices play an important role in rotations because they leave the origin fixed and preserve all angles and distances. Consequently, an object's geometric integrity is maintained after a rotation, which is why an orthogonal transform is known as a *rigid motion* transform.

A rotation transform also preserves orientations, which means that left-handed and right-handed axial systems (frames) remain unaltered after a rotation. Such changes in orientation will occur with a reflection transform.

4.12 Diagonal Matrix

A *diagonal matrix* is a square matrix whose elements are zero, apart from its diagonal:

$$\mathbf{A} = \begin{bmatrix} a_{11} & 0 & \dots & 0 \\ 0 & a_{22} & \dots & 0 \\ \vdots & \vdots & \ddots & \vdots \\ 0 & 0 & \dots & a_{nn} \end{bmatrix}.$$

The determinant of a diagonal matrix must be

$$\det \mathbf{A} = a_{11} \times a_{22} \times \cdots \times a_{nn}.$$

Here is a diagonal matrix with its determinant

$$\mathbf{A} = \begin{bmatrix} 2 & 0 & 0 \\ 0 & 3 & 0 \\ 0 & 0 & 4 \end{bmatrix}$$
$$|\mathbf{A}| = 2 \times 3 \times 4 = 24.$$

The identity matrix **I** is a diagonal matrix with a determinant of 1.

4.13 Trace

The *trace* of a square matrix **A** is the sum of its diagonal elements and written as Tr (**A**). For example:

$$\mathbf{A} = \begin{bmatrix} 1 & 2 & 3 & 4 \\ 2 & 3 & 4 & 5 \\ 3 & 4 & 5 & 6 \\ 4 & 5 & 6 & 7 \end{bmatrix}$$
$$\mathrm{Tr}\,(\mathbf{A}) = 1 + 3 + 5 + 7 = 16.$$

In Chap. 9 we use the trace of a square matrix to reveal the angle of rotation associated with a rotation matrix. And as we will be using the product of two or more rotation transforms we require to establish that

$$\mathrm{Tr}\,(\mathbf{AB}) = \mathrm{Tr}\,(\mathbf{BA})$$

to reassure ourselves that the trace operation is not sensitive to transform order, and is readily proved as follows.

Given two square matrices **A** and **B**:

$$\mathbf{A} = \begin{bmatrix} a_{11} & \cdots & \cdots & a_{1n} \\ \cdots & a_{22} & \cdots & a_{2n} \\ \cdots & \cdots & \cdots & \cdots \\ a_{n1} & \cdots & \cdots & a_{nn} \end{bmatrix}, \quad \mathbf{B} = \begin{bmatrix} b_{11} & \cdots & \cdots & b_{1n} \\ \cdots & b_{22} & \cdots & b_{2n} \\ \cdots & \cdots & \cdots & \cdots \\ b_{n1} & \cdots & \cdots & b_{nn} \end{bmatrix}$$

then,

$$\mathbf{AB} = \begin{bmatrix} a_{11} & \cdots & \cdots & a_{1n} \\ \cdots & a_{22} & \cdots & a_{2n} \\ \cdots & \cdots & \cdots & \cdots \\ a_{n1} & \cdots & \cdots & a_{nn} \end{bmatrix} \begin{bmatrix} b_{11} & \cdots & \cdots & b_{1n} \\ \cdots & b_{22} & \cdots & b_{2n} \\ \cdots & \cdots & \cdots & \cdots \\ b_{n1} & \cdots & \cdots & b_{nn} \end{bmatrix}$$

$$\mathbf{AB} = \begin{bmatrix} a_{11}b_{11} & \cdots & \cdots & a_{1n} \\ \cdots & a_{22}b_{22} & \cdots & a_{2n} \\ \cdots & \cdots & \cdots & \cdots \\ a_{n1} & \cdots & \cdots & a_{nn}b_{nn} \end{bmatrix}$$

and $\mathrm{Tr}\,(\mathbf{AB}) = a_{11}b_{11} + a_{22}b_{22} + \cdots + a_{nn}b_{nn}$.

Hopefully, it is obvious that reversing the matrix sequence to **BA** only reverses the a and b scalar elements on the diagonal, and therefore does not affect the trace operation.

4.14 Symmetric Matrix

It is worth exploring two types of matrices called *symmetric* and *antisymmetric* matrices, as we refer to them in later chapters. A symmetric matrix is a matrix which equals its own transpose:

$$\mathbf{A} = \mathbf{A}^{\mathrm{T}}.$$

For example, the following matrix is symmetric:

$$\mathbf{A} = \begin{bmatrix} 1 & 3 & 4 \\ 3 & 2 & 4 \\ 4 & 4 & 3 \end{bmatrix}.$$

The symmetric part of any square matrix can be isolated as follows. Given a matrix **A** and its transpose $\mathbf{A}^{\mathrm{T}}$

$$\mathbf{A} = \begin{bmatrix} a_{11} & a_{12} & \dots & a_{1n} \\ a_{21} & a_{22} & \dots & a_{2n} \\ \vdots & \vdots & \ddots & \vdots \\ a_{n1} & a_{n2} & \dots & a_{nn} \end{bmatrix}, \quad \mathbf{A}^{\mathrm{T}} = \begin{bmatrix} a_{11} & a_{21} & \dots & a_{n1} \\ a_{12} & a_{22} & \dots & a_{n2} \\ \vdots & \vdots & \ddots & \vdots \\ a_{1n} & a_{2n} & \dots & a_{nn} \end{bmatrix}$$

their sum is

$$\mathbf{A} + \mathbf{A}^{\mathrm{T}} = \begin{bmatrix} 2a_{11} & a_{12} + a_{21} & \dots & a_{1n} + a_{n1} \\ a_{12} + a_{21} & 2a_{22} & \dots & a_{2n} + a_{n2} \\ \vdots & \vdots & \ddots & \vdots \\ a_{1n} + a_{n1} & a_{2n} + a_{n2} & \dots & 2a_{nn} \end{bmatrix}.$$

By inspection, $\mathbf{A} + \mathbf{A}^{\mathrm{T}}$ is symmetric, and if we divide throughout by 2 we have

$$\mathbf{S} = \frac{1}{2}\left(\mathbf{A} + \mathbf{A}^{\mathrm{T}}\right)$$

which is defined as the symmetric part of **A**. For example, given

$$\mathbf{A} = \begin{bmatrix} a_{11} & a_{12} & a_{13} \\ a_{21} & a_{22} & a_{23} \\ a_{31} & a_{32} & a_{33} \end{bmatrix}, \quad \mathbf{A}^{\mathrm{T}} = \begin{bmatrix} a_{11} & a_{21} & a_{31} \\ a_{12} & a_{22} & a_{32} \\ a_{13} & a_{23} & a_{33} \end{bmatrix}$$

then

$$\begin{aligned} \mathbf{S} &= \frac{1}{2}\left(\mathbf{A} + \mathbf{A}^{\mathrm{T}}\right) \\ &= \begin{bmatrix} a_{11} & (a_{12} + a_{21})/2 & (a_{13} + a_{31})/2 \\ (a_{12} + a_{21})/2 & a_{22} & a_{23} + a_{32} \\ (a_{13} + a_{31})/2 & (a_{23} + a_{32})/2 & a_{33} \end{bmatrix} \end{aligned}$$

$$= \begin{bmatrix} a_{11} & s_3/2 & s_2/2 \\ s_3/2 & a_{22} & s_1/2 \\ s_2/2 & s_1/2 & a_{33} \end{bmatrix}$$

where

$$s_1 = a_{23} + a_{32}$$
$$s_2 = a_{13} + a_{31}$$
$$s_3 = a_{12} + a_{21}.$$

Using a real example:

$$\mathbf{A} = \begin{bmatrix} 0 & 1 & 4 \\ 3 & 1 & 4 \\ 4 & 2 & 6 \end{bmatrix}, \quad \mathbf{A}^{\mathrm{T}} = \begin{bmatrix} 0 & 3 & 4 \\ 1 & 1 & 2 \\ 4 & 4 & 6 \end{bmatrix}$$

$$\mathbf{S} = \begin{bmatrix} 0 & 2 & 4 \\ 2 & 1 & 3 \\ 4 & 3 & 6 \end{bmatrix}$$

which equals its own transpose.

4.15 Antisymmetric Matrix

An *antisymmetric matrix* is a matrix whose transpose is its own negative:

$$\mathbf{A}^{\mathrm{T}} = -\mathbf{A}$$

and is also known as a *skew symmetric matrix.*

As the elements of $\mathbf{A}$ and $\mathbf{A}^{\mathrm{T}}$ are related by

$$a_{row,col} = -a_{col,row}.$$

When $k = row = col$:

$$a_{k,k} = -a_{k,k}$$

which implies that the diagonal elements must be zero. For example, this is an antisymmetric matrix

$$\begin{bmatrix} 0 & 6 & 2 \\ -6 & 0 & -4 \\ -2 & 4 & 0 \end{bmatrix}.$$

In general, we have

$$\mathbf{A} = \begin{bmatrix} a_{11} & a_{12} & \dots & a_{1n} \\ a_{21} & a_{22} & \dots & a_{2n} \\ \vdots & \vdots & \ddots & \vdots \\ a_{n1} & a_{n2} & \dots & a_{nn} \end{bmatrix}, \quad \mathbf{A}^{\mathrm{T}} = \begin{bmatrix} a_{11} & a_{21} & \dots & a_{n1} \\ a_{12} & a_{22} & \dots & a_{n2} \\ \vdots & \vdots & \ddots & \vdots \\ a_{1n} & a_{2n} & \dots & a_{nn} \end{bmatrix}$$

and their difference is

$$\mathbf{A}-\mathbf{A}^{\mathrm{T}}=\begin{bmatrix} 0 & a_{12}-a_{21} & \dots & a_{1n}-a_{n1} \\ -(a_{12}-a_{21}) & 0 & \dots & a_{2n}-a_{n2} \\ \vdots & \vdots & \ddots & \vdots \\ -(a_{1n}-a_{n1}) & -(a_{2n}-a_{n2}) & \dots & 0 \end{bmatrix}.$$

It is clear that $\mathbf{A}-\mathbf{A}^{\mathrm{T}}$ is antisymmetric, and if we divide throughout by 2 we have

$$\mathbf{Q}=\frac{1}{2}\left(\mathbf{A}-\mathbf{A}^{\mathrm{T}}\right).$$

For example:

$$\mathbf{A}=\begin{bmatrix} a_{11} & a_{12} & a_{13} \\ a_{21} & a_{22} & a_{23} \\ a_{31} & a_{32} & a_{33} \end{bmatrix}, \quad \mathbf{A}^{\mathrm{T}}=\begin{bmatrix} a_{11} & a_{21} & a_{31} \\ a_{12} & a_{22} & a_{32} \\ a_{13} & a_{23} & a_{33} \end{bmatrix}$$

$$\mathbf{Q}=\begin{bmatrix} 0 & (a_{12}-a_{21})/2 & (a_{13}-a_{31})/2 \\ (a_{21}-a_{12})/2 & 0 & (a_{23}-a_{32})/2 \\ (a_{31}-a_{13})/2 & (a_{32}-a_{23})/2 & 0 \end{bmatrix}$$

and if we maintain some symmetry with the subscripts, we have

$$\begin{aligned} \mathbf{Q}&=\begin{bmatrix} 0 & (a_{12}-a_{21})/2 & -(a_{31}-a_{13})/2 \\ -(a_{12}-a_{21})/2 & 0 & (a_{23}-a_{32})/2 \\ (a_{31}-a_{13})/2 & -(a_{23}-a_{32})/2 & 0 \end{bmatrix} \\ &=\begin{bmatrix} 0 & q_3/2 & -q_2/2 \\ -q_3/2 & 0 & q_1/2 \\ q_2/2 & -q_1/2 & 0 \end{bmatrix} \end{aligned}$$

where

$$\begin{aligned} q_1&=a_{23}-a_{32} \\ q_2&=a_{31}-a_{13} \\ q_3&=a_{12}-a_{21}. \end{aligned}$$

Using a real example:

$$\mathbf{A}=\begin{bmatrix} 0 & 1 & 4 \\ 3 & 1 & 4 \\ 4 & 2 & 6 \end{bmatrix}, \quad \mathbf{A}^{\mathrm{T}}=\begin{bmatrix} 0 & 3 & 4 \\ 1 & 1 & 2 \\ 4 & 4 & 6 \end{bmatrix}$$

$$\mathbf{Q}=\begin{bmatrix} 0 & -1 & 0 \\ 1 & 0 & 1 \\ 0 & -1 & 0 \end{bmatrix}.$$

Furthermore, we have already computed

$$\mathbf{S}=\begin{bmatrix} 0 & 2 & 4 \\ 2 & 1 & 3 \\ 4 & 3 & 6 \end{bmatrix}$$

and

$$\mathbf{S}+\mathbf{Q}=\begin{bmatrix}0 & 1 & 4\\ 3 & 1 & 4\\ 4 & 2 & 6\end{bmatrix}=\mathbf{A}.$$

4.16 Inverting a Pair of Matrices

In later chapters we form the products of two or more matrices, and in some cases require to find their inverse. In anticipation of this requirement, let's compute the inverse of a pair of matrices.

Given two transforms $\mathbf{T}$ and $\mathbf{R}$, the product $\mathbf{TR}$ and its inverse $(\mathbf{TR})^{-1}$ must equal the identity matrix $\mathbf{I}$:

$$(\mathbf{TR})(\mathbf{TR})^{-1}=\mathbf{I}$$

and multiplying throughout by $\mathbf{T}^{-1}$ we have

$$\begin{aligned}\mathbf{T}^{-1}\mathbf{TR}(\mathbf{TR})^{-1}&=\mathbf{T}^{-1}\\ \mathbf{R}(\mathbf{TR})^{-1}&=\mathbf{T}^{-1}.\end{aligned}$$

Multiplying throughout by $\mathbf{R}^{-1}$ we have

$$\begin{aligned}\mathbf{R}^{-1}\mathbf{R}(\mathbf{TR})^{-1}&=\mathbf{R}^{-1}\mathbf{T}^{-1}\\ (\mathbf{TR})^{-1}&=\mathbf{R}^{-1}\mathbf{T}^{-1}.\end{aligned}$$

Therefore, if $\mathbf{T}$ and $\mathbf{R}$ are invertible, then

$$(\mathbf{TR})^{-1}=\mathbf{R}^{-1}\mathbf{T}^{-1}.$$

Generalising this result to a triple product such as $\mathbf{STR}$ we can reason that

$$(\mathbf{STR})^{-1}=\mathbf{R}^{-1}\mathbf{T}^{-1}\mathbf{S}^{-1}.$$

4.17 Eigenvectors and Eigenvalues

Matrices represent linear transforms that scale, translate, shear, reflect or rotate points, whilst leaving the origin untouched. For example, the following 2D transform

$$\begin{bmatrix}4 & 1\\ 1 & 4\end{bmatrix}\begin{bmatrix}x\\ y\end{bmatrix}=\begin{bmatrix}x'\\ y'\end{bmatrix}$$

transforms the points on four unit squares as shown in Fig. 4.1 where we see a pronounced stretching in the first and third quadrants, and reduced stretching in the second and fourth quadrants.

It should be clear from Fig. 4.1 that any point (k,k) is transformed to another point $(5k,5k)$, and that its mirror point $(-k,-k)$ is transformed to $(-5k,-5k)$.

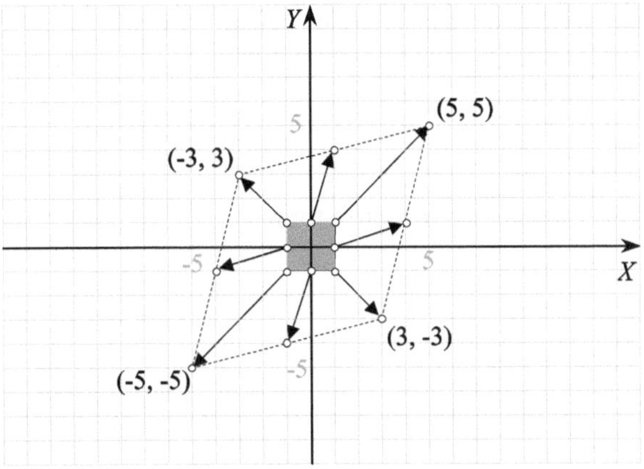

Fig. 4.1 Transforming points on four unit squares

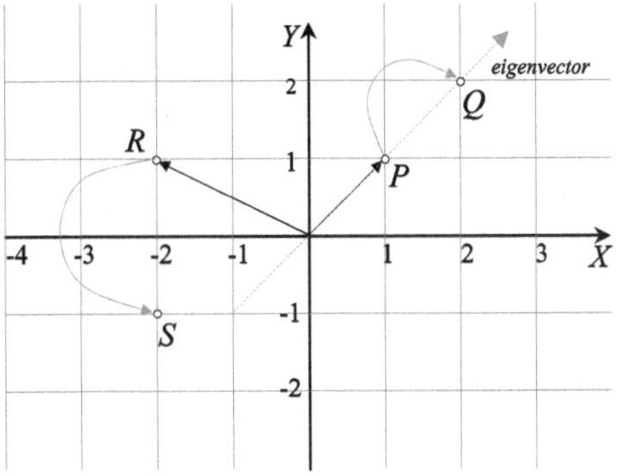

Fig. 4.2 How a transform reacts to different points

Similarly, any point $(-k, k)$ is transformed to another point $(-3k, 3k)$, and its mirror point $(k, -k)$ is transformed to $(3k, -3k)$. Thus the transform shows a particular bias towards points lying on vectors $[k \quad k]^T$ and $[-k \quad k]^T$, where $k \neq 0$.

These vectors are called *eigenvectors* and the scaling factor is its *eigenvalue*. Figure 4.2 shows a scenario where a transform t moves point R to S, whilst the same transform moves P – which lies on one of t's eigenvectors, to Q – which also lies on the same eigenvector.

We can define an eigenvector and its eigenvalue as follows. Given a square matrix $\mathbf{A}$, a non-zero vector $\mathbf{v}$ is an eigenvector, and λ is the corresponding eigenvalue if

$$\mathbf{A}\mathbf{v} = \lambda\mathbf{v}$$

where λ is a scalar.

The German word *eigen* means *characteristic*, *own*, *latent* or *special*, and eigenvector means a special vector associated with a transform. The equation that determines the existence of any eigenvectors is called the *characteristic equation* of a square matrix, and is given by

$$\det(\mathbf{A} - \lambda\mathbf{I}) = 0. \tag{4.18}$$

Let's derive the characteristic equation (4.18).

Consider the 2D transform t that maps the point (x, y) to another point $(ax + by, cx + dy)$:

$$t\,(x, y) \mapsto (ax + by, cx + dy)\,.$$

This is expressed in matrix form as

$$t : \mathbf{v} \mapsto \mathbf{A}\mathbf{v}$$

or

$$\begin{bmatrix} x' \\ y' \end{bmatrix} = \begin{bmatrix} a & b \\ c & d \end{bmatrix} \begin{bmatrix} x \\ y \end{bmatrix}$$

where

$$\mathbf{A} = \begin{bmatrix} a & b \\ c & d \end{bmatrix}, \quad \mathbf{v} = \begin{bmatrix} x \\ y \end{bmatrix}.$$

Therefore, if $\mathbf{v}$ is an eigenvector of t, and λ its associated eigenvalue, then

$$\mathbf{A}\mathbf{v} = \lambda\mathbf{v}$$
$$\begin{bmatrix} a & b \\ c & d \end{bmatrix} \begin{bmatrix} x \\ y \end{bmatrix} = \lambda \begin{bmatrix} x \\ y \end{bmatrix}$$

or in equation terms:

$$ax + by = \lambda x$$
$$cx + dy = \lambda y.$$

Rearranging, we have

$$(a - \lambda)\,x + by = 0$$
$$cx + (d - \lambda)\,y = 0$$

or back in matrix form:

$$\begin{bmatrix} a - \lambda & b \\ c & d - \lambda \end{bmatrix} \begin{bmatrix} x \\ y \end{bmatrix} = \begin{bmatrix} 0 \\ 0 \end{bmatrix}.$$

For a non-zero $[x \quad y]^{\mathrm{T}}$ to exist, we must have

$$\det \begin{bmatrix} a - \lambda & b \\ c & d - \lambda \end{bmatrix} = 0$$

which is called the *characteristic equation*. Let's use this on the transform

$$\begin{bmatrix} 4 & 1 \\ 1 & 4 \end{bmatrix} \begin{bmatrix} x \\ y \end{bmatrix} = \begin{bmatrix} x' \\ y' \end{bmatrix}.$$

Then

$$\begin{vmatrix} 4-\lambda & 1 \\ 1 & 4-\lambda \end{vmatrix} = 0$$

$$\begin{aligned} (4-\lambda)^2 - 1 &= 0 \\ \lambda^2 - 8\lambda + 16 - 1 &= 0 \\ \lambda^2 - 8\lambda + 15 &= 0 \\ (\lambda - 5)(\lambda - 3) &= 0. \end{aligned}$$

Thus $\lambda = 5$ and $\lambda = 3$, are the two eigenvalues we observed in Fig. 4.1. Next, we substitute the two values of λ in

$$\begin{bmatrix} 4-\lambda & 1 \\ 1 & 4-\lambda \end{bmatrix} \begin{bmatrix} x \\ y \end{bmatrix} = \begin{bmatrix} 0 \\ 0 \end{bmatrix}$$

to extract the eigenvectors. Let's start with $\lambda = 5$:

$$\begin{bmatrix} -1 & 1 \\ 1 & -1 \end{bmatrix} \begin{bmatrix} x \\ y \end{bmatrix} = \begin{bmatrix} 0 \\ 0 \end{bmatrix}$$

which represents the equation $y = x$ or the vector $[k \quad k]^{\mathrm{T}}$. Next, we substitute $\lambda = 3$:

$$\begin{bmatrix} 1 & 1 \\ 1 & 1 \end{bmatrix} \begin{bmatrix} x \\ y \end{bmatrix} = \begin{bmatrix} 0 \\ 0 \end{bmatrix}$$

which represents the equation $y = -x$ or the vector $[-k \quad k]^{\mathrm{T}}$.

Thus we have discovered that the transform possesses two eigenvectors $[k \quad k]^{\mathrm{T}}$ and $[-k \quad k]^{\mathrm{T}}$ and their respective eigenvalues $\lambda = 5$ and $\lambda = 3$, as predicted.

The characteristic equation may have real or complex solutions, and if they are complex, there are no real eigenvectors. For example, we have already come across the 2D transform for rotating points about the origin:

$$\mathbf{A} = \begin{bmatrix} \cos\beta & -\sin\beta \\ \sin\beta & \cos\beta \end{bmatrix}$$

and we would not expect this to have any real eigenvectors, as this would imply that it shows a rotational preference to certain points. Let's explore this transform to see how the characteristic equation behaves.

The characteristic equation is

$$\begin{vmatrix} \cos\beta - \lambda & -\sin\beta \\ \sin\beta & \cos\beta - \lambda \end{vmatrix} = 0$$

where β is the angle of rotation. Therefore,

$$\begin{aligned} (\cos\beta - \lambda)^2 + \sin^2\beta &= 0 \\ \lambda^2 - 2\lambda\cos\beta + \cos^2\beta + \sin^2\beta &= 0 \\ \lambda^2 - 2\lambda\cos\beta + 1 &= 0. \end{aligned}$$

This quadratic in λ is solved using

$$\lambda = \frac{-b \pm \sqrt{b^2 - 4ac}}{2a}$$

where $a = 1$, $b = -2\cos\beta$, $c = 1$:

$$\begin{aligned}
\lambda &= \frac{2\cos\beta \pm \sqrt{4\cos^2\beta - 4}}{2} \\
&= \cos\beta \pm \sqrt{\cos^2\beta - 1} \\
&= \cos\beta \pm \sqrt{-\sin^2\beta} \\
\lambda_1 &= \cos\beta + i\sin\beta \\
\lambda_2 &= \cos\beta - i\sin\beta
\end{aligned}$$

which are complex numbers.

The corresponding complex eigenvectors are

$$\begin{aligned}
\mathbf{v}_1 &= \begin{bmatrix} 1 \\ i \end{bmatrix} \\
\mathbf{v}_2 &= \begin{bmatrix} 1 \\ -i \end{bmatrix}.
\end{aligned}$$

Now let's investigate the eigenvectors associated with a 3D transform. We start with the arbitrary transform

$$\mathbf{A} = \begin{bmatrix} 3 & 0 & 1 \\ -1 & 3 & 3 \\ 1 & 0 & 3 \end{bmatrix}$$

and its characteristic equation is

$$\begin{vmatrix} 3-\lambda & 0 & 1 \\ -1 & 3-\lambda & 3 \\ 1 & 0 & 3-\lambda \end{vmatrix} = 0.$$

Expanding the determinant using the top row we have

$$\begin{aligned}
(3-\lambda)\begin{vmatrix} 3-\lambda & 3 \\ 0 & 3-\lambda \end{vmatrix} - 0 + \begin{vmatrix} -1 & 3-\lambda \\ 1 & 0 \end{vmatrix} &= 0 \\
(3-\lambda)(3-\lambda)^2 - (3-\lambda) &= 0 \\
(3-\lambda)\left[(3-\lambda)^2 - 1\right] &= 0 \\
(3-\lambda)\left(\lambda^2 - 6\lambda + 8\right) &= 0 \\
(3-\lambda)(\lambda-4)(\lambda-2) &= 0
\end{aligned}$$

which has solutions $\lambda = 2, 3, 4$. Let's substitute these values of λ in the original equations to reveal the eigenvectors:

$$\begin{cases}
(3-\lambda)x \qquad\qquad\quad + \qquad\quad z = 0 \\
\quad -x + (3-\lambda)y + \qquad 3z = 0 \\
\qquad x \qquad\qquad\quad + (3-\lambda)z = 0.
\end{cases}$$

With $\lambda = 2$ we have $z = -x$ from the 1st equation. Substituting this in the 2nd equation we have $y = 4x$, which permits us to state that the associated eigenvector is of the form $[k \quad 4k \quad -k]^{\mathrm{T}}$.

With $\lambda = 3$ we have $z = 0$ from the 1st equation, and $x = 0$ from the 3rd equation, which permits us to state that the associated eigenvector is of the form $[0 \quad k \quad 0]^{\mathrm{T}}$.

With $\lambda = 4$ we have $z = x$ from the 1st equation. Substituting this in the 2nd equation we have $y = 2x$, which permits us to state that the associated eigenvector is of the form $[k \quad 2k \quad k]^{\mathrm{T}}$.

Therefore, the eigenvectors and eigenvalues are

$$
\begin{aligned}
[k \quad 4k \quad -k]^{\mathrm{T}} \quad & \lambda = 2\\
[0 \quad k \quad 0]^{\mathrm{T}} \quad & \lambda = 3\\
[k \quad 2k \quad k]^{\mathrm{T}} \quad & \lambda = 4
\end{aligned}
$$

where $k \neq 0$.

The major problem with the above technique is that it requires careful analysis to untangle the eigenvector, and ideally, we require a deterministic algorithm to reveal the result. We will discover that such a technique is available in Chap. 9.

4.18 Vector Products

Vectors are regarded as single column or single row matrices, which permits us to express their products neatly. For example, given two vectors

$$
\mathbf{v} = \begin{bmatrix} a \\ b \\ c \end{bmatrix}, \quad \mathbf{w} = \begin{bmatrix} x \\ y \\ z \end{bmatrix}
$$

then

$$
\begin{aligned}
\mathbf{v} \cdot \mathbf{w} &= \mathbf{v}^{\mathrm{T}} \mathbf{w}\\
&= \begin{bmatrix} a & b & c \end{bmatrix} \begin{bmatrix} x \\ y \\ z \end{bmatrix} = ax + by + cz.
\end{aligned}
$$

Similarly, the vector cross product is written

$$
\begin{aligned}
\mathbf{v} \times \mathbf{w} = \begin{bmatrix} a \\ b \\ c \end{bmatrix} \times \begin{bmatrix} x \\ y \\ z \end{bmatrix} &= \begin{bmatrix} \mathbf{i} & \mathbf{j} & \mathbf{k} \\ a & b & c \\ x & y & z \end{bmatrix}\\
&= (bz - cy)\mathbf{i} - (az - xc)\mathbf{j} + (ay - bx)\mathbf{k}\\
&= \begin{bmatrix} bz - cy \\ -az + xc \\ ay - bx \end{bmatrix}.
\end{aligned}
$$

4.19 Summary

Matrices play an important role in representing rotations, especially orthogonal matrices, which is why they have been reviewed in this chapter. The inverse matrix is also an important concept to grasp as this provides the mechanism for reversing a rotation or change of frame. We will also come across eigenvectors in later chapters, which is why they were explained in some detail.

4.19.1 Summary of Matrix Operations

Matrix (2×2)

$$\mathbf{A} = \begin{bmatrix} a & b \\ c & d \end{bmatrix}.$$

Matrix (3×3)

$$\mathbf{B} = \begin{bmatrix} a & b & c \\ d & e & f \\ g & h & i \end{bmatrix}.$$

Transpose

$$\mathbf{A}^{\mathrm{T}} = \begin{bmatrix} a & c \\ b & d \end{bmatrix}, \quad \mathbf{B}^{\mathrm{T}} = \begin{bmatrix} a & d & g \\ b & e & h \\ c & f & i \end{bmatrix}.$$

Identity

$$\mathbf{I} = \begin{bmatrix} 1 & 0 \\ 0 & 1 \end{bmatrix}, \quad \mathbf{I} = \begin{bmatrix} 1 & 0 & 0 \\ 0 & 1 & 0 \\ 0 & 0 & 1 \end{bmatrix}.$$

Adding and subtracting

$$\mathbf{M} \pm \mathbf{N} = [m_{row,col} \pm n_{row,col}].$$

Multiplying by a scalar

$$\pm\lambda\mathbf{M} = [\pm\lambda m_{row,col}].$$

Product transpose

$$[\mathbf{MN}]^{\mathrm{T}} = \mathbf{N}^{\mathrm{T}}\mathbf{M}^{\mathrm{T}}.$$

Sum/difference transpose

$$[\mathbf{M} \pm \mathbf{N}]^{\mathrm{T}} = \mathbf{M}^{\mathrm{T}} \pm \mathbf{N}^{\mathrm{T}}.$$

Determinant

$$\det \mathbf{A} = |\mathbf{A}| = ad - bc$$
$$\det \mathbf{B} = |\mathbf{B}| = aei + bfg + cdh - ceg - afh - bdi.$$

Inverse

$$\mathbf{M}^{-1} = \frac{(\text{cofactor matrix of } \mathbf{M})^{\mathrm{T}}}{\det \mathbf{M}}$$
$$[\mathbf{MN}]^{-1} = \mathbf{N}^{-1}\mathbf{M}^{-1}.$$

Orthogonal

$$\mathbf{M} \text{ is orthogonal if } \mathbf{M}^{\mathrm{T}} = \mathbf{M}^{-1}.$$

Trace

$$\mathrm{Tr}(\mathbf{A}) = a + d$$
$$\mathrm{Tr}(\mathbf{B}) = a + e + i$$
$$\mathrm{Tr}(\mathbf{MN}) = \mathrm{Tr}(\mathbf{NM}).$$

Symmetric

$$\mathbf{M} \text{ is symmetric if } \mathbf{M} = \mathbf{M}^{\mathrm{T}}.$$

Symmetric part S

$$\mathbf{S} = \frac{1}{2}\left(\mathbf{M} + \mathbf{M}^{\mathrm{T}}\right).$$

Antisymmetric

$$\mathbf{M} \text{ is antisymmetric if } \mathbf{M} = -\mathbf{M}^{\mathrm{T}}.$$

Antisymmetric part Q

$$\mathbf{Q} = \frac{1}{2}\left(\mathbf{M} - \mathbf{M}^{\mathrm{T}}\right).$$

Eigenvector

$$\mathbf{v} \text{ is the eigenvector of } \mathbf{M} \text{ if } \mathbf{Mv} = \lambda\mathbf{v}.$$

Eigenvalue

$$\lambda \text{ is the eigenvalue of } \mathbf{M} \text{ if } \mathbf{Mv} = \lambda\mathbf{v}.$$

Chapter 5
Quaternions

5.1 Introduction

As mentioned earlier, quaternions were invented by Sir William Rowan Hamilton in 1843. Sir William was looking to generalise complex numbers in higher dimensions, and it took 14 years of toil before he stumbled upon the idea of using a 4D notation – hence the name '*quaternion*'.

5.2 Definition

The definition and associated rules for a quaternion are:

$$\mathbf{q} = a + bi + cj + dk$$

where a, b, c and d are scalars, and i, j and k are imaginary and obey the following rules:

$$\begin{aligned}
&i^2 = -1, \quad && j^2 = -1, \quad && k^2 = -1, \quad && ijk = -1\\
&ij = k, && jk = i, && ki = j\\
&ji = -k, && kj = -i, && ik = -j.
\end{aligned}$$

Although quaternions had some enthusiastic supporters, there were many mathematicians and scientists who were suspicious of the need to involve so many imaginary terms. Towards the end of the nineteenth century Josiah Gibbs resolved the problem by declaring that the three imaginary quantities could be viewed as a 3D vector and changed the original $bi + cj + dk$ into $b\mathbf{i} + c\mathbf{j} + d\mathbf{k}$, where $\mathbf{i}$, $\mathbf{j}$ and $\mathbf{k}$ are unit Cartesian vectors. Today, it is convenient in computer graphics to write a quaternion in two ways:

$$\mathbf{q} = s, \mathbf{v} \tag{5.1}$$

$$\mathbf{q} = s + \mathbf{v} \tag{5.2}$$

where s is a scalar, and $\mathbf{v}$ is a 3D vector.

J. Vince, *Rotation Transforms for Computer Graphics*,
DOI 10.1007/978-0-85729-154-7_5,

The difference is rather subtle: in (5.1) the scalar and vector are separated by a comma, whereas in (5.2) a '+' sign is used as in complex numbers. Although the idea of adding a scalar to a vector seems strange, this notation is used in this book as it helps us understand the ideas behind multivectors, which are covered in the next chapter. Since Hamilton's invention, mathematicians have successfully applied quaternions to rotate points about an arbitrary axis, which is why we are interested in them.

A quaternion then, is the combination of a scalar and a vector:

$$\mathbf{q} = s + \mathbf{v}$$

where s is a scalar and $\mathbf{v}$ is a 3D vector. If we express the vector $\mathbf{v}$ in terms of its components, we have

$$\mathbf{q} = s + x\mathbf{i} + y\mathbf{j} + z\mathbf{k} \quad \text{where } s, x, y, z \text{ are all scalars.}$$

Later on we will discover that in the context of a rotation transform, $\mathbf{v}$ is used to represent the axis of rotation, and the scalar s encodes the angle of rotation.

5.2.1 Axioms

Quaternions share the same axioms as complex numbers apart from multiplication, where they do not commute.

Addition:

$$\begin{aligned} \text{Commutative} \quad & \mathbf{q}_1 + \mathbf{q}_2 = \mathbf{q}_2 + \mathbf{q}_1 \\ \text{Associative} \quad & (\mathbf{q}_1 + \mathbf{q}_2) + \mathbf{q}_3 = \mathbf{q}_1 + (\mathbf{q}_2 + \mathbf{q}_3). \end{aligned}$$

Multiplication:

$$\begin{aligned} \text{Associative} \quad & (\mathbf{q}_1\mathbf{q}_2)\mathbf{q}_3 = \mathbf{q}_1(\mathbf{q}_2\mathbf{q}_3) \\ \text{Non-commutative} \quad & \mathbf{q}_1\mathbf{q}_2 \neq \mathbf{q}_2\mathbf{q}_1. \end{aligned}$$

5.3 Adding and Subtracting Quaternions

Two quaternions $\mathbf{q}_1$ and $\mathbf{q}_2$

$$\begin{aligned} \mathbf{q}_1 &= s_1 + x_1\mathbf{i} + y_1\mathbf{j} + z_1\mathbf{k} \\ \mathbf{q}_2 &= s_2 + x_2\mathbf{i} + y_2\mathbf{j} + z_2\mathbf{k} \end{aligned}$$

are equal if, and only if, their corresponding terms are equal. Furthermore, like vectors, they can be added and subtracted as follows:

$$\mathbf{q}_1 \pm \mathbf{q}_2 = (s_1 \pm s_2) + (x_1 \pm x_2)\mathbf{i} + (y_1 \pm y_2)\mathbf{j} + (z_1 \pm z_2)\mathbf{k}.$$

For example:

$$\begin{aligned}
\mathbf{q}_1 &= 0.6 + 2\mathbf{i} + 4\mathbf{j} - 3\mathbf{k} \\
\mathbf{q}_2 &= 0.2 + 3\mathbf{i} + 5\mathbf{j} + 7\mathbf{k} \\
\mathbf{q}_1 + \mathbf{q}_2 &= 0.8 + 5\mathbf{i} + 9\mathbf{j} + 4\mathbf{k} \\
\mathbf{q}_1 - \mathbf{q}_2 &= 0.4 - \mathbf{i} - \mathbf{j} - 10\mathbf{k}.
\end{aligned}$$

5.4 Multiplying Quaternions

When multiplying quaternions we must employ the following rules:

$$\begin{aligned}
&\mathbf{i}^2 = -1, \quad \mathbf{j}^2 = -1, \quad \mathbf{k}^2 = -1, \quad \mathbf{ijk} = -1 \\
&\mathbf{ij} = \mathbf{k}, \quad \mathbf{jk} = \mathbf{i}, \quad \mathbf{ki} = \mathbf{j} \\
&\mathbf{ji} = -\mathbf{k}, \quad \mathbf{kj} = -\mathbf{i}, \quad \mathbf{ik} = -\mathbf{j}.
\end{aligned}$$

Note that quaternion addition is commutative, however, the rules make quaternion products non-commutative. For example:

$$\begin{aligned}
\mathbf{q}_1 &= s_1 + \mathbf{v}_1 = s_1 + x_1\mathbf{i} + y_1\mathbf{j} + z_1\mathbf{k} \\
\mathbf{q}_2 &= s_2 + \mathbf{v}_2 = s_2 + x_2\mathbf{i} + y_2\mathbf{j} + z_2\mathbf{k} \\
\mathbf{q}_1\mathbf{q}_2 &= (s_1s_2 - x_1x_2 - y_1y_2 - z_1z_2) + (s_1x_2 + s_2x_1 + y_1z_2 - y_2z_1)\mathbf{i} \\
&\quad + (s_1y_2 + s_2y_1 + z_1x_2 - z_2x_1)\mathbf{j} + (s_1z_2 + s_2z_1 + x_1y_2 - x_2y_1)\mathbf{k} \\
&= s_1s_2 - (x_1x_2 + y_1y_2 + z_1z_2) + s_1(x_2\mathbf{i} + y_2\mathbf{j} + z_2\mathbf{k}) + s_2(x_1\mathbf{i} + y_1\mathbf{j} + z_1\mathbf{k}) \\
&\quad + (y_1z_2 - y_2z_1)\mathbf{i} + (z_1x_2 - z_2x_1)\mathbf{j} + (x_1y_2 - x_2y_1)\mathbf{k}
\end{aligned}$$

which can be rewritten using the dot and cross product notation as

$$\mathbf{q}_1\mathbf{q}_2 = s_1s_2 - \mathbf{v}_1 \cdot \mathbf{v}_2 + s_1\mathbf{v}_2 + s_2\mathbf{v}_1 + \mathbf{v}_1 \times \mathbf{v}_2$$

where

$$\begin{aligned}
s_1s_2 - \mathbf{v}_1 \cdot \mathbf{v}_2 &\quad \text{is a scalar, and} \\
s_1\mathbf{v}_2 + s_2\mathbf{v}_1 + \mathbf{v}_1 \times \mathbf{v}_2 &\quad \text{is a vector.}
\end{aligned}$$

For example:

$$\begin{aligned}
\mathbf{q}_1 &= 1 + 2\mathbf{i} + 3\mathbf{j} + 4\mathbf{k} \\
\mathbf{q}_2 &= 2 - \mathbf{i} + 5\mathbf{j} - 2\mathbf{k} \\
\mathbf{q}_1\mathbf{q}_2 &= (1 + 2\mathbf{i} + 3\mathbf{j} + 4\mathbf{k})(2 - \mathbf{i} + 5\mathbf{j} - 2\mathbf{k}) \\
\mathbf{q}_1\mathbf{q}_2 &= \big(1 \times 2 - 2 \times (-1) + 3 \times 5 + 4 \times (-2)\big) \\
&\quad + 1(-\mathbf{i} + 5\mathbf{j} - 2\mathbf{k}) + 2(2\mathbf{i} + 3\mathbf{j} + 4\mathbf{k}) \\
&\quad + \big(3 \times (-2) - 4 \times 5\big)\mathbf{i} - \big(2 \times (-2) - 4 \times (-1)\big)\mathbf{j} + \big(2 \times 5 - 3 \times (-1)\big)\mathbf{k} \\
&= -3 + 3\mathbf{i} + 11\mathbf{j} + 6\mathbf{k} - 26\mathbf{i} + 13\mathbf{k} \\
&= -3 - 23\mathbf{i} + 11\mathbf{j} + 19\mathbf{k}
\end{aligned}$$

which is another quaternion. You may wish to evaluate $\mathbf{q}_2\mathbf{q}_1$ and show that $\mathbf{q}_1\mathbf{q}_2 \neq \mathbf{q}_2\mathbf{q}_1$.

5.5 Pure Quaternion

A pure quaternion has a zero scalar term:

$$\mathbf{q} = 0 + \mathbf{v}$$

which is a vector. Therefore,

$$\begin{aligned}\mathbf{q}_1 &= 0 + \mathbf{v}_1\\ \mathbf{q}_2 &= 0 + \mathbf{v}_2\\ \mathbf{q}_1\mathbf{q}_2 &= -\mathbf{v}_1 \cdot \mathbf{v}_2 + \mathbf{v}_1 \times \mathbf{v}_2\end{aligned}$$

which leads to a rather strange result for the square of a pure quaternion:

$$\begin{aligned}\mathbf{q}\mathbf{q} &= -\mathbf{v} \cdot \mathbf{v} + \mathbf{v} \times \mathbf{v}\\ &= -\mathbf{v} \cdot \mathbf{v}\\ &= -|\mathbf{v}|^2\end{aligned}$$

a negative real number! In Hamilton's day, physicists found this result difficult to accept, and on top of all the imaginary terms refused to adopt quaternions and embraced the vector analysis proposed by Gibbs *et al.*

5.6 Magnitude of a Quaternion

The *magnitude*, *norm* or *modulus* of a quaternion is written $|\mathbf{q}|$ and equals

$$\begin{aligned}\mathbf{q} &= s + x\mathbf{i} + y\mathbf{j} + z\mathbf{k}\\ |\mathbf{q}| &= \sqrt{s^2 + x^2 + y^2 + z^2}.\end{aligned}$$

For example:

$$\begin{aligned}\mathbf{q} &= 1 + 2\mathbf{i} + 4\mathbf{j} - 3\mathbf{k}\\ |\mathbf{q}| &= \sqrt{1^2 + 2^2 + 4^2 + (-3)^2} = \sqrt{30}.\end{aligned}$$

5.7 Unit Quaternion

A unit quaternion has a magnitude equal to 1:

$$|\mathbf{q}| = \sqrt{s^2 + x^2 + y^2 + z^2} = 1.$$

Any quaternion $\mathbf{q}$ can be normalised to a unit quaternion $\hat{\mathbf{q}}$ by dividing by its magnitude:

$$\hat{\mathbf{q}} = \frac{\mathbf{q}}{|\mathbf{q}|}.$$

5.8 The Quaternion Conjugate

We have already discovered that the complex conjugate of a complex number $z = a + bi$ is given by

$$z^* = a - bi$$

and is very useful in computing the inverse of z. The *quaternion conjugate* plays a similar role in computing the inverse of a quaternion. Therefore, given

$$\begin{aligned}\mathbf{q} &= s + \mathbf{v}\\ &= s + x\mathbf{i} + y\mathbf{j} + z\mathbf{k}\end{aligned}$$

its conjugate is defined as

$$\begin{aligned}\mathbf{q}^* &= s - \mathbf{v}\\ &= s - x\mathbf{i} - y\mathbf{j} - z\mathbf{k}.\end{aligned}$$

If we compute the product $\mathbf{q}\mathbf{q}^*$ we obtain

$$\begin{aligned}\mathbf{q}\mathbf{q}^* &= (s + \mathbf{v})(s - \mathbf{v})\\ &= s^2 + \mathbf{v}\cdot\mathbf{v} + s\mathbf{v} - s\mathbf{v} + \mathbf{v}\times(-\mathbf{v}) = s^2 + \mathbf{v}\cdot\mathbf{v}\\ &= s^2 + x^2 + y^2 + z^2\end{aligned}$$

which is a scalar and implies that

$$\mathbf{q}\mathbf{q}^* = |\mathbf{q}|^2$$

or

$$|\mathbf{q}| = \sqrt{\mathbf{q}\mathbf{q}^*}.$$

Similarly, we can show that $\mathbf{q}\mathbf{q}^* = \mathbf{q}^*\mathbf{q}$.

Now let's show that $(\mathbf{q}_1\mathbf{q}_2)^* = \mathbf{q}_2^*\mathbf{q}_1^*$. We start with quaternions $\mathbf{q}_1$ and $\mathbf{q}_2$:

$$\begin{aligned}\mathbf{q}_1 &= s_1 + \mathbf{v}_1\\ \mathbf{q}_2 &= s_2 + \mathbf{v}_2\\ \mathbf{q}_1\mathbf{q}_2 &= (s_1 + \mathbf{v}_1)(s_2 + \mathbf{v}_2)\\ &= s_1s_2 - \mathbf{v}_1\cdot\mathbf{v}_2 + s_1\mathbf{v}_2 + s_2\mathbf{v}_1 + \mathbf{v}_1\times\mathbf{v}_2\\ (\mathbf{q}_1\mathbf{q}_2)^* &= s_1s_2 - \mathbf{v}_1\cdot\mathbf{v}_2 - s_1\mathbf{v}_2 - s_2\mathbf{v}_1 - \mathbf{v}_1\times\mathbf{v}_2. \qquad (5.3)\end{aligned}$$

Next, we compute $\mathbf{q}_2^*\mathbf{q}_1^*$

$$\begin{aligned}
\mathbf{q}_1^* &= s_1 - \mathbf{v}_1 \\
\mathbf{q}_2^* &= s_2 - \mathbf{v}_2 \\
\mathbf{q}_2^*\mathbf{q}_1^* &= (s_2 - \mathbf{v}_2)(s_1 - \mathbf{v}_1) \\
&= s_1 s_2 - \mathbf{v}_1 \cdot \mathbf{v}_2 + s_2(-\mathbf{v}_1) + s_1(-\mathbf{v}_2) + (-\mathbf{v}_2) \times (-\mathbf{v}_1) \\
&= s_1 s_2 - \mathbf{v}_1 \cdot \mathbf{v}_2 - s_1\mathbf{v}_2 - s_2\mathbf{v}_1 - \mathbf{v}_1 \times \mathbf{v}_2
\end{aligned} \tag{5.4}$$

and as (5.3) equals (5.4), $(\mathbf{q}_1\mathbf{q}_2)^* = \mathbf{q}_2^*\mathbf{q}_1^*$.

5.9 The Inverse Quaternion

Given a quaternion $\mathbf{q}$ we can compute its inverse $\mathbf{q}^{-1}$ as follows.

By definition, we require that

$$\mathbf{q}\mathbf{q}^{-1} = \mathbf{q}^{-1}\mathbf{q} = 1. \tag{5.5}$$

First, we multiply (5.5) by $\mathbf{q}^*$

$$\mathbf{q}^*\mathbf{q}\mathbf{q}^{-1} = \mathbf{q}^*\mathbf{q}^{-1}\mathbf{q} = \mathbf{q}^* \tag{5.6}$$

and from (5.6) we can write

$$\mathbf{q}^{-1} = \frac{\mathbf{q}^*}{\mathbf{q}^*\mathbf{q}} = \frac{\mathbf{q}^*}{|\mathbf{q}|^2}.$$

If $\mathbf{q}$ is a unit quaternion, then $\mathbf{q}^{-1} = \mathbf{q}^*$, which is useful when reversing a rotational sequence. Therefore, as

$$(\mathbf{q}_1\mathbf{q}_2)^* = \mathbf{q}_2^*\mathbf{q}_1^*$$

then

$$(\mathbf{q}_1\mathbf{q}_2)^{-1} = \mathbf{q}_2^{-1}\mathbf{q}_1^{-1}.$$

For completeness let's evaluate the inverse of $\mathbf{q}$ where

$$\begin{aligned}
\mathbf{q} &= 1 + \frac{1}{\sqrt{3}}\mathbf{i} + \frac{1}{\sqrt{3}}\mathbf{j} + \frac{1}{\sqrt{3}}\mathbf{k} \\
\mathbf{q}^* &= 1 - \frac{1}{\sqrt{3}}\mathbf{i} - \frac{1}{\sqrt{3}}\mathbf{j} - \frac{1}{\sqrt{3}}\mathbf{k} \\
|\mathbf{q}|^2 &= 1 + \frac{1}{3} + \frac{1}{3} + \frac{1}{3} = 2 \\
\mathbf{q}^{-1} &= \frac{1}{2}\left(1 - \frac{1}{\sqrt{3}}\mathbf{i} - \frac{1}{\sqrt{3}}\mathbf{j} - \frac{1}{\sqrt{3}}\mathbf{k}\right) \\
&= \frac{1}{2} - \frac{1}{\sqrt{12}}\mathbf{i} - \frac{1}{\sqrt{12}}\mathbf{j} - \frac{1}{\sqrt{12}}\mathbf{k}.
\end{aligned}$$

5.10 Summary

Quaternions offer a powerful algebra for rotating points about an arbitrary axis and it is important that they are fully understood before proceeding. We have yet to see how quaternions actually perform this rotational task, which is covered in Chap. 11.

5.10.1 Summary of Quaternion Operations

$$\begin{aligned}\mathbf{q} &= s + \mathbf{v}\\ &= s + x\mathbf{i} + y\mathbf{j} + z\mathbf{k}\end{aligned}$$

where s, x, y, z are scalars, and

$$\begin{aligned}&\mathbf{i}^2 = -1, \quad \mathbf{j}^2 = -1, \quad \mathbf{k}^2 = -1, \quad \mathbf{ijk} = -1\\ &\mathbf{ij} = \mathbf{k}, \quad \mathbf{jk} = \mathbf{i}, \quad \mathbf{ki} = \mathbf{j}\\ &\mathbf{ji} = -\mathbf{k}, \quad \mathbf{kj} = -\mathbf{i}, \quad \mathbf{ik} = -\mathbf{j}.\end{aligned}$$

Addition and subtraction

$$\mathbf{q}_1 \pm \mathbf{q}_2 = (s_1 \pm s_2) + (x_1 \pm x_2)\mathbf{i} + (y_1 \pm y_2)\mathbf{j} + (z_1 \pm z_2)\mathbf{k}.$$

Product

$$\begin{aligned}\mathbf{q}_1\mathbf{q}_2 &= (s_1s_2 - x_1x_2 - y_1y_2 - z_1z_2) + (s_1x_2 + s_2x_1 + y_1z_2 - y_2z_1)\mathbf{i}\\ &\quad + (s_1y_2 + s_2y_1 + z_1x_2 - z_2x_1)\mathbf{j} + (s_1z_2 + s_2z_1 + x_1y_2 - x_2y_1)\mathbf{k}\\ \mathbf{q}_1\mathbf{q}_2 &= (s_1s_2 - \mathbf{v}_1 \cdot \mathbf{v}_2) + s_1\mathbf{v}_2 + s_2\mathbf{v}_1 + \mathbf{v}_1 \times \mathbf{v}_2.\end{aligned}$$

Pure

$$\mathbf{q} = 0 + \mathbf{v}.$$

Magnitude

$$|\mathbf{q}| = \sqrt{s^2 + x^2 + y^2 + z^2}.$$

Unit

$$|\mathbf{q}| = \sqrt{s^2 + x^2 + y^2 + z^2} = 1.$$

Quaternion conjugate

$$\begin{aligned}\mathbf{q}^* &= s - \mathbf{v} = s - x\mathbf{i} - y\mathbf{j} - z\mathbf{k}\\ (\mathbf{q}_1\mathbf{q}_2)^* &= \mathbf{q}_2^*\mathbf{q}_1^*.\end{aligned}$$

Inverse

$$\begin{aligned}\mathbf{q}^{-1} &= \frac{\mathbf{q}^*}{\mathbf{q}^*\mathbf{q}} = \frac{\mathbf{q}^*}{|\mathbf{q}|^2}\\ (\mathbf{q}_1\mathbf{q}_2)^{-1} &= \mathbf{q}_2^{-1}\mathbf{q}_1^{-1}.\end{aligned}$$

Chapter 6
Multivectors

6.1 Introduction

This is a brief introduction to *multivectors* of *geometric algebra* and we only explore those elements associated with rotations. Those readers who wish to pursue the subject further may wish to consult the author's books: *Geometric Algebra for Computer Graphics* [5] or *Geometric Algebra: An Algebraic System for Computer Games and Animation* [6].

We regard vectors as *directed* lines or *oriented* lines, but if *they* exist, why shouldn't oriented planes and oriented volumes exist? Well they do, which is what geometric algebra is about. Unfortunately when vectors were invented, the work of the German mathematician, Hermann Grassmann (1809–1877), was not understood and consequently ignored. In retrospect this was unfortunate, as Grassmann had invented an exceedingly powerful algebra for geometry, and it has taken a further century for it to emerge through the work of William Kingdon Clifford (1845–1879) and David Hestenes. So let's explore an exciting algebra that offers new ways of handling rotations.

6.2 Symmetric and Antisymmetric Functions

Symmetric (*even*) and *antisymmetric* (*odd*) functions play an important role in understanding multivectors. For example, $f(\beta)$ is a symmetric function if

$$f(-\beta) = f(\beta)$$

an example being $\cos\beta$ where $\cos(-\beta) = \cos\beta$. Whereas, $f(\beta)$ is an antisymmetric function if

$$f(-\beta) = -f(\beta)$$

an example being $\sin\beta$ where $\sin(-\beta) = -\sin\beta$.

J. Vince, *Rotation Transforms for Computer Graphics*,
DOI 10.1007/978-0-85729-154-7_6, © Springer-Verlag London Limited 2011

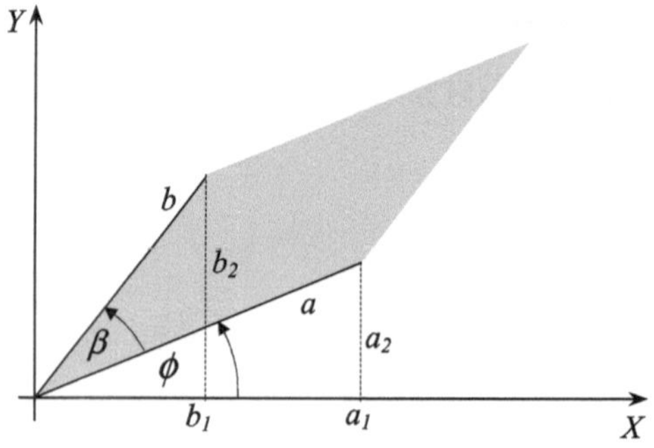

Fig. 6.1 Two line segments a and b separated by $+\beta$

6.3 Trigonometric Foundations

Figure 6.1 shows two line segments a and b with coordinates (a_1, a_2) and (b_1, b_2) respectively. The lines are separated by an angle β, and it is a trivial exercise to show that

$$ab \sin\beta = a_1 b_2 - a_2 b_1$$

which equals the area of the parallelogram formed by a and b. What is interesting is that reversing the relative orientation of the lines such that b is rotated $-\beta$ relative to a makes

$$ab \sin\beta = -(a_1 b_2 - a_2 b_1)$$

which means that this is antisymmetric due to the sine function.

We know from the definition of the scalar product of vectors that

$$ab \cos\beta = a_1 b_1 + a_2 b_2$$

which remains unaltered if the relative orientation of the lines is reversed, which means that this is symmetric due to the cosine function.

6.4 Vectorial Foundations

If we form the algebraic product of two 2D vectors **a** and **b** we have:

$$\begin{aligned}
\mathbf{a} &= a_1\mathbf{i} + a_2\mathbf{j} \\
\mathbf{b} &= b_1\mathbf{i} + b_2\mathbf{j} \\
\mathbf{ab} &= a_1 b_1 \mathbf{i}^2 + a_2 b_2 \mathbf{j}^2 + a_1 b_2 \mathbf{ij} + a_2 b_1 \mathbf{ji}.
\end{aligned} \tag{6.1}$$

It is clear from (6.1) that $a_1 b_1 \mathbf{i}^2 + a_2 b_2 \mathbf{j}^2$ has something to do with $ab\cos\beta$, and $a_1 b_2 \mathbf{ij} + a_2 b_1 \mathbf{ji}$ has something to do with $ab \sin\beta$. The product **ab** creates the terms $\mathbf{i}^2$, $\mathbf{j}^2$, **ij** and **ji**, which are resolved as follows.

6.5 Inner and Outer Products

Let's assume that the products **ij** and **ji** in (6.1) anticommute: $\mathbf{ji} = -\mathbf{ij}$. Therefore,

$$\mathbf{ab} = a_1b_1\mathbf{i}^2 + a_2b_2\mathbf{j}^2 + (a_1b_2 - a_2b_1)\mathbf{ij} \tag{6.2}$$

and if we reverse the product to **ba** we obtain

$$\mathbf{ba} = a_1b_1\mathbf{i}^2 + a_2b_2\mathbf{j}^2 - (a_1b_2 - a_2b_1)\mathbf{ij}. \tag{6.3}$$

From (6.2) and (6.3) we see that the product of two vectors contains a symmetric component

$$a_1b_1\mathbf{i}^2 + a_2b_2\mathbf{j}^2$$

and an antisymmetric component

$$(a_1b_2 - a_2b_1)\mathbf{ij}.$$

Geometric algebra defines the product **ab** as the sum of two other products called the *inner* and *outer* products. The inner product has the form

$$\mathbf{a}\cdot\mathbf{b} = |\mathbf{a}||\mathbf{b}|\cos\beta \tag{6.4}$$

where

$$\begin{aligned}\mathbf{a}\cdot\mathbf{b} &= (a_1\mathbf{i} + a_2\mathbf{j})\cdot(b_1\mathbf{i} + b_2\mathbf{j})\\ &= a_1b_1\mathbf{i}\cdot\mathbf{i} + a_1b_2\mathbf{i}\cdot\mathbf{j} + a_2b_1\mathbf{j}\cdot\mathbf{i} + a_2b_2\mathbf{j}\cdot\mathbf{j}\\ &= a_1b_1 + a_2b_2\end{aligned}$$

which is the familiar scalar product. The outer product uses the wedge '$\wedge$' symbol, which is why it is also called the *wedge product* and has the form

$$\mathbf{a}\wedge\mathbf{b} = |\mathbf{a}||\mathbf{b}|\sin\beta\mathbf{i}\wedge\mathbf{j} \tag{6.5}$$

where

$$\begin{aligned}\mathbf{a}\wedge\mathbf{b} &= (a_1\mathbf{i} + a_2\mathbf{j})\wedge(b_1\mathbf{i} + b_2\mathbf{j})\\ &= a_1b_1\mathbf{i}\wedge\mathbf{i} + a_1b_2\mathbf{i}\wedge\mathbf{j} + a_2b_1\mathbf{j}\wedge\mathbf{i} + a_2b_2\mathbf{j}\wedge\mathbf{j}\\ &= (a_1b_2 - a_2b_1)\mathbf{i}\wedge\mathbf{j}\end{aligned}$$

which enables us to write

$$\mathbf{ab} = \mathbf{a}\cdot\mathbf{b} + \mathbf{a}\wedge\mathbf{b} \tag{6.6}$$

$$\mathbf{ab} = |\mathbf{a}||\mathbf{b}|\cos\beta + |\mathbf{a}||\mathbf{b}|\sin\beta\mathbf{i}\wedge\mathbf{j}. \tag{6.7}$$

6.6 The Geometric Product in 2D

Clifford named the sum of the two products the *geometric product*, which means that (6.6) reads: The geometric product **ab** is the sum of the inner product "**a** dot **b**" and the outer product "**a** wedge **b**".

Given the definition of the geometric product, let's evaluate $\mathbf{i}^2$:

$$\mathbf{ii} = \mathbf{i} \cdot \mathbf{i} + \mathbf{i} \wedge \mathbf{i}.$$

Using the definition for the inner product (6.4) we have

$$\mathbf{i} \cdot \mathbf{i} = 1 \times 1 \times \cos 0^\circ = 1$$

whereas using the definition of the outer product (6.5) we have

$$\mathbf{i} \wedge \mathbf{i} = 1 \times 1 \times \sin 0^\circ \mathbf{i} \wedge \mathbf{i} = 0.$$

Thus $\mathbf{i}^2 = 1$ and $\mathbf{j}^2 = 1$, and $\mathbf{aa} = |\mathbf{a}|^2$:

$$\begin{aligned} \mathbf{aa} &= \mathbf{a} \cdot \mathbf{a} + \mathbf{a} \wedge \mathbf{a} \\ &= |\mathbf{a}||\mathbf{a}| \cos 0^\circ + |\mathbf{a}||\mathbf{a}| \sin 0^\circ \mathbf{i} \wedge \mathbf{j} \\ \mathbf{aa} &= |\mathbf{a}|^2. \end{aligned}$$

This result is much more satisfying than the square of a pure quaternion $\mathbf{q}$:

$$\mathbf{qq} = -|\mathbf{q}|^2.$$

Now let's evaluate $\mathbf{ij}$:

$$\mathbf{ij} = \mathbf{i} \cdot \mathbf{j} + \mathbf{i} \wedge \mathbf{j}.$$

Using the definition for the inner product (6.4) we have

$$\mathbf{i} \cdot \mathbf{j} = 1 \times 1 \times \cos 90^\circ = 0$$

whereas using the definition of the outer product (6.5) we have

$$\mathbf{i} \wedge \mathbf{j} = 1 \times 1 \times \sin 90^\circ \mathbf{i} \wedge \mathbf{j} = \mathbf{i} \wedge \mathbf{j}.$$

Thus $\mathbf{ij} = \mathbf{i} \wedge \mathbf{j}$. But what is $\mathbf{i} \wedge \mathbf{j}$? Well, it is a new object called a *bivector* and defines the orientation of the plane containing $\mathbf{i}$ and $\mathbf{j}$. As the order of the vectors is from $\mathbf{i}$ to $\mathbf{j}$, the angle is $+90^\circ$ and $\sin(+90)^\circ = 1$. Whereas, if the order is from $\mathbf{j}$ to $\mathbf{i}$ the angle is -90° and $\sin(-90^\circ) = -1$. Consequently,

$$\begin{aligned} \mathbf{ji} &= \mathbf{j} \cdot \mathbf{i} + \mathbf{j} \wedge \mathbf{i} \\ &= 0 + 1 \times 1 \times \sin(-90^\circ) \mathbf{i} \wedge \mathbf{j} \\ &= -\mathbf{i} \wedge \mathbf{j}. \end{aligned}$$

A useful way of visualising the bivector $\mathbf{i} \wedge \mathbf{j}$ is to imagine moving along the vector $\mathbf{i}$ and then along the vector $\mathbf{j}$, which creates an anticlockwise rotation. Conversely, for the bivector $\mathbf{j} \wedge \mathbf{i}$, imagine moving along the vector $\mathbf{j}$ followed by vector $\mathbf{i}$, which creates a clockwise rotation. Another useful picture is to sweep vector $\mathbf{j}$ along vector $\mathbf{i}$ to create an anticlockwise rotation, and vice versa for $\mathbf{j} \wedge \mathbf{i}$. These ideas are shown in Fig. 6.2.

The following equation

$$\mathbf{ab} = 9 + 12\mathbf{i} \wedge \mathbf{j}$$

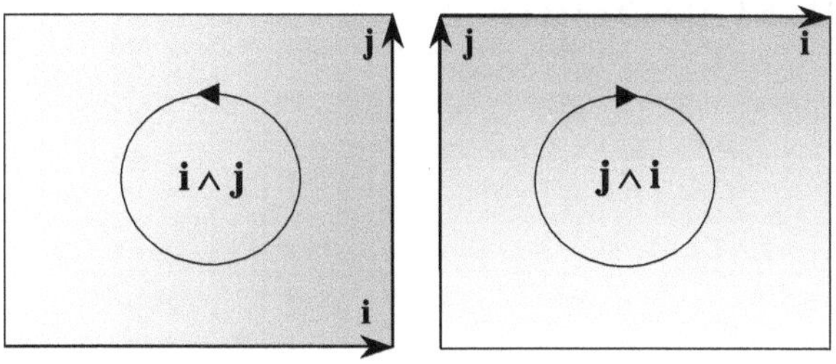

Fig. 6.2 An anticlockwise and clockwise bivector

means that the geometric product of two vectors **a** and **b** creates a scalar, inner product of 9, and an outer product of 12 on the **i**–**j** plane. For example, given

$$\begin{aligned}\mathbf{a} &= 3\mathbf{i}\\ \mathbf{b} &= 3\mathbf{i}+4\mathbf{j}\\ \mathbf{ab} &= (3\mathbf{i})\cdot(3\mathbf{i}+4\mathbf{j})+(3\mathbf{i})\wedge(3\mathbf{i}+4\mathbf{j})\\ &= 9+9\mathbf{i}\wedge\mathbf{i}+12\mathbf{i}\wedge\mathbf{j}\\ &= 9+12\mathbf{i}\wedge\mathbf{j}.\end{aligned}$$

The 9 represents $|\mathbf{a}||\mathbf{b}|\cos\beta$, whereas the 12 represents an area $|\mathbf{a}||\mathbf{b}|\sin\beta$ on the **i**–**j** plane. The angle between the two vectors β is given by

$$\beta = \cos^{-1}(3/5).$$

However, reversing the product, we obtain

$$\begin{aligned}\mathbf{ba} &= (3\mathbf{i}+4\mathbf{j})\cdot(3\mathbf{i})+(3\mathbf{i}+4\mathbf{j})\wedge(3\mathbf{i})\\ &= 9+9\mathbf{i}\wedge\mathbf{i}+12\mathbf{j}\wedge\mathbf{i}\\ &= 9-12\mathbf{i}\wedge\mathbf{j}\end{aligned}$$

where the sign of the outer (wedge) product has flipped to reflect the new orientation of the vectors relative to the accepted orientation of the basis bivector.

So the geometric product combines the scalar and wedge products into a single product, where the scalar product is the symmetric component and the wedge product is the antisymmetric component. Now let's see how these products behave in 3D.

6.7 The Geometric Product in 3D

Before we begin let's introduce some new notation to simplify future algebraic expressions. Rather than use **i**, **j** and **k** to represent the unit basis vectors let's employ $\mathbf{e}_1$, $\mathbf{e}_2$ and $\mathbf{e}_3$ respectively. This implies that (6.7) can be written

$$\mathbf{ab} = |\mathbf{a}||\mathbf{b}|\cos\beta + |\mathbf{a}||\mathbf{b}|\sin\beta\mathbf{e}_1\wedge\mathbf{e}_2.$$

We also remind ourselves that we are working with a right-handed axial system, where, using our right-hand, the thumb aligns with $\mathbf{e}_1$, the x-axis, the first finger

Fig. 6.3 The 3D bivectors

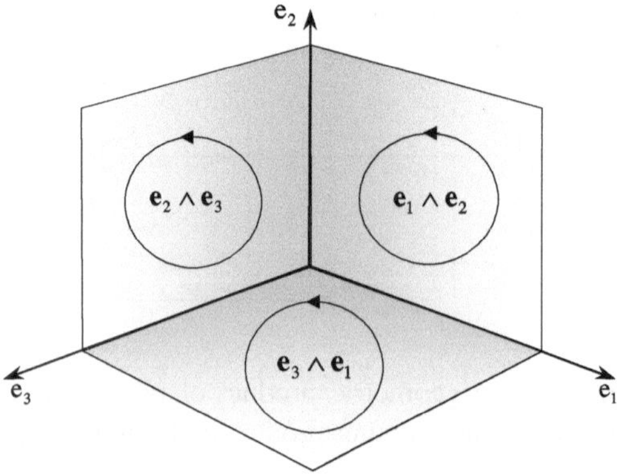

aligns with $\mathbf{e}_2$, the y-axis, and the middle finger aligns with $\mathbf{e}_3$, the z-axis. We begin with two 3D vectors:

$$\begin{aligned}\mathbf{a} &= a_1\mathbf{e}_1 + a_2\mathbf{e}_2 + a_3\mathbf{e}_3\\ \mathbf{b} &= b_1\mathbf{e}_1 + b_2\mathbf{e}_2 + b_3\mathbf{e}_3\end{aligned}$$

and their inner product is

$$\begin{aligned}\mathbf{a}\cdot\mathbf{b} &= (a_1\mathbf{e}_1 + a_2\mathbf{e}_2 + a_3\mathbf{e}_3)\cdot(b_1\mathbf{e}_1 + b_2\mathbf{e}_2 + b_3\mathbf{e}_3)\\ &= a_1b_1 + a_2b_2 + a_3b_3\end{aligned}$$

and their outer product is

$$\begin{aligned}\mathbf{a}\wedge\mathbf{b} &= (a_1\mathbf{e}_1 + a_2\mathbf{e}_2 + a_3\mathbf{e}_3)\wedge(b_1\mathbf{e}_1 + b_2\mathbf{e}_2 + b_3\mathbf{e}_3)\\ &= a_1b_2\mathbf{e}_1\wedge\mathbf{e}_2 + a_1b_3\mathbf{e}_1\wedge\mathbf{e}_3 + a_2b_1\mathbf{e}_2\wedge\mathbf{e}_1\\ &\quad + a_2b_3\mathbf{e}_2\wedge\mathbf{e}_3 + a_3b_1\mathbf{e}_3\wedge\mathbf{e}_1 + a_3b_2\mathbf{e}_3\wedge\mathbf{e}_2\\ &= (a_1b_2 - a_2b_1)\mathbf{e}_1\wedge\mathbf{e}_2 + (a_2b_3 - a_3b_2)\mathbf{e}_2\wedge\mathbf{e}_3 + (a_3b_1 - a_1b_3)\mathbf{e}_3\wedge\mathbf{e}_1.\end{aligned} \tag{6.8}$$

This time we have three unit basis bivectors: $\mathbf{e}_1\wedge\mathbf{e}_2$, $\mathbf{e}_2\wedge\mathbf{e}_3$, $\mathbf{e}_3\wedge\mathbf{e}_1$, and three associated scalar multipliers: $(a_1b_2 - a_2b_1)$, $(a_2b_3 - a_3b_2)$, $(a_3b_1 - a_1b_3)$ respectively. These bivectors are the basis for a right-handed oriented axial system.

Continuing with the idea described in the previous section, the three bivectors represent the three planes containing the respective vectors as shown in Fig. 6.3, and the scalar multipliers are projections of the area of the vector parallelogram onto the three bivectors as shown in Fig. 6.4. Note that this is the accepted definition for a right-handed space. The orientation of the vectors **a** and **b** determine whether the projected areas are positive or negative.

Equation (6.8) should look familiar as it looks similar to the cross product $\mathbf{a}\times\mathbf{b}$:

$$\mathbf{a}\times\mathbf{b} = (a_1b_2 - a_2b_1)\mathbf{e}_3 + (a_2b_3 - a_3b_2)\mathbf{e}_1 + (a_3b_1 - a_1b_3)\mathbf{e}_2. \tag{6.9}$$

This similarity is no accident, for when Hamilton invented quaternions he did not recognise the possibility of bivectors, and invented some rules which eventually

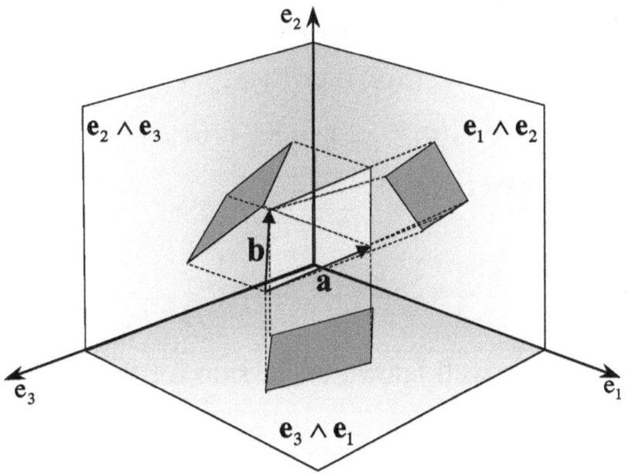

Fig. 6.4 The projections on the three bivectors

became the vector product. In fact, we show later that quaternions are bivectors in disguise. We can see that a simple relationship exists between (6.8) and (6.9):

$$\mathbf{e}_1 \wedge \mathbf{e}_2 \text{ and } \mathbf{e}_3$$
$$\mathbf{e}_2 \wedge \mathbf{e}_3 \text{ and } \mathbf{e}_1$$
$$\mathbf{e}_3 \wedge \mathbf{e}_1 \text{ and } \mathbf{e}_2.$$

The wedge product bivectors are perpendicular to the vector components of the cross product. So the wedge product is just another way of representing the cross product. However, the wedge product introduces a very important bonus – it works in space of any dimension; whereas, the cross product is only comfortable in 3D. Not only that, the wedge (outer) product is a product that creates volumes, hypervolumes, and can also be applied to vectors, bivectors, trivectors, etc.

6.8 The Outer Product of Three 3D Vectors

Having seen that the outer product of two 3D vectors is represented by areal projections onto the three basis bivectors, let's compute the outer product of three 3D vectors:

$$\mathbf{a} = a_1\mathbf{e}_1 + a_2\mathbf{e}_2 + a_3\mathbf{e}_3$$
$$\mathbf{b} = b_1\mathbf{e}_1 + b_2\mathbf{e}_2 + b_3\mathbf{e}_3$$
$$\mathbf{c} = c_1\mathbf{e}_1 + c_2\mathbf{e}_2 + c_3\mathbf{e}_3$$
$$\begin{aligned}\mathbf{a} \wedge \mathbf{b} \wedge \mathbf{c} &= (a_1\mathbf{e}_1 + a_2\mathbf{e}_2 + a_3\mathbf{e}_3) \wedge (b_1\mathbf{e}_1 + b_2\mathbf{e}_2 + b_3\mathbf{e}_3) \wedge (c_1\mathbf{e}_1 + c_2\mathbf{e}_2 + c_3\mathbf{e}_3)\\ &= ((a_1b_2 - a_2b_1)\mathbf{e}_1 \wedge \mathbf{e}_2 + (a_2b_3 - a_3b_2)\mathbf{e}_2 \wedge \mathbf{e}_3 + (a_3b_1 - a_1b_3)\mathbf{e}_3 \wedge \mathbf{e}_1)\\ &\quad \wedge (c_1\mathbf{e}_1 + c_2\mathbf{e}_2 + c_3\mathbf{e}_3).\end{aligned}$$

At this stage we introduce another axiom: the outer product is associative. This means that $\mathbf{a} \wedge (\mathbf{b} \wedge \mathbf{c}) = (\mathbf{a} \wedge \mathbf{b}) \wedge \mathbf{c}$. Therefore, knowing that $\mathbf{a} \wedge \mathbf{a} = 0$:

$$\begin{aligned}\mathbf{a}\wedge\mathbf{b}\wedge\mathbf{c} &= c_3(a_1b_2-a_2b_1)\mathbf{e}_1\wedge\mathbf{e}_2\wedge\mathbf{e}_3 + c_1(a_2b_3-a_3b_2)\mathbf{e}_2\wedge\mathbf{e}_3\wedge\mathbf{e}_1\\ &\quad + c_2(a_3b_1-a_1b_3)\mathbf{e}_3\wedge\mathbf{e}_1\wedge\mathbf{e}_2\\ &= (c_3(a_1b_2-a_2b_1)+c_1(a_2b_3-a_3b_2)+c_2(a_3b_1-a_1b_3))\mathbf{e}_1\wedge\mathbf{e}_2\wedge\mathbf{e}_3\end{aligned}$$

or using a determinant:

$$\mathbf{a}\wedge\mathbf{b}\wedge\mathbf{c} = \begin{vmatrix} a_1 & b_1 & c_1\\ a_2 & b_2 & c_2\\ a_3 & b_3 & c_3\end{vmatrix}\mathbf{e}_1\wedge\mathbf{e}_2\wedge\mathbf{e}_3$$

which is the well-known expression for the volume of a parallelpiped formed by three vectors.

The term $\mathbf{e}_1\wedge\mathbf{e}_2\wedge\mathbf{e}_3$ is a *trivector* and implies that the volume is oriented. If the sign of the determinant is positive, the original three vectors possess the same orientation of the three basis vectors. If the sign of the determinant is negative, the three vectors oppose the orientation of the basis trivector.

6.9 Axioms

One of the features of geometric algebra is that it behaves very similar to the everyday algebra of reals:

Axiom 6.1 The associative rule:

$$\mathbf{a}(\mathbf{bc}) = (\mathbf{ab})\mathbf{c}.$$

Axiom 6.2 The left and right distributive rules:

$$\mathbf{a}(\mathbf{b}+\mathbf{c}) = \mathbf{ab}+\mathbf{ac}$$
$$(\mathbf{b}+\mathbf{c})\mathbf{a} = \mathbf{ba}+\mathbf{ca}.$$

The next four axioms describe how vectors interact with a scalar λ:

Axiom 6.3

$$(\lambda\mathbf{a})\mathbf{b} = \lambda(\mathbf{ab}) = \lambda\mathbf{ab}.$$

Axiom 6.4

$$\lambda(\phi\mathbf{a}) = (\lambda\phi)\mathbf{a}.$$

Axiom 6.5

$$\lambda(\mathbf{a}+\mathbf{b}) = \lambda\mathbf{a}+\lambda\mathbf{b}.$$

Axiom 6.6

$$(\lambda+\phi)\mathbf{a} = \lambda\mathbf{a}+\phi\mathbf{a}.$$

Axiom 6.7

$$\mathbf{a}^2 = |\mathbf{a}|^2.$$

6.10 Notation

Having abandoned $\mathbf{i}, \mathbf{j}, \mathbf{k}$ for $\mathbf{e}_1, \mathbf{e}_2, \mathbf{e}_3$, it is convenient to convert geometric products $\mathbf{e}_1\mathbf{e}_2\cdots\mathbf{e}_n$ to $\mathbf{e}_{12\cdots n}$. For example, $\mathbf{e}_1\mathbf{e}_2\mathbf{e}_3 \equiv \mathbf{e}_{123}$. Furthermore, we must get used to the following substitutions:

$$\begin{aligned}
\mathbf{e}_i\mathbf{e}_i\mathbf{e}_j &= \mathbf{e}_j \\
\mathbf{e}_{21} &= -\mathbf{e}_{12} \\
\mathbf{e}_{312} &= \mathbf{e}_{123} \\
\mathbf{e}_{112} &= \mathbf{e}_2 \\
\mathbf{e}_{121} &= -\mathbf{e}_2.
\end{aligned}$$

6.11 Grades, Pseudoscalars and Multivectors

As geometric algebra embraces such a wide range of objects, it is convenient to *grade* them as follows: scalars are grade 0, vectors are grade 1, bivectors are grade 2, and trivectors are grade 3, and so on for higher dimensions. In such a graded algebra it is traditional to call the highest grade element a *pseudoscalar*. Thus in 2D the pseudoscalar is $\mathbf{e}_{12}$ and in 3D the pseudoscalar is $\mathbf{e}_{123}$.

One very powerful feature of geometric algebra is the idea of a *multivector*, which is a linear combination of a scalar, vector, bivector, trivector, or any other higher dimensional object. For example the following are multivectors:

$$\begin{aligned}
\mathbf{A} &= 3 + (2\mathbf{e}_1 + 3\mathbf{e}_2 + 4\mathbf{e}_3) + (5\mathbf{e}_{12} + 6\mathbf{e}_{23} + 7\mathbf{e}_{31}) + 8\mathbf{e}_{123} \\
\mathbf{B} &= 2 + (2\mathbf{e}_1 + 2\mathbf{e}_2 + 3\mathbf{e}_3) + (4\mathbf{e}_{12} + 5\mathbf{e}_{23} + 6\mathbf{e}_{31}) + 7\mathbf{e}_{123}
\end{aligned}$$

and we can form their sum:

$$\mathbf{A} + \mathbf{B} = 5 + (4\mathbf{e}_1 + 5\mathbf{e}_2 + 7\mathbf{e}_3) + (9\mathbf{e}_{12} + 11\mathbf{e}_{23} + 13\mathbf{e}_{31}) + 15\mathbf{e}_{123}$$

or their difference:

$$\mathbf{A} - \mathbf{B} = 1 + (\mathbf{e}_2 + \mathbf{e}_3) + (\mathbf{e}_{12} + \mathbf{e}_{23} + \mathbf{e}_{31}) + \mathbf{e}_{123}.$$

We can even form their product **AB**.

We can isolate any grade of a multivector using the following notation:

$$\langle multivector \rangle_g$$

where g identifies a particular grade. For example, say we have the following multivector:

$$2 + 3\mathbf{e}_1 + 2\mathbf{e}_2 - 5\mathbf{e}_{12} + 6\mathbf{e}_{123}$$

we extract the scalar term using:

$$\langle 2 + 3\mathbf{e}_1 + 2\mathbf{e}_2 - 5\mathbf{e}_{12} + 6\mathbf{e}_{123} \rangle_0 = 2$$

the vector term using

$$\langle 2 + 3\mathbf{e}_1 + 2\mathbf{e}_2 - 5\mathbf{e}_{12} + 6\mathbf{e}_{123} \rangle_1 = 3\mathbf{e}_1 + 2\mathbf{e}_2$$

the bivector term using:

$$\langle 2 + 3\mathbf{e}_1 + 2\mathbf{e}_2 - 5\mathbf{e}_{12} + 6\mathbf{e}_{123} \rangle_2 = -5\mathbf{e}_{12}$$

and the trivector term using:

$$\langle 2 + 3\mathbf{e}_1 + 2\mathbf{e}_2 - 5\mathbf{e}_{12} + 6\mathbf{e}_{123} \rangle_3 = 6\mathbf{e}_{123}.$$

It is also worth pointing out that the inner vector product converts two grade 1 elements, i.e. vectors, into a grade 0 element, i.e. a scalar, whereas the outer vector product converts two grade 1 elements into a grade 2 element, i.e. a bivector. Thus the inner product is a grade lowering operation, while the outer product is a grade raising operation. These qualities of the inner and outer products are associated with higher grade elements in the algebra. This is why the scalar product is renamed as the inner product, because the scalar product is synonymous with transforming vectors into scalars. Whereas, the inner product transforms two elements of grade n into a grade $n-1$ element.

6.12 Redefining the Inner and Outer Products

As the geometric product is defined in terms of the inner and outer products, it seems only natural to expect that similar functions exist relating the inner and outer products in terms of the geometric product. Such functions do exist and emerge when we combine the following two equations:

$$\mathbf{ab} = \mathbf{a} \cdot \mathbf{b} + \mathbf{a} \wedge \mathbf{b} \tag{6.10}$$

$$\mathbf{ba} = \mathbf{a} \cdot \mathbf{b} - \mathbf{a} \wedge \mathbf{b}. \tag{6.11}$$

Adding and subtracting (6.10) and (6.11) we have

$$\mathbf{a} \cdot \mathbf{b} = \frac{1}{2}(\mathbf{ab} + \mathbf{ba}) \tag{6.12}$$

$$\mathbf{a} \wedge \mathbf{b} = \frac{1}{2}(\mathbf{ab} - \mathbf{ba}). \tag{6.13}$$

Equations (6.12) and (6.13) are used frequently to define the products between different grade elements.

6.13 The Inverse of a Vector

In traditional vector analysis we accept that it is impossible to divide by a vector, but that is not so in geometric algebra. In fact, we don't actually divide a multivector by another vector but find a way of representing the inverse of a vector. For example, we know that a unit vector $\hat{\mathbf{a}}$ is defined as

$$\hat{\mathbf{a}} = \frac{\mathbf{a}}{|\mathbf{a}|}$$

and using the geometric product

$$\hat{\mathbf{a}}^2 = \frac{\mathbf{a}^2}{|\mathbf{a}|^2} = 1$$

therefore,

$$\mathbf{b} = \frac{\mathbf{a}^2\mathbf{b}}{|\mathbf{a}|^2}$$

and exploiting the associative nature of the geometric product we have

$$\mathbf{b} = \frac{\mathbf{a}(\mathbf{ab})}{|\mathbf{a}|^2}. \tag{6.14}$$

Equation (6.14) is effectively stating that, given the geometric product $\mathbf{ab}$ we can recover the vector $\mathbf{b}$ by pre-multiplying by $\mathbf{a}^{-1}$:

$$\mathbf{a}^{-1} = \frac{\mathbf{a}}{|\mathbf{a}|^2}.$$

Similarly, we can recover the vector $\mathbf{a}$ as follows by post-multiplying by $\mathbf{b}^{-1}$:

$$\mathbf{a} = \frac{(\mathbf{ab})\mathbf{b}}{|\mathbf{b}|^2}.$$

For example:

$$\mathbf{a} = \mathbf{e}_1 + 2\mathbf{e}_2$$
$$\mathbf{b} = 3\mathbf{e}_1 + 2\mathbf{e}_2$$

their geometric product is

$$\mathbf{ab} = 7 - 4\mathbf{e}_{12}.$$

Therefore, given $\mathbf{ab}$ and $\mathbf{a}$, we can recover $\mathbf{b}$ as follows:

$$\begin{aligned}
\mathbf{b} &= \left(\frac{\mathbf{e}_1 + 2\mathbf{e}_2}{5}\right)(7 - 4\mathbf{e}_{12})\\
&= \frac{1}{5}(7\mathbf{e}_1 - 4\mathbf{e}_{112} + 14\mathbf{e}_2 - 8\mathbf{e}_{212})\\
&= \frac{1}{5}(7\mathbf{e}_1 - 4\mathbf{e}_2 + 14\mathbf{e}_2 + 8\mathbf{e}_1)\\
&= 3\mathbf{e}_1 + 2\mathbf{e}_2.
\end{aligned}$$

Similarly, given **ab** and **b**, **a** is recovered as follows:

$$\begin{aligned}
\mathbf{a} &= (7 - 4\mathbf{e}_{12})\left(\frac{3\mathbf{e}_1 + 2\mathbf{e}_2}{13}\right) \\
&= \frac{1}{13}(21\mathbf{e}_1 + 14\mathbf{e}_2 - 12\mathbf{e}_{121} - 8\mathbf{e}_{122}) \\
&= \frac{1}{13}(21\mathbf{e}_1 + 14\mathbf{e}_2 + 12\mathbf{e}_2 - 8\mathbf{e}_1) \\
&= \mathbf{e}_1 + 2\mathbf{e}_2.
\end{aligned}$$

Note that the inverse of a unit vector is the original vector:

$$\hat{\mathbf{a}}^{-1} = \frac{\hat{\mathbf{a}}}{|\hat{\mathbf{a}}|^2} = \hat{\mathbf{a}}.$$

6.14 The Imaginary Properties of the Outer Product

So far we know that the outer product of two vectors is represented by one or more unit basis vectors, such as

$$\mathbf{a} \wedge \mathbf{b} = \lambda_1 \mathbf{e}_{12} + \lambda_2 \mathbf{e}_{23} + \lambda_3 \mathbf{e}_{31}$$

where, in this case, the λ terms represent areas projected onto their respective unit basis bivectors. But what has not emerged is that the outer product is an imaginary quantity, which is revealed by expanding $\mathbf{e}_{12}^2$:

$$\mathbf{e}_{12}^2 = \mathbf{e}_{1212}$$

but as

$$\mathbf{e}_{21} = -\mathbf{e}_{12}$$

then

$$\begin{aligned}
\mathbf{e}_{1(21)2} &= -\mathbf{e}_{1(12)2} \\
&= -\mathbf{e}_1^2\mathbf{e}_2^2 \\
\mathbf{e}_{12}^2 &= -1.
\end{aligned}$$

Consequently, the geometric product effectively creates a complex number! Thus in a 2D scenario, given:

$$\begin{aligned}
\mathbf{a} &= a_1\mathbf{e}_1 + a_2\mathbf{e}_2 \\
\mathbf{b} &= b_1\mathbf{e}_1 + b_2\mathbf{e}_2
\end{aligned}$$

their geometric product is

$$\mathbf{ab} = (a_1b_1 + a_2b_2) + (a_1b_2 - a_2b_1)\mathbf{e}_{12}$$

and knowing that $\mathbf{e}_{12} = i$, then we can write **ab** as

$$\mathbf{ab} = (a_1b_1 + a_2b_2) + (a_1b_2 - a_2b_1)i. \tag{6.15}$$

However, this notation is not generally adopted by the geometric algebra community. The reason being that i is normally only associated with a scalar, with which it commutes. Whereas in 2D, $\mathbf{e}_{12}$ is associated with scalars and vectors, and although scalars present no problem, under some conditions, it anticommutes with vectors. Consequently, an upper-case I is used so that there is no confusion between the two elements. Thus (6.15) is written as

$$\mathbf{ab} = (a_1b_1 + a_2b_2) + (a_1b_2 - a_2b_1)I$$

where

$$I^2 = -1.$$

It goes without saying that the 3D unit basis bivectors are also imaginary quantities, so too, is $\mathbf{e}_{123}$.

Multiplying a complex number by i rotates it 90° on the complex plane. Therefore, it should be no surprise that multiplying a 2D vector by $\mathbf{e}_{12}$ rotates it by 90°. However, because vectors are sensitive to their product partners, we must remember that pre-multiplying a vector by $\mathbf{e}_{12}$ rotates a vector clockwise and post-multiplying rotates a vector anticlockwise. For instance, post-multiplying $\mathbf{e}_1$ by $\mathbf{e}_{12}$ creates $\mathbf{e}_2$, which is an anticlockwise rotation, whereas, pre-multiplying $\mathbf{e}_1$ by $\mathbf{e}_{12}$ creates $-\mathbf{e}_2$, which is a clockwise rotation.

Whilst on the subject of rotations, let's consider what happens in 3D. We begin with a 3D vector

$$\mathbf{a} = a_1\mathbf{e}_1 + a_2\mathbf{e}_2 + a_3\mathbf{e}_3$$

and the unit basis bivector $\mathbf{e}_{12}$ as shown in Fig. 6.5. Next we construct their geometric product

$$\begin{aligned}
\mathbf{e}_{12}\mathbf{a} &= a_1\mathbf{e}_{12}\mathbf{e}_1 + a_2\mathbf{e}_{12}\mathbf{e}_2 + a_3\mathbf{e}_{12}\mathbf{e}_3 \\
&= a_1\mathbf{e}_{121} + a_2\mathbf{e}_{122} + a_3\mathbf{e}_{123} \\
&= -a_1\mathbf{e}_2 + a_2\mathbf{e}_1 + a_3\mathbf{e}_{123} \\
&= a_2\mathbf{e}_1 - a_1\mathbf{e}_2 + a_3\mathbf{e}_{123}
\end{aligned}$$

which contains two parts: a vector $(a_2\mathbf{e}_1 - a_1\mathbf{e}_2)$ and a volume $a_3\mathbf{e}_{123}$.

Figure 6.5 shows how the projection of vector $\mathbf{a}$ is rotated anticlockwise on the bivector $\mathbf{e}_{12}$. A volume is also created perpendicular to the bivector. This enables us to predict that if the vector is coplanar with the bivector, the entire vector is rotated 90° and the volume component is zero.

Post-multiplying $\mathbf{a}$ by $\mathbf{e}_{12}$ creates

$$\mathbf{ae}_{12} = -a_2\mathbf{e}_1 + a_1\mathbf{e}_2 + a_3\mathbf{e}_{123}$$

which shows that while the volumetric element has remained the same, the projected vector is rotated anticlockwise. You may wish to show that the same happens with the other two bivectors.

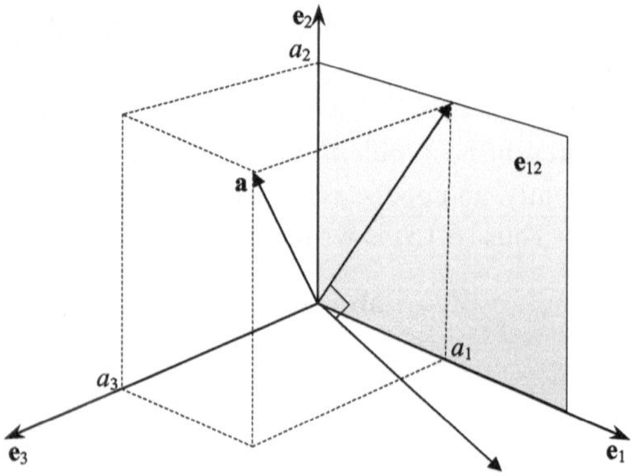

Fig. 6.5 The effect of multiplying a vector by a bivector

6.15 Duality

The ability to exchange pairs of geometric elements such as lines and planes involves a *dual* operation, which in geometric algebra is relatively easy to define. For example, given a multivector $\mathbf{A}$ its dual $\mathbf{A}^*$ is defined as

$$\mathbf{A}^* = I\mathbf{A}$$

where I is the local pseudoscalar. For 2D this is $\mathbf{e}_{12}$ and for 3D it is $\mathbf{e}_{123}$. Therefore, given:

$$\mathbf{a} = a_1\mathbf{e}_1 + a_2\mathbf{e}_2$$

its dual is

$$\begin{aligned}
\mathbf{a}^* &= \mathbf{e}_{12}(a_1\mathbf{e}_1 + a_2\mathbf{e}_2)\\
&= a_1\mathbf{e}_{121} + a_2\mathbf{e}_{122}\\
&= a_2\mathbf{e}_1 - a_1\mathbf{e}_2
\end{aligned}$$

which is another vector rotated 90° anticlockwise.

It is easy to show that $(\mathbf{a}^*)^* = -\mathbf{a}$, and two further dual operations return the vector back to $\mathbf{a}$.

In 3D the dual of a vector $\mathbf{e}_1$ is

$$\mathbf{e}_{123}\mathbf{e}_1 = \mathbf{e}_{1231} = \mathbf{e}_{23}$$

which is the perpendicular bivector. Similarly, the dual of $\mathbf{e}_2$ is $\mathbf{e}_{31}$ and the dual of $\mathbf{e}_3$ is $\mathbf{e}_{12}$.

For a general vector $a_1\mathbf{e}_1 + a_2\mathbf{e}_2 + a_3\mathbf{e}_3$ its dual is

$$\begin{aligned}
\mathbf{e}_{123}(a_1\mathbf{e}_1 + a_2\mathbf{e}_2 + a_3\mathbf{e}_3) &= a_1\mathbf{e}_{1231} + a_2\mathbf{e}_{1232} + a_3\mathbf{e}_{1233}\\
&= a_3\mathbf{e}_{12} + a_1\mathbf{e}_{23} + a_2\mathbf{e}_{31}.
\end{aligned}$$

The duals of the 3D basis bivectors are:

$$\begin{aligned}
\mathbf{e}_{123}\mathbf{e}_{12} &= \mathbf{e}_{12312} = -\mathbf{e}_3\\
\mathbf{e}_{123}\mathbf{e}_{23} &= \mathbf{e}_{12323} = -\mathbf{e}_1\\
\mathbf{e}_{123}\mathbf{e}_{31} &= \mathbf{e}_{12331} = -\mathbf{e}_2.
\end{aligned}$$

Table 6.1 Hamilton's quaternion product rules

	i	j	k
i	-1	k	$-j$
j	$-k$	-1	i
k	j	$-i$	-1

6.16 The Relationship Between the Vector Product and the Outer Product

We have already discovered that there is a very close relationship between the vector product and the outer product, and we will see what happens when we form the cross and wedge product of two 3D vectors:

$$\mathbf{a} = a_1\mathbf{e}_1 + a_2\mathbf{e}_2 + a_3\mathbf{e}_3$$

$$\mathbf{b} = b_1\mathbf{e}_1 + b_2\mathbf{e}_2 + b_3\mathbf{e}_3$$

$$\mathbf{a} \times \mathbf{b} = (a_2b_3 - a_3b_2)\mathbf{e}_1 + (a_3b_1 - a_1b_3)\mathbf{e}_2 + (a_1b_2 - a_2b_1)\mathbf{e}_3 \tag{6.16}$$

$$\begin{aligned}\mathbf{a} \wedge \mathbf{b} &= (a_2b_3 - a_3b_2)\mathbf{e}_2 \wedge \mathbf{e}_3 + (a_3b_1 - a_1b_3)\mathbf{e}_3 \wedge \mathbf{e}_1 + (a_1b_2 - a_2b_1)\mathbf{e}_1 \wedge \mathbf{e}_2\\ &= (a_2b_3 - a_3b_2)\mathbf{e}_{23} + (a_3b_1 - a_1b_3)\mathbf{e}_{31} + (a_1b_2 - a_2b_1)\mathbf{e}_{12}.\end{aligned} \tag{6.17}$$

Multiplying (6.17) by I_{123} we obtain

$$\begin{aligned}I_{123}(\mathbf{a} \wedge \mathbf{b}) &= (a_2b_3 - a_3b_2)\mathbf{e}_{123}\mathbf{e}_{23} + (a_3b_1 - a_1b_3)\mathbf{e}_{123}\mathbf{e}_{31}\\ &\quad + (a_1b_2 - a_2b_1)\mathbf{e}_{123}\mathbf{e}_{12}\\ &= -((a_2b_3 - a_3b_2)\mathbf{e}_1 + (a_3b_1 - a_1b_3)\mathbf{e}_2 + (a_1b_2 - a_2b_1)\mathbf{e}_3)\end{aligned}$$

which is identical to the cross product (6.9) apart from its sign. Therefore, we can state:

$$\mathbf{a} \times \mathbf{b} = -I_{123}(\mathbf{a} \wedge \mathbf{b}).$$

6.17 The Relationship Between Quaternions and Bivectors

Hamilton's rules for the imaginaries i, j and k are shown in Table 6.1, whilst Table 6.2 shows the rules for 3D bivector products. Although there is some agreement between the table entries, there is a sign reversal in some of them. However, if we switch to a left-handed axial system the bivectors become $\mathbf{e}_{32}$, $\mathbf{e}_{13}$, $\mathbf{e}_{21}$ and their products are as shown in Table 6.3. If we now create a one-to-one correspondence (isomorphism) between the two systems:

$$i \leftrightarrow \mathbf{e}_{32}, \quad j \leftrightarrow \mathbf{e}_{13}, \quad k \leftrightarrow \mathbf{e}_{21}$$

there is a true correspondence between quaternions and a left-handed set of bivectors.

Table 6.2 3D bivector product rules

	$\mathbf{e}_{23}$	$\mathbf{e}_{31}$	$\mathbf{e}_{12}$
$\mathbf{e}_{23}$	-1	$-\mathbf{e}_{12}$	$\mathbf{e}_{31}$
$\mathbf{e}_{31}$	$\mathbf{e}_{12}$	-1	$-\mathbf{e}_{23}$
$\mathbf{e}_{12}$	$-\mathbf{e}_{31}$	$\mathbf{e}_{23}$	-1

Table 6.3 Left-handed 3D bivector product rules

	$\mathbf{e}_{32}$	$\mathbf{e}_{13}$	$\mathbf{e}_{21}$
$\mathbf{e}_{32}$	-1	$\mathbf{e}_{21}$	$-\mathbf{e}_{13}$
$\mathbf{e}_{13}$	$-\mathbf{e}_{21}$	-1	$\mathbf{e}_{32}$
$\mathbf{e}_{21}$	$\mathbf{e}_{13}$	$-\mathbf{e}_{32}$	-1

6.18 Reverse of a Multivector

You will have noticed how sensitive geometric algebra is to the sequence of vectors, and it should not be too much of a surprise to learn that a special function exists to reverse sequences of elements. For example, given

$$\mathbf{A} = \mathbf{ab}$$

the reverse of $\mathbf{A}$ is denoted using the dagger symbol $\mathbf{A}^{\dagger}$

$$\mathbf{A}^{\dagger} = \mathbf{ba}$$

or the tilde symbol $\tilde{\mathbf{A}}$

$$\tilde{\mathbf{A}} = \mathbf{ba}.$$

The dagger symbol is used in this book.

Obviously, scalars are unaffected by reversion, neither are vectors. However, bivectors and trivectors flip their signs:

$$(\mathbf{e}_1\mathbf{e}_2)^{\dagger} = \mathbf{e}_2\mathbf{e}_1 = -\mathbf{e}_1\mathbf{e}_2$$

and

$$(\mathbf{e}_1\mathbf{e}_2\mathbf{e}_3)^{\dagger} = \mathbf{e}_3\mathbf{e}_2\mathbf{e}_1 = -\mathbf{e}_1\mathbf{e}_2\mathbf{e}_3.$$

When reversing a multivector containing terms up to a trivector, it's only the bivector and trivector terms that are reversed. For example, given a multivector $\mathbf{A}$

$$\mathbf{A} = \lambda + \mathbf{v} + \mathbf{B} + \mathbf{T}$$

where

λ is a scalar

$\mathbf{v}$ is a vector

$\mathbf{B}$ is a bivector, and

$\mathbf{T}$ is a trivector

then

$$\mathbf{A}^{\dagger} = \lambda + \mathbf{v} - \mathbf{B} - \mathbf{T}.$$

Let's illustrate this reversion process with an example.

Given three vectors

$$\mathbf{a} = 2\mathbf{e}_1 + 3\mathbf{e}_2$$
$$\mathbf{b} = 4\mathbf{e}_1 - 2\mathbf{e}_2$$
$$\mathbf{c} = \mathbf{e}_1 + \mathbf{e}_2$$

the products **ab** and **ba** are

$$\mathbf{ab} = (2\mathbf{e}_1 + 3\mathbf{e}_2)(4\mathbf{e}_1 - 2\mathbf{e}_2) = 2 - 16\mathbf{e}_{12}$$
$$\mathbf{ba} = (4\mathbf{e}_1 - 2\mathbf{e}_2)(2\mathbf{e}_1 + 3\mathbf{e}_2) = 2 + 16\mathbf{e}_{12}.$$

Thus

$$(\mathbf{ab})^{\dagger} = \mathbf{ba}.$$

Furthermore, the products **abc** and **cba** are

$$\mathbf{abc} = (2\mathbf{e}_1 + 3\mathbf{e}_2)(4\mathbf{e}_1 - 2\mathbf{e}_2)(\mathbf{e}_1 + \mathbf{e}_2) = -14\mathbf{e}_1 + 18\mathbf{e}_2$$
$$\mathbf{cba} = (\mathbf{e}_1 + \mathbf{e}_2)(4\mathbf{e}_1 - 2\mathbf{e}_2)(2\mathbf{e}_1 + 3\mathbf{e}_2) = -14\mathbf{e}_1 + 18\mathbf{e}_2.$$

And as there are only vectors terms there are no sign changes.

Reversion plays an important role in rotors and we will meet them again in the next chapter.

6.19 Summary

This chapter has covered the basic ideas behind geometric algebra which offers an algebraic framework for oriented lines (vectors), oriented planes (bivectors) and oriented volumes (trivectors), not to mention higher dimensional objects. We have yet to discover how they offer an alternative way of rotating points in the plane and in 3D space.

6.19.1 Summary of Multivector Operations

Inner product: 2D vectors

$$\mathbf{a} = a_1\mathbf{e}_1 + a_2\mathbf{e}_2$$
$$\mathbf{b} = b_1\mathbf{e}_1 + b_2\mathbf{e}_2$$
$$\mathbf{a} \cdot \mathbf{b} = |\mathbf{a}||\mathbf{b}|\cos\beta = a_1b_1 + a_2b_2.$$

Inner product: 3D vectors

$$
\begin{aligned}
\mathbf{a} &= a_1\mathbf{e}_1 + a_2\mathbf{e}_2 + a_3\mathbf{e}_3\\
\mathbf{b} &= b_1\mathbf{e}_1 + b_2\mathbf{e}_2 + b_3\mathbf{e}_3\\
\mathbf{a}\cdot\mathbf{b} &= |\mathbf{a}||\mathbf{b}|\cos\beta = a_1b_1 + a_2b_2 + a_3b_3.
\end{aligned}
$$

Outer product: 2D vectors

$$
\begin{aligned}
\mathbf{a} &= a_1\mathbf{e}_1 + a_2\mathbf{e}_2\\
\mathbf{b} &= b_1\mathbf{e}_1 + b_2\mathbf{e}_2\\
\mathbf{a}\wedge\mathbf{b} &= \begin{vmatrix} a_1 & a_2\\ b_1 & b_2 \end{vmatrix}\mathbf{e}_1\wedge\mathbf{e}_2.
\end{aligned}
$$

Outer product: 3D vectors

$$
\begin{aligned}
\mathbf{a} &= a_1\mathbf{e}_1 + a_2\mathbf{e}_2 + a_3\mathbf{e}_3\\
\mathbf{b} &= b_1\mathbf{e}_1 + b_2\mathbf{e}_2 + b_3\mathbf{e}_3\\
\mathbf{c} &= c_1\mathbf{e}_1 + c_2\mathbf{e}_2 + c_3\mathbf{e}_3\\
\mathbf{a}\wedge\mathbf{b} &= \begin{vmatrix} a_1 & a_2\\ b_1 & b_2 \end{vmatrix}\mathbf{e}_1\wedge\mathbf{e}_2 + \begin{vmatrix} a_2 & a_3\\ b_2 & b_3 \end{vmatrix}\mathbf{e}_2\wedge\mathbf{e}_3 + \begin{vmatrix} a_3 & a_1\\ b_3 & b_1 \end{vmatrix}\mathbf{e}_3\wedge\mathbf{e}_1\\
\mathbf{a}\wedge\mathbf{b}\wedge\mathbf{c} &= \begin{vmatrix} a_1 & a_2 & a_3\\ b_1 & b_2 & b_3\\ c_1 & c_2 & c_3 \end{vmatrix}\mathbf{e}_1\wedge\mathbf{e}_2\wedge\mathbf{e}_3\\
|\mathbf{a}\wedge\mathbf{b}| &= |\mathbf{a}||\mathbf{b}|\sin\beta.
\end{aligned}
$$

Geometric product

$$
\begin{aligned}
\mathbf{ab} &= \mathbf{a}\cdot\mathbf{b} + \mathbf{a}\wedge\mathbf{b}\\
\mathbf{ba} &= \mathbf{a}\cdot\mathbf{b} - \mathbf{a}\wedge\mathbf{b}\\
\mathbf{a}\cdot\mathbf{b} &= \frac{1}{2}(\mathbf{ab} + \mathbf{ba})\\
\mathbf{a}\wedge\mathbf{b} &= \frac{1}{2}(\mathbf{ab} - \mathbf{ba}).
\end{aligned}
$$

Inverse of a vector

$$
\mathbf{a}^{-1} = \frac{\mathbf{a}}{|\mathbf{a}|^2}.
$$

Duality

$$
\begin{aligned}
\mathbf{A}^{\star} &= I\mathbf{A}\\
I &= \text{the local pseudoscalar.}
\end{aligned}
$$

Reverse of a multivector

$$
\begin{aligned}
\mathbf{A} &= \mathbf{ab}\\
\mathbf{A}^{\dagger} &= \mathbf{ba}.
\end{aligned}
$$

Chapter 7
Rotation Transforms in the Plane

7.1 Introduction

In this chapter we investigate the techniques for rotating points in the plane using geometry, matrices and multivectors. These ideas will be developed further in the next chapter where we consider rotating frames of reference in the plane.

7.2 2D Transforms

2D transforms are used to scale, translate, rotate, reflect and shear shapes. For example, a point $P(x, y)$ is transformed into $P'(x', y')$ by modifying x and y using

$$\begin{aligned} x' &= ax + by \\ y' &= cx + dy. \end{aligned}$$

By using different values for a, b, c and d we can scale, shear, reflect or rotate a point about the origin. However, this transform cannot effect a translation as we need to increment both x and y by values which are spatial offsets. To achieve this we employ homogeneous coordinates.

7.2.1 Homogeneous Coordinates

Homogeneous coordinates are used to define a point in the plane using three coordinates instead of two. This means that for a point (x, y) there exists a homogeneous point (xt, yt, t) where t is an arbitrary number. The values of x and y are found by dividing xt and yt by t. For example, the point $(2, 3)$ has homogeneous coordinates $(4, 6, 2)$, because $(4/2, 6/2, 2/2) = (2, 3, 1)$. But the homogeneous point $(4, 6, 2)$ is not unique to $(2, 3)$ – $(8, 12, 4)$, $(10, 15, 5)$ and $(200, 300, 100)$ are all possible homogeneous coordinates for $(2, 3)$.

J. Vince, *Rotation Transforms for Computer Graphics*,
DOI 10.1007/978-0-85729-154-7_7,

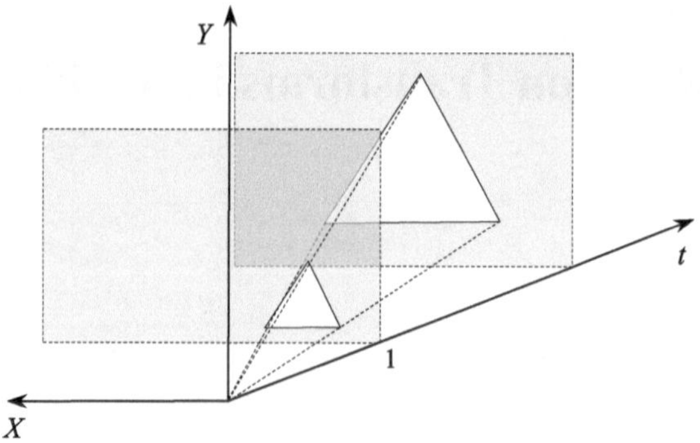

Fig. 7.1 2D homogeneous coordinates can be visualised as a plane in 3D space where $t = 1$, for convenience

For our purposes we can imagine that a collection of homogeneous points of the form (xt, yt, t) exist on an xy-plane where t is the z-coordinate as illustrated in Fig. 7.1. The figure shows a triangle on the $t = 1$ plane, and a similar, larger triangle on a more distant plane. Thus, instead of working in two dimensions, we can work on an arbitrary xy-plane in three dimensions. The t-coordinate of the plane is immaterial because the x- and y-coordinates are eventually scaled by t. To keep things simple it is convenient to choose $t = 1$, which means that the point (x, y) has homogeneous coordinates $(x, y, 1)$ making scaling unnecessary.

If we substitute 3D homogeneous coordinates for traditional 2D Cartesian coordinates we must attach 1 to every (x, y) pair. When a point $(x, y, 1)$ is transformed, it emerges as $(x', y', 1)$, and we discard the 1. This may seem a futile exercise, but it resolves the problem of creating a translation transform.

Consider the following transform on the homogeneous point $(x, y, 1)$:

$$x' = ax + by + e$$
$$y' = cx + dy + f$$

this is represented in matrix form as

$$\begin{bmatrix} x' \\ y' \\ 1 \end{bmatrix} = \begin{bmatrix} a & b & e \\ c & d & f \\ 0 & 0 & 1 \end{bmatrix} \begin{bmatrix} x \\ y \\ 1 \end{bmatrix}$$

and resolves the problem of translation by adding e and f to x' and y' respectively. However, this has to be paid for in terms of extra memory required to store the larger matrix.

7.3 Matrix Transforms

In this section we investigate strategies for designing matrix transforms to translate and rotate points about the origin and an arbitrary point. We will also see how the inverse transform is used to establish new frames of reference.

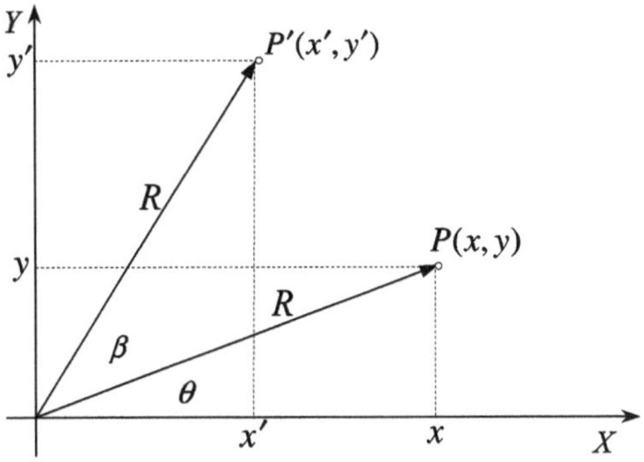

Fig. 7.2 The point $P(x, y)$ is rotated through an angle β to $P'(x', y')$

7.3.1 Translate a Point

Perhaps the simplest transform is that of point translation. For example, to translate a point $P(x, y)$ by (t_x, t_y), we only require

$$\begin{aligned} x' &= x + t_x \\ y' &= y + t_y \end{aligned}$$

which is represented by this homogeneous matrix

$$\begin{bmatrix} x' \\ y' \\ 1 \end{bmatrix} = \begin{bmatrix} 1 & 0 & t_x \\ 0 & 1 & t_y \\ 0 & 0 & 1 \end{bmatrix} \begin{bmatrix} x \\ y \\ 1 \end{bmatrix}.$$

We will refer to this translate matrix as $\mathbf{T}_{t_x,t_y}$.

As an example, let's translate the point $P(2, 3)$ by $(4, 5)$, which moves it to $P'(6, 8)$:

$$\begin{bmatrix} 6 \\ 8 \\ 1 \end{bmatrix} = \begin{bmatrix} 1 & 0 & 4 \\ 0 & 1 & 5 \\ 0 & 0 & 1 \end{bmatrix} \begin{bmatrix} 2 \\ 3 \\ 1 \end{bmatrix}.$$

7.3.2 Rotate a Point About the Origin

Figure 7.2 shows a point $P(x, y)$ which is rotated an angle β about the origin to $P'(x', y')$, and as we are dealing with a pure rotation, both P' and P are distance R from the origin.

From Fig. 7.2 it can be seen that

$$\begin{aligned} \cos\theta &= x/R \\ \sin\theta &= y/R \\ x' &= R\cos(\theta + \beta) \\ y' &= R\sin(\theta + \beta) \end{aligned}$$

and substituting the identities for $\cos(\theta+\beta)$ and $\sin(\theta+\beta)$ we have

$$\begin{aligned}
x' &= R(\cos\theta\cos\beta - \sin\theta\sin\beta)\\
&= R\left(\frac{x}{R}\cos\beta - \frac{y}{R}\sin\beta\right)\\
&= x\cos\beta - y\sin\beta\\
y' &= R(\sin\theta\cos\beta + \cos\theta\sin\beta)\\
&= R\left(\frac{y}{R}\cos\beta + \frac{x}{R}\sin\beta\right)\\
&= x\sin\beta + y\cos\beta
\end{aligned}$$

or in matrix form

$$\begin{bmatrix} x' \\ y' \\ 1 \end{bmatrix} = \begin{bmatrix} \cos\beta & -\sin\beta & 0 \\ \sin\beta & \cos\beta & 0 \\ 0 & 0 & 1 \end{bmatrix} \begin{bmatrix} x \\ y \\ 1 \end{bmatrix}$$

and is the homogeneous transform for rotating points about the origin. For example, to rotate a point 90° about the origin the transform becomes

$$\begin{bmatrix} 0 \\ 1 \\ 1 \end{bmatrix} = \begin{bmatrix} 0 & -1 & 0 \\ 1 & 0 & 0 \\ 0 & 0 & 1 \end{bmatrix} \begin{bmatrix} 1 \\ 0 \\ 1 \end{bmatrix}$$

where we see the point $(1, 0, 1)$ becomes $(0, 1, 1)$ and we ignore the homogeneous scaling factor of 1.

Rotating a point 360° about the origin the transform becomes the identity matrix:

$$\begin{bmatrix} x' \\ y' \\ 1 \end{bmatrix} = \begin{bmatrix} 1 & 0 & 0 \\ 0 & 1 & 0 \\ 0 & 0 & 1 \end{bmatrix} \begin{bmatrix} x \\ y \\ 1 \end{bmatrix}.$$

The following observations can be made about the rotation matrix $\mathbf{R}_\beta$:

$$\mathbf{R}_\beta = \begin{bmatrix} \cos\beta & -\sin\beta & 0 \\ \sin\beta & \cos\beta & 0 \\ 0 & 0 & 1 \end{bmatrix}.$$

Its determinant equals 1:

$$\det\mathbf{R}_\beta = \cos^2\beta + \sin^2\beta = 1.$$

Its transpose is

$$\mathbf{R}_\beta^{\mathrm{T}} = \begin{bmatrix} \cos\beta & \sin\beta & 0 \\ -\sin\beta & \cos\beta & 0 \\ 0 & 0 & 1 \end{bmatrix}.$$

The product $\mathbf{R}_\beta\mathbf{R}_\beta^{\mathrm{T}} = \mathbf{I}$:

$$\mathbf{R}_\beta\mathbf{R}_\beta^{\mathrm{T}} = \begin{bmatrix} \cos\beta & -\sin\beta & 0 \\ \sin\beta & \cos\beta & 0 \\ 0 & 0 & 1 \end{bmatrix} \begin{bmatrix} \cos\beta & \sin\beta & 0 \\ -\sin\beta & \cos\beta & 0 \\ 0 & 0 & 1 \end{bmatrix} = \begin{bmatrix} 1 & 0 & 0 \\ 0 & 1 & 0 \\ 0 & 0 & 1 \end{bmatrix}$$

and because $\mathbf{R}_\beta \mathbf{R}_\beta^{\mathrm{T}}$ equals the identity matrix, $\mathbf{R}_\beta^{-1} = \mathbf{R}_\beta^{\mathrm{T}}$:

$$\mathbf{R}_\beta^{-1} = \begin{bmatrix} \cos\beta & \sin\beta & 0 \\ -\sin\beta & \cos\beta & 0 \\ 0 & 0 & 1 \end{bmatrix}$$

confirms that $\mathbf{R}_\beta$ is orthogonal.

7.3.3 Rotate a Point About an Arbitrary Point

Now let's see how to rotate a point (x, y) about an arbitrary point (t_x, t_y). The strategy involves making the point of rotation a temporary origin, which is achieved by subtracting (t_x, t_y) from the coordinates (x, y) respectively. Next, we perform a rotation about the temporary origin, and finally, we add (t_x, t_y) back to the rotated point to compensate for the original subtraction. Here are the steps:
1. Subtract (t_x, t_y) to create a new temporary origin:

$$\begin{aligned} x_1 &= x - t_x \\ y_1 &= y - t_y. \end{aligned}$$

2. Rotate (x_1, y_1) about the temporary origin by β:

$$\begin{aligned} x_2 &= (x - t_x)\cos\beta - (y - t_y)\sin\beta \\ y_2 &= (x - t_x)\sin\beta + (y - t_y)\cos\beta. \end{aligned}$$

3. Add (t_x, t_y) to the rotated point (x_2, y_2) to return to the original origin:

$$\begin{aligned} x' &= x_2 + t_x \\ y' &= y_2 + t_y \\ x' &= (x - t_x)\cos\beta - (y - t_y)\sin\beta + t_x \\ y' &= (x - t_x)\sin\beta + (y - t_y)\cos\beta + t_y. \end{aligned}$$

Simplifying, we obtain

$$\begin{aligned} x' &= x\cos\beta - y\sin\beta + t_x(1 - \cos\beta) + t_y\sin\beta \\ y' &= x\sin\beta + y\cos\beta + t_y(1 - \cos\beta) - t_x\sin\beta \end{aligned}$$

and in matrix form we have

$$\begin{bmatrix} x' \\ y' \\ 1 \end{bmatrix} = \begin{bmatrix} \cos\beta & -\sin\beta & t_x(1 - \cos\beta) + t_y\sin\beta \\ \sin\beta & \cos\beta & t_y(1 - \cos\beta) - t_x\sin\beta \\ 0 & 0 & 1 \end{bmatrix} \begin{bmatrix} x \\ y \\ 1 \end{bmatrix}. \qquad (7.1)$$

For example, if we rotate the point (2, 1), 90° about the point (1, 1) the transform (7.1) becomes

$$\begin{bmatrix} 1 \\ 2 \\ 1 \end{bmatrix} = \begin{bmatrix} 0 & -1 & 2 \\ 1 & 0 & 0 \\ 0 & 0 & 1 \end{bmatrix} \begin{bmatrix} 2 \\ 1 \\ 1 \end{bmatrix}.$$

The above algebraic approach to derive the rotation transform is relatively easy. However, it is also possible to use matrices to derive composite transforms, such as a reflection relative to an arbitrary line or scaling and rotation relative to an arbitrary point. All of these linear transforms are called *affine* transforms, as parallel lines remain parallel after being transformed. Furthermore, the word 'affine' is used to imply that there is a strong geometric *affinity* between the original and transformed shape. One cannot always guarantee that angles and lengths are preserved, as these can change when different scaling factors are used. For completeness, let's derive the above transform using matrices.

A transform for rotating a point β about the origin is given by

$$\mathbf{R}_\beta = \begin{bmatrix} \cos\beta & -\sin\beta & 0 \\ \sin\beta & \cos\beta & 0 \\ 0 & 0 & 1 \end{bmatrix}$$

and a transform for translating a point (t_x, t_y) relative to the origin is given by

$$\mathbf{T}_{t_x,t_y} = \begin{bmatrix} 1 & 0 & t_x \\ 0 & 1 & t_y \\ 0 & 0 & 1 \end{bmatrix}.$$

We can use $\mathbf{R}_\beta$ and $\mathbf{T}_{t_x,t_y}$ to develop a composite transform for rotating a point about an arbitrary point (t_x, t_y) as follows:

$$\begin{bmatrix} x' \\ y' \\ 1 \end{bmatrix} = \mathbf{T}_{t_x,t_y}\mathbf{R}_\beta\mathbf{T}_{-t_x,-t_y}\begin{bmatrix} x \\ y \\ 1 \end{bmatrix} \tag{7.2}$$

where

$\mathbf{T}_{-t_x,-t_y}$	creates a temporary origin
$\mathbf{R}_\beta$	rotates β about the origin
$\mathbf{T}_{t_x,t_y}$	returns to the original position.

Note that the transform sequence starts on the right next to the original coordinates, working leftwards.

Equation (7.2) expands to

$$\begin{bmatrix} x' \\ y' \\ 1 \end{bmatrix} = \begin{bmatrix} 1 & 0 & t_x \\ 0 & 1 & t_y \\ 0 & 0 & 1 \end{bmatrix}\begin{bmatrix} \cos\beta & -\sin\beta & 0 \\ \sin\beta & \cos\beta & 0 \\ 0 & 0 & 1 \end{bmatrix}\begin{bmatrix} 1 & 0 & -t_x \\ 0 & 1 & -t_y \\ 0 & 0 & 1 \end{bmatrix}\begin{bmatrix} x \\ y \\ 1 \end{bmatrix}.$$

Next, we multiply these matrices together to form a single matrix. Let's begin by multiplying the $\mathbf{R}_\beta$ and $\mathbf{T}_{-t_x,-t_y}$ matrices, which produces

$$\begin{bmatrix} x' \\ y' \\ 1 \end{bmatrix} = \begin{bmatrix} 1 & 0 & t_x \\ 0 & 1 & t_y \\ 0 & 0 & 1 \end{bmatrix}\begin{bmatrix} \cos\beta & -\sin\beta & -t_x\cos\beta + t_y\sin\beta \\ \sin\beta & \cos\beta & -t_x\sin\beta - t_y\cos\beta \\ 0 & 0 & 1 \end{bmatrix}\begin{bmatrix} x \\ y \\ 1 \end{bmatrix}$$

and finally we obtain

$$\begin{bmatrix} x' \\ y' \\ 1 \end{bmatrix} = \begin{bmatrix} \cos\beta & -\sin\beta & t_x(1-\cos\beta) + t_y\sin\beta \\ \sin\beta & \cos\beta & t_y(1-\cos\beta) - t_x\sin\beta \\ 0 & 0 & 1 \end{bmatrix}\begin{bmatrix} x \\ y \\ 1 \end{bmatrix}$$

which is the same as the previous transform (7.1).

7.3.4 Rotate and Translate a Point

There are two ways we can combine the rotate and translate transforms into a single transform. The first method starts by translating a point $P(x, y)$ using $\mathbf{T}(t_x, t_y)$ to an intermediate point, which is then rotated using $\mathbf{R}_\beta$. The problem with this strategy is that the radius of rotation becomes large and subjects the point to a large circular motion. The normal way is to first subject the point to a rotation about the origin and then translate it:

$$P' = \mathbf{T}_{t_x,t_y}\mathbf{R}_\beta P$$

$$\begin{bmatrix} x' \\ y' \\ 1 \end{bmatrix} = \begin{bmatrix} 1 & 0 & t_x \\ 0 & 1 & t_y \\ 0 & 0 & 1 \end{bmatrix} \begin{bmatrix} \cos\beta & -\sin\beta & 0 \\ \sin\beta & \cos\beta & 0 \\ 0 & 0 & 1 \end{bmatrix} \begin{bmatrix} x \\ y \\ 1 \end{bmatrix}$$

$$\begin{bmatrix} x' \\ y' \\ 1 \end{bmatrix} = \begin{bmatrix} \cos\beta & -\sin\beta & t_x \\ \sin\beta & \cos\beta & t_y \\ 0 & 0 & 1 \end{bmatrix} \begin{bmatrix} x \\ y \\ 1 \end{bmatrix}.$$

For example, consider rotating the point $P(1, 0)$, 90° and then translating it by $(1, 0)$. The rotation moves P to (0, 1) and the translation moves it to $(1, 1)$. This is confirmed by the above transform:

$$\begin{bmatrix} 1 \\ 1 \\ 1 \end{bmatrix} = \begin{bmatrix} 0 & -1 & 1 \\ 1 & 0 & 0 \\ 0 & 0 & 1 \end{bmatrix} \begin{bmatrix} 1 \\ 0 \\ 1 \end{bmatrix}.$$

7.3.5 Composite Rotations

It is worth confirming that if we rotate a point β about the origin, and follow this by a rotation of θ, this is equivalent to a single rotation of $\theta + \beta$, so $\mathbf{R}_\theta\mathbf{R}_\beta = \mathbf{R}_{\theta+\beta}$. Let's start with the transforms $\mathbf{R}_\beta$ and $\mathbf{R}_\theta$:

$$\mathbf{R}_\beta = \begin{bmatrix} \cos\beta & -\sin\beta & 0 \\ \sin\beta & \cos\beta & 0 \\ 0 & 0 & 1 \end{bmatrix}$$

$$\mathbf{R}_\theta = \begin{bmatrix} \cos\theta & -\sin\theta & 0 \\ \sin\theta & \cos\theta & 0 \\ 0 & 0 & 1 \end{bmatrix}.$$

We can represent the double rotation by the product $\mathbf{R}_\theta\mathbf{R}_\beta$:

$$\mathbf{R}_\theta\mathbf{R}_\beta = \begin{bmatrix} \cos\theta & -\sin\theta & 0 \\ \sin\theta & \cos\theta & 0 \\ 0 & 0 & 1 \end{bmatrix} \begin{bmatrix} \cos\beta & -\sin\beta & 0 \\ \sin\beta & \cos\beta & 0 \\ 0 & 0 & 1 \end{bmatrix}$$

$$= \begin{bmatrix} \cos\theta\cos\beta - \sin\theta\sin\beta & -\cos\theta\sin\beta - \sin\theta\cos\beta & 0 \\ \sin\theta\cos\beta + \cos\theta\sin\beta & -\sin\theta\sin\beta + \cos\theta\cos\beta & 0 \\ 0 & 0 & 1 \end{bmatrix}$$

$$= \begin{bmatrix} \cos(\theta+\beta) & -\sin(\theta+\beta) & 0 \\ \sin(\theta+\beta) & \cos(\theta+\beta) & 0 \\ 0 & 0 & 1 \end{bmatrix}$$

which confirms that the composite rotation is equivalent to a single rotation of $\theta+\beta$.

7.4 Inverse Transforms

Given a transform $\mathbf{A}$, its inverse $\mathbf{A}^{-1}$ is defined as such that

$$\mathbf{A}\mathbf{A}^{-1} = \mathbf{A}^{-1}\mathbf{A} = \mathbf{I}$$

where $\mathbf{I}$ is the identity matrix. So let's identify the inverse translation and rotation transforms.

We know that the translation matrix is given by

$$\mathbf{T}_{t_x,t_y} = \begin{bmatrix} 1 & 0 & t_x \\ 0 & 1 & t_y \\ 0 & 0 & 1 \end{bmatrix}$$

and we could reason that the inverse of $\mathbf{T}_{t_x,t_y}$ must be a translation in the opposite direction:

$$\mathbf{T}^{-1}_{t_x,t_y} = \begin{bmatrix} 1 & 0 & -t_x \\ 0 & 1 & -t_y \\ 0 & 0 & 1 \end{bmatrix}.$$

We can confirm this by computing $\mathbf{T}^{-1}_{t_x,t_y}$ from the cofactor matrix of $\mathbf{T}_{t_x,t_y}$, transposing it and dividing by its determinant:

$$\text{cofactor matrix of } \mathbf{T}_{t_x,t_y} = \begin{bmatrix} 1 & 0 & 0 \\ 0 & 1 & 0 \\ -t_x & -t_y & 1 \end{bmatrix}$$

$$\mathbf{T}^{\mathrm{T}}_{t_x,t_y} = \begin{bmatrix} 1 & 0 & -t_x \\ 0 & 1 & -t_y \\ 0 & 0 & 1 \end{bmatrix}$$

and as $\det \mathbf{T}_{t_x,t_y} = 1$, we can write

$$\mathbf{T}^{-1}_{t_x,t_y} = \begin{bmatrix} 1 & 0 & -t_x \\ 0 & 1 & -t_y \\ 0 & 0 & 1 \end{bmatrix}.$$

So our reasoning is correct. Furthermore, $\mathbf{T}_{t_x,t_y}\mathbf{T}^{-1}_{t_x,t_y} = \mathbf{I}$:

$$\mathbf{T}_{t_x,t_y}\mathbf{T}^{-1}_{t_x,t_y} = \begin{bmatrix} 1 & 0 & t_x \\ 0 & 1 & t_y \\ 0 & 0 & 1 \end{bmatrix} \begin{bmatrix} 1 & 0 & -t_x \\ 0 & 1 & -t_y \\ 0 & 0 & 1 \end{bmatrix} = \begin{bmatrix} 1 & 0 & 0 \\ 0 & 1 & 0 \\ 0 & 0 & 1 \end{bmatrix}.$$

Similarly, we know that the rotation transform is given by

$$\mathbf{R}_\beta = \begin{bmatrix} \cos\beta & -\sin\beta & 0 \\ \sin\beta & \cos\beta & 0 \\ 0 & 0 & 1 \end{bmatrix}$$

and we can reason that the inverse of $\mathbf{R}_\beta$ must be a rotation in the opposite direction, i.e. a rotation of $-\beta$:

$$\mathbf{R}_\beta^{-1} = \begin{bmatrix} \cos\beta & \sin\beta & 0 \\ -\sin\beta & \cos\beta & 0 \\ 0 & 0 & 1 \end{bmatrix}.$$

We can also compute $\mathbf{R}_\beta^{-1}$ by forming the cofactor matrix of $\mathbf{R}_\beta$, transposing it and dividing by its determinant:

$$\text{cofactor matrix of } \mathbf{R}_\beta = \begin{bmatrix} \cos\beta & -\sin\beta & 0 \\ \sin\beta & \cos\beta & 0 \\ 0 & 0 & 1 \end{bmatrix}$$

$$\mathbf{R}_\beta^{\mathrm{T}} = \begin{bmatrix} \cos\beta & \sin\beta & 0 \\ -\sin\beta & \cos\beta & 0 \\ 0 & 0 & 1 \end{bmatrix}$$

and as $\det \mathbf{R}_\beta = 1$, we can write

$$\mathbf{R}_\beta^{-1} = \begin{bmatrix} \cos\beta & \sin\beta & 0 \\ -\sin\beta & \cos\beta & 0 \\ 0 & 0 & 1 \end{bmatrix}.$$

So our reasoning is correct. Furthermore, $\mathbf{R}_\beta \mathbf{R}_\beta^{-1} = \mathbf{I}$:

$$\mathbf{R}_\beta \mathbf{R}_\beta^{-1} = \begin{bmatrix} \cos\beta & -\sin\beta & 0 \\ \sin\beta & \cos\beta & 0 \\ 0 & 0 & 1 \end{bmatrix} \begin{bmatrix} \cos\beta & \sin\beta & 0 \\ -\sin\beta & \cos\beta & 0 \\ 0 & 0 & 1 \end{bmatrix} = \begin{bmatrix} 1 & 0 & 0 \\ 0 & 1 & 0 \\ 0 & 0 & 1 \end{bmatrix}.$$

7.5 Multivector Transforms

Multivectors are linear combinations of vectors, bivectors, trivectors, etc., plus a scalar. They possess imaginary qualities and consequently have the ability to rotate vectors. Although it is unusual to employ multivectors in 2D computer graphics, they have been included to introduce their rotational qualities.

7.5.1 Translate a Point

Figure 7.3 shows a point $P(x, y)$ with position vector $\mathbf{p}$, and is translated by (t_x, t_y) using

$$\mathbf{p}' = \mathbf{p} + \mathbf{t}$$

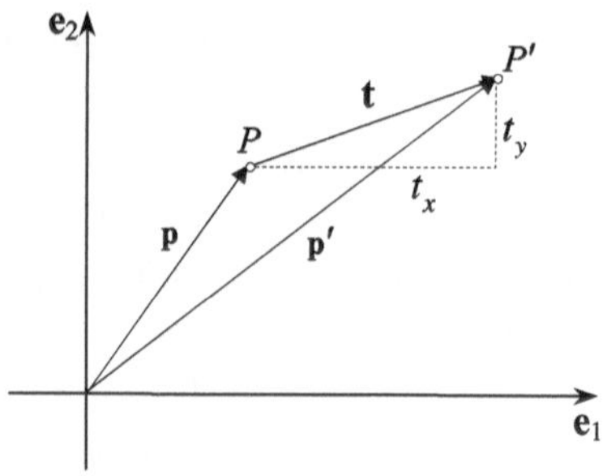

Fig. 7.3 Translate a point by (t_x, t_y)

where $\mathbf{t} = [t_x \quad t_y]^{\mathrm{T}}$. The position vector $\mathbf{p}'$ points to the translated point P' with components:

$$\mathbf{p}' = (x + t_x)\,\mathbf{e}_1 + \left(y + t_y\right)\mathbf{e}_2$$

and agrees with the translation matrix

$$\begin{bmatrix} x' \\ y' \\ 1 \end{bmatrix} = \begin{bmatrix} 1 & 0 & t_x \\ 0 & 1 & t_y \\ 0 & 0 & 1 \end{bmatrix} \begin{bmatrix} x \\ y \\ 1 \end{bmatrix}.$$

7.5.2 Rotational Qualities of the Unit Bivector

We know from Chap. 2 that multiplying a complex number by imaginary i rotates that complex number by 90°. In geometric algebra the 2D pseudoscalar $\mathbf{e}_{12}$ is also imaginary in that $\mathbf{e}_{12}^2 = -1$, and has similar rotational properties, but has the extra feature of controlling the direction of rotation. For example, Fig. 7.4 shows $\mathbf{p}\mathbf{e}_{12}$ which rotates $\mathbf{p}$, 90°:

$$\begin{aligned} \mathbf{p} &= p_1\mathbf{e}_1 + p_2\mathbf{e}_2 \\ \mathbf{p}\mathbf{e}_{12} &= (p_1\mathbf{e}_1 + p_2\mathbf{e}_2)\,\mathbf{e}_{12} \\ &= p_1\mathbf{e}_2 - p_2\mathbf{e}_1 \\ &= -p_2\mathbf{e}_1 + p_1\mathbf{e}_2. \end{aligned}$$

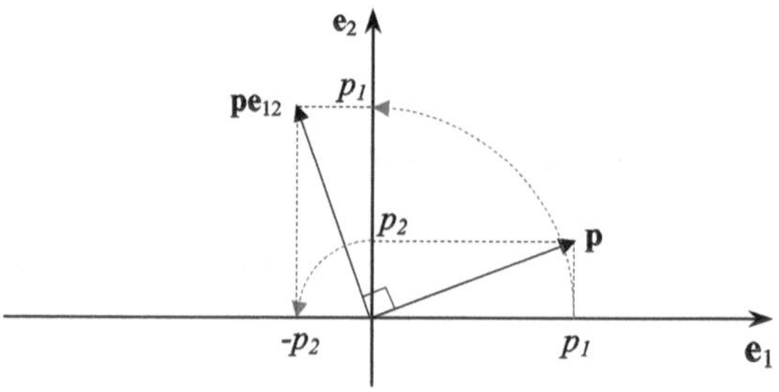

Fig. 7.4 $\mathbf{p}\mathbf{e}_{12}$ rotates $\mathbf{p}$, 90°

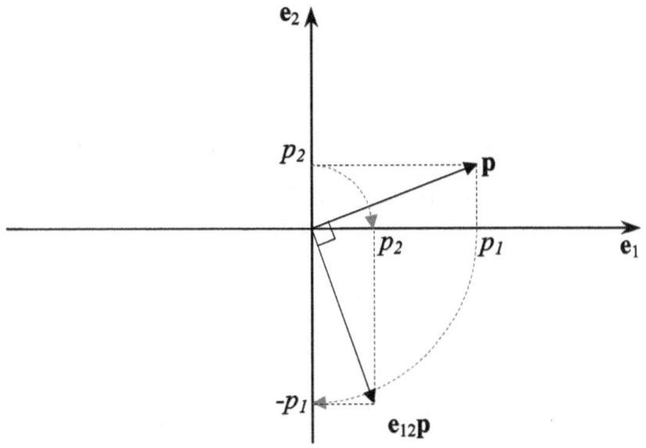

Fig. 7.5 $\mathbf{e}_{12}\mathbf{p}$ rotates $\mathbf{p}$, $-90°$

However, the reverse product $\mathbf{e}_{12}\mathbf{p}$ rotates $\mathbf{p}$, $-90°$:

$$\begin{aligned}\mathbf{p} &= p_1\mathbf{e}_1 + p_2\mathbf{e}_2\\ \mathbf{p}\mathbf{e}_{12} &= \mathbf{e}_{12}\left(p_1\mathbf{e}_1 + p_2\mathbf{e}_2\right)\\ &= -p_1\mathbf{e}_2 + p_2\mathbf{e}_1\\ &= p_2\mathbf{e}_1 - p_1\mathbf{e}_2\end{aligned}$$

as shown in Fig. 7.5.

We also discovered in Chap. 2 that a complex number $z = a + bi$ can be represented in exponential form as

$$z = |z|\mathrm{e}^{i\beta} = |z|\left(\cos\beta + i\sin\beta\right)$$

which, if used to multiply another complex number, scales it by $|z|$ and rotates it β.

Figure 7.6 shows a plane defined by $\mathbf{m} \wedge \mathbf{n}$ and the vectors $\mathbf{n}$ and $\mathbf{m}$ such that $\mathbf{n}$ is rotated β further than $\mathbf{m}$:

$$\begin{aligned}\mathbf{n} &= n_1\mathbf{e}_1 + n_2\mathbf{e}_2\\ \mathbf{m} &= m_1\mathbf{e}_1 + m_2\mathbf{e}_2\\ \mathbf{n}\mathbf{m} &= \mathbf{n}\cdot\mathbf{m} - \mathbf{m}\wedge\mathbf{n}\\ &= |\mathbf{n}||\mathbf{m}|\cos\beta - |\mathbf{m}||\mathbf{n}|\sin\beta\mathbf{e}_{12}\\ &= |\mathbf{n}||\mathbf{m}|\left(\cos\beta - \sin\beta\mathbf{e}_{12}\right).\end{aligned}$$

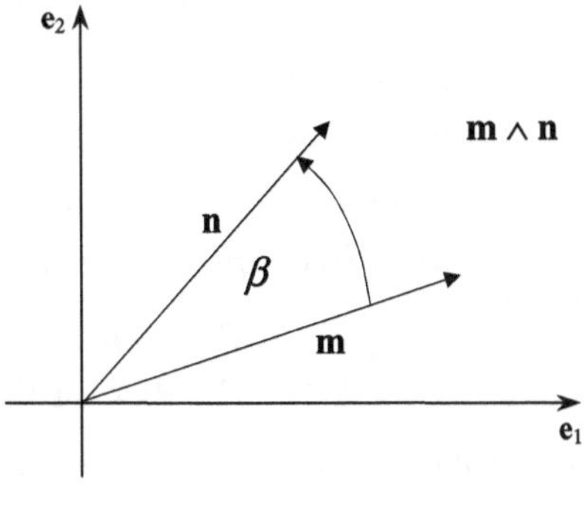

Fig. 7.6 The bivector $\mathbf{m} \wedge \mathbf{n}$

Pre-multiplying a vector $\mathbf{p}$ by the product $\mathbf{nm}$ creates $\mathbf{p}'$ such that given:

$$\begin{aligned}
\mathbf{p} &= p_1\mathbf{e}_1 + p_2\mathbf{e}_2 \\
\mathbf{nmp} &= |\mathbf{n}||\mathbf{m}|\,(\cos\beta - \sin\beta\mathbf{e}_{12})\,(p_1\mathbf{e}_1 + p_2\mathbf{e}_2) \\
&= |\mathbf{n}||\mathbf{m}|\,(\cos\beta p_1\mathbf{e}_1 + \cos\beta p_2\mathbf{e}_2 + \sin\beta p_1\mathbf{e}_2 - \sin\beta p_2\mathbf{e}_1) \\
\mathbf{p}' &= |\mathbf{n}||\mathbf{m}|\big[(\cos\beta p_1 - \sin\beta p_2)\,\mathbf{e}_1 + (\sin\beta p_1 + \cos\beta p_2)\,\mathbf{e}_2\big] \\
\begin{bmatrix} p'_1 \\ p'_2 \end{bmatrix} &= |\mathbf{n}||\mathbf{m}| \begin{bmatrix} \cos\beta & -\sin\beta \\ \sin\beta & \cos\beta \end{bmatrix} \begin{bmatrix} p_1 \\ p_2 \end{bmatrix}
\end{aligned}$$

and confirms that the vector $\mathbf{p}$ is rotated β and scaled by $|\mathbf{n}||\mathbf{m}|$.

Post-multiplying a vector $\mathbf{p}$ by the product $\mathbf{nm}$ creates $\mathbf{p}'$ such that given:

$$\begin{aligned}
\mathbf{p} &= p_1\mathbf{e}_1 + p_2\mathbf{e}_2 \\
\mathbf{pnm} &= (p_1\mathbf{e}_1 + p_2\mathbf{e}_2)\,|\mathbf{n}||\mathbf{m}|\,(\cos\beta - \sin\beta\mathbf{e}_{12}) \\
&= |\mathbf{n}||\mathbf{m}|\,(p_1\mathbf{e}_1\cos\beta - p_1\mathbf{e}_2\sin\beta + p_2\mathbf{e}_2\cos\beta + p_2\mathbf{e}_1\sin\beta) \\
\mathbf{p}' &= |\mathbf{n}||\mathbf{m}|\big((p_1\cos\beta + p_2\sin\beta)\,\mathbf{e}_1 + (-p_1\sin\beta + p_2\cos\beta)\,\mathbf{e}_2\big) \\
\begin{bmatrix} p'_1 \\ p'_2 \end{bmatrix} &= |\mathbf{n}||\mathbf{m}| \begin{bmatrix} \cos\beta & \sin\beta \\ -\sin\beta & \cos\beta \end{bmatrix} \begin{bmatrix} p_1 \\ p_2 \end{bmatrix}
\end{aligned}$$

and confirms that the vector $\mathbf{p}$ is rotated $-\beta$ and scaled by $|\mathbf{n}||\mathbf{m}|$. By making $\mathbf{n}$ and $\mathbf{m}$ unit vectors, the product $\mathbf{nm}$ rotates a vector without scaling, which is an essential quality for a rotation transform.

Before proceeding, we should clarify the effect of reversing the product $\mathbf{nm}$ to $\mathbf{mn}$. Therefore, assuming that vectors $\mathbf{n}$ and $\mathbf{m}$ remain unchanged, the product $\mathbf{mn}$ is given by:

$$\begin{aligned}
\mathbf{n} &= n_1\mathbf{e}_1 + n_2\mathbf{e}_2 \\
\mathbf{m} &= m_1\mathbf{e}_1 + m_2\mathbf{e}_2 \\
\mathbf{mn} &= \mathbf{m}\cdot\mathbf{n} + \mathbf{m}\wedge\mathbf{n} \\
&= |\mathbf{n}||\mathbf{m}|\cos\beta + |\mathbf{m}||\mathbf{n}|\sin\beta\mathbf{e}_{12} \\
&= |\mathbf{n}||\mathbf{m}|\,(\cos\beta + \sin\beta\mathbf{e}_{12})\,.
\end{aligned}$$

Pre-multiplying a vector $\mathbf{p}$ by the product $\mathbf{mn}$ creates $\mathbf{p}'$ such that given:

$$\begin{aligned}
\mathbf{p} &= p_1\mathbf{e}_1 + p_2\mathbf{e}_2 \\
\mathbf{mnp} &= |\mathbf{n}||\mathbf{m}|\,(\cos\beta + \sin\beta\mathbf{e}_{12})\,(p_1\mathbf{e}_1 + p_2\mathbf{e}_2) \\
&= |\mathbf{n}||\mathbf{m}|\,(\cos\beta p_1\mathbf{e}_1 + \cos\beta p_2\mathbf{e}_2 - \sin\beta p_1\mathbf{e}_2 + \sin\beta p_2\mathbf{e}_1) \\
\mathbf{p}' &= |\mathbf{n}||\mathbf{m}|\big((\cos\beta p_1 + \sin\beta p_2)\,\mathbf{e}_1 + (-\sin\beta p_1 + \cos\beta p_2)\,\mathbf{e}_2\big) \\
\begin{bmatrix} p'_1 \\ p'_2 \end{bmatrix} &= |\mathbf{n}||\mathbf{m}| \begin{bmatrix} \cos\beta & \sin\beta \\ -\sin\beta & \cos\beta \end{bmatrix} \begin{bmatrix} p_1 \\ p_2 \end{bmatrix}
\end{aligned}$$

and confirms that the vector $\mathbf{p}$ is rotated $-\beta$ and scaled by $|\mathbf{n}||\mathbf{m}|$.

It should be no surprise that post-multiplying a vector **p** by the product **mn** rotates it β.

The above results are summarised as follows:

$$\begin{aligned} \mathbf{nmp} = \mathbf{pmn} &\quad \text{rotates } \mathbf{p}, \quad \beta \\ \mathbf{pnm} = \mathbf{mnp} &\quad \text{rotates } \mathbf{p}, \ -\beta. \end{aligned}$$

7.5.3 Rotate a Point About the Origin

In Chap. 9 on 3D rotations we show the origins of rotors in geometric algebra using double reflections. The plane containing the vectors **n** and **m** is defined by the wedge product $\mathbf{m} \wedge \mathbf{n}$, which means we can write the product **mn** as

$$\mathbf{mn} = \mathbf{m} \cdot \mathbf{n} + \mathbf{m} \wedge \mathbf{n}$$

and the product **nm** in the same plane as

$$\mathbf{nm} = \mathbf{m} \cdot \mathbf{n} - \mathbf{m} \wedge \mathbf{n}$$

which accounts for the negative sign in the following bivector term

$$\mathbf{nm} = |\mathbf{n}||\mathbf{m}| \left(\cos\beta - \sin\beta \mathbf{e}_{12} \right).$$

Furthermore, if we make **n** and **m** unit vectors we can replace them by a rotor $\mathbf{R}_\beta$ whose magnitude is 1 because

$$|\mathbf{R}_\beta| = \sqrt{\cos^2\beta + \sin^2\beta} = 1$$

therefore no scaling occurs, which means that

$$\begin{aligned} \mathbf{R}_\beta \mathbf{p} &\quad \text{rotates } \mathbf{p}, \beta \quad \text{i.e. anticlockwise, and} \\ \mathbf{p}\mathbf{R}_\beta &\quad \text{rotates } \mathbf{p}, -\beta \quad \text{i.e. clockwise.} \end{aligned}$$

The effect of this rotor is illustrated as follows:

$$\begin{aligned} \mathbf{p} &= \mathbf{e}_1 + \mathbf{e}_2 \\ \mathbf{R}_{45^\circ} &= \cos 45^\circ - \sin 45^\circ \mathbf{e}_{12} \\ &= \frac{\sqrt{2}}{2} - \frac{\sqrt{2}}{2}\mathbf{e}_{12} \\ \mathbf{p}' = \mathbf{R}_{45^\circ}\mathbf{p} &= \left(\frac{\sqrt{2}}{2} - \frac{\sqrt{2}}{2}\mathbf{e}_{12} \right)(\mathbf{e}_1 + \mathbf{e}_2) \\ &= \frac{\sqrt{2}}{2}\mathbf{e}_1 + \frac{\sqrt{2}}{2}\mathbf{e}_2 + \frac{\sqrt{2}}{2}\mathbf{e}_2 - \frac{\sqrt{2}}{2}\mathbf{e}_1 \\ &= \sqrt{2}\mathbf{e}_2 \end{aligned}$$

as shown in Fig. 7.7.

Fig. 7.7 The effect of rotor $\mathbf{R}_{45^\circ}$ on vector $\mathbf{p}$

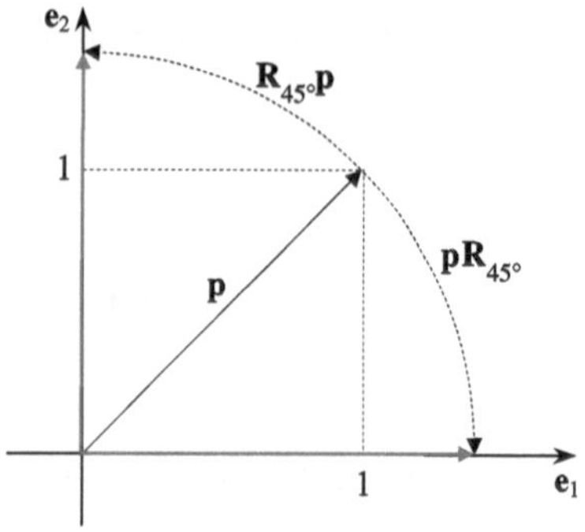

Similarly, reversing the product to $\mathbf{pR}_{45^\circ}$ we obtain

$$\begin{aligned}\mathbf{p}' = \mathbf{pR}_{45^\circ} &= (\mathbf{e}_1 + \mathbf{e}_2)\left(\frac{\sqrt{2}}{2} - \frac{\sqrt{2}}{2}\mathbf{e}_{12}\right)\\ &= \frac{\sqrt{2}}{2}\mathbf{e}_1 - \frac{\sqrt{2}}{2}\mathbf{e}_2 + \frac{\sqrt{2}}{2}\mathbf{e}_2 + \frac{\sqrt{2}}{2}\mathbf{e}_1\\ &= \sqrt{2}\mathbf{e}_1\end{aligned}$$

as shown in Fig. 7.7.

Geometric algebra also employs a reversion function which reverses the sequence of elements in a multivector by switching the signs of bivector and trivector elements. Instead of reversing the sequence of $\mathbf{p}$ and $\mathbf{R}_\beta$, we can reverse $\mathbf{R}_\beta$ using $\mathbf{R}_\beta^\dagger$:

$$\begin{aligned}\mathbf{R}_\beta &= \cos\beta - \sin\beta\mathbf{e}_{12}\\ \mathbf{R}_\beta^\dagger &= \cos\beta + \sin\beta\mathbf{e}_{12}\end{aligned}$$

therefore,

$$\begin{aligned}\mathbf{p}' = \mathbf{R}_{45^\circ}^\dagger\mathbf{p} &= \left(\frac{\sqrt{2}}{2} + \frac{\sqrt{2}}{2}\mathbf{e}_{12}\right)(\mathbf{e}_1 + \mathbf{e}_2)\\ &= \frac{\sqrt{2}}{2}\mathbf{e}_1 + \frac{\sqrt{2}}{2}\mathbf{e}_2 - \frac{\sqrt{2}}{2}\mathbf{e}_2 + \frac{\sqrt{2}}{2}\mathbf{e}_1\\ &= \sqrt{2}\mathbf{e}_1\end{aligned}$$

and

$$\begin{aligned}\mathbf{p}' = \mathbf{pR}_{45^\circ}^\dagger &= (\mathbf{e}_1 + \mathbf{e}_2)\left(\frac{\sqrt{2}}{2} + \frac{\sqrt{2}}{2}\mathbf{e}_{12}\right)\\ &= \frac{\sqrt{2}}{2}\mathbf{e}_1 + \frac{\sqrt{2}}{2}\mathbf{e}_2 + \frac{\sqrt{2}}{2}\mathbf{e}_2 - \frac{\sqrt{2}}{2}\mathbf{e}_1\\ &= \sqrt{2}\mathbf{e}_2\end{aligned}$$

which means that

$$\mathbf{R}_\beta^\dagger \mathbf{p} \quad \text{rotates } \mathbf{p}, -\beta \quad \text{i.e. clockwise}$$
$$\mathbf{p}\mathbf{R}_\beta^\dagger \quad \text{rotates } \mathbf{p}, \beta \quad \text{i.e. anticlockwise}$$

and

$$\mathbf{R}_\beta \mathbf{p} = \mathbf{p}\mathbf{R}_\beta^\dagger$$
$$\mathbf{p}\mathbf{R}_\beta = \mathbf{R}_\beta^\dagger \mathbf{p}.$$

Using the rotor $\mathbf{R}_\beta$ in a single-sided transformation only works for vectors in the plane of rotation, which satisfies everything we do in 2D. However, in 3D we have to employ a double-sided, half-angle formula of the form $\mathbf{R}_\beta \mathbf{p} \mathbf{R}_\beta^\dagger$, which is covered in Chap. 11.

7.5.4 Rotate a Point About an Arbitrary Point

Earlier in this chapter we developed a transform for rotating a point about an arbitrary point. Let's show how we can approach the same problem using geometric algebra. Figure 7.8 shows the geometry describing how the point P is rotated β about T to P', and by inspection we can write

$$\mathbf{p}' = \mathbf{t} + \mathbf{R}_\beta (\mathbf{p} - \mathbf{t}).$$

Using the previous example, where $T = (1, 1)$, $P = (2, 1)$ and $\beta = 90°$ we have

$$\begin{aligned}
\mathbf{R}_{90°} &= \cos 90° - \sin 90° \mathbf{e}_{12} = -\mathbf{e}_{12} \\
\mathbf{t} &= \mathbf{e}_1 + \mathbf{e}_2 \\
\mathbf{p} &= 2\mathbf{e}_1 + \mathbf{e}_2 \\
\mathbf{p}' &= \mathbf{e}_1 + \mathbf{e}_2 - \mathbf{e}_{12} (2\mathbf{e}_1 + \mathbf{e}_2 - \mathbf{e}_1 - \mathbf{e}_2) \\
&= \mathbf{e}_1 + \mathbf{e}_2 - \mathbf{e}_{12}\mathbf{e}_1 \\
&= \mathbf{e}_1 + \mathbf{e}_2 + \mathbf{e}_2 \\
&= \mathbf{e}_1 + 2\mathbf{e}_2
\end{aligned}$$

which is correct.

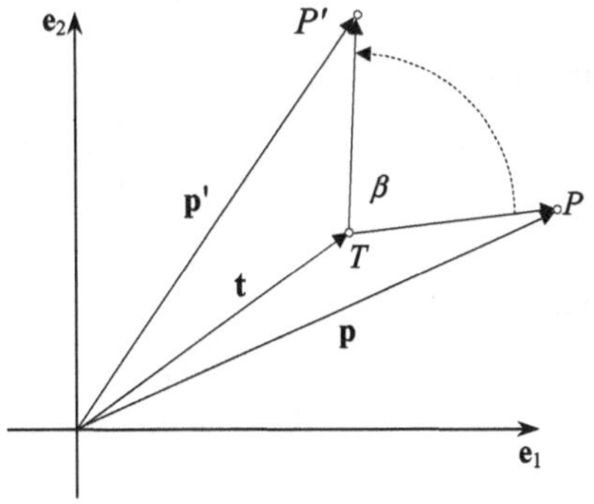

Fig. 7.8 Using a rotor $\mathbf{R}_\beta$ to rotate P about T

If we expand the definition of $\mathbf{p}'$ we obtain:

$$\begin{aligned}
\mathbf{p}' &= \mathbf{t} + \mathbf{R}_\beta (\mathbf{p} - \mathbf{t}) \\
&= t_x \mathbf{e}_1 + t_y \mathbf{e}_2 + (\cos\beta - \sin\beta \mathbf{e}_{12}) \big((x - t_x)\mathbf{e}_1 + (y - t_y)\mathbf{e}_2\big) \\
&= t_x \mathbf{e}_1 + t_y \mathbf{e}_2 + \big(x\cos\beta - y\sin\beta - t_x\cos\beta + t_y\sin\beta\big)\mathbf{e}_1 \\
&\quad + (x\sin\beta + y\cos\beta - t_x\sin\beta - t_y\cos\beta)\mathbf{e}_2 \\
&= \big(x\cos\beta - y\sin\beta + t_x(1 - \cos\beta) + t_y\sin\beta\big)\mathbf{e}_1 \\
&\quad + \big(x\sin\beta + y\cos\beta + t_y(1 - \cos\beta) - t_x\sin\beta\big)\mathbf{e}_2
\end{aligned}$$

which in matrix form is

$$\begin{bmatrix} x' \\ y' \\ 1 \end{bmatrix} = \begin{bmatrix} \cos\beta & -\sin\beta & t_x(1 - \cos\beta) + t_y\sin\beta \\ \sin\beta & \cos\beta & t_y(1 - \cos\beta) - t_x\sin\beta \\ 0 & 0 & 1 \end{bmatrix} \begin{bmatrix} x \\ y \\ 1 \end{bmatrix}$$

and agrees with the original transform (7.1).

7.6 Summary

In this chapter we have seen how the translation and rotation transforms are used to rotate points about the origin and arbitrary points. We have also seen how the inverse transforms translate and rotate in the opposite directions which will be used in the next chapter to relate points in different frames of reference.

We have also seen how multivectors provide an alternative approach based upon vectors, bivectors and rotors, and can undertake the same tasks. However, we have discovered that fundamentally they are matrix transforms in disguise, albeit, an effective one.

In order to show the patterns that exist between these two mathematical approaches, all the commands are summarised.

7.6.1 Summary of Matrix Transforms

Translate a point

$$\mathbf{T}_{t_x,t_y} = \begin{bmatrix} 1 & 0 & t_x \\ 0 & 1 & t_y \\ 0 & 0 & 1 \end{bmatrix}.$$

Rotate a point

$$\mathbf{R}_\beta = \begin{bmatrix} \cos\beta & -\sin\beta & 0 \\ \sin\beta & \cos\beta & 0 \\ 0 & 0 & 1 \end{bmatrix}.$$

Rotate a point about (t_x, t_y)

$$\mathbf{T}_{t_x,t_y}\mathbf{R}_\beta\mathbf{T}_{-t_x,-t_y} = \begin{bmatrix} \cos\beta & -\sin\beta & t_x(1-\cos\beta)+t_y\sin\beta \\ \sin\beta & \cos\beta & t_y(1-\cos\beta)-t_x\sin\beta \\ 0 & 0 & 1 \end{bmatrix}.$$

7.6.2 Summary of Multivector Transforms

Rotor definition

$$\mathbf{R}_\beta = \cos\beta - \sin\beta\mathbf{e}_{12}$$
$$\mathbf{R}_\beta^\dagger = \cos\beta + \sin\beta\mathbf{e}_{12}.$$

Translate a point

$$\mathbf{p}' = \mathbf{p} + \mathbf{t} = (x + t_x)\mathbf{e}_1 + (y + t_y)\mathbf{e}_2.$$

Rotate a point

$$\mathbf{p}' = \mathbf{R}_\beta\mathbf{p} = (\cos\beta - \sin\beta\mathbf{e}_{12})(x\mathbf{e}_1 + y\mathbf{e}_2)$$
$$\mathbf{p}' = \mathbf{p}\mathbf{R}_\beta^\dagger = (x\mathbf{e}_1 + y\mathbf{e}_2)(\cos\beta + \sin\beta\mathbf{e}_{12}).$$

Rotate a point about (t_x, t_y)

$$\mathbf{p}' = \mathbf{t} + \mathbf{R}_\beta(\mathbf{p} - \mathbf{t}) = t_x\mathbf{e}_1 + t_y\mathbf{e}_2 + (\cos\beta - \sin\beta\mathbf{e}_{12})\big((x - t_x)\mathbf{e}_1 + (y - t_y)\mathbf{e}_2\big)$$
$$\mathbf{p}' = \mathbf{t} + (\mathbf{p} - \mathbf{t})\mathbf{R}_\beta^\dagger = t_x\mathbf{e}_1 + t_y\mathbf{e}_2 + \big((x - t_x)\mathbf{e}_1 + (y - t_y)\mathbf{e}_2\big)(\cos\beta + \sin\beta\mathbf{e}_{12}).$$

Chapter 8
Frames of Reference in the Plane

8.1 Introduction

In the previous chapter we covered the transforms for rotating points in the plane with respect to a *fixed* frame of reference. In this chapter we investigate the transforms for computing the coordinates of points in rotated frames of reference using geometry, matrices and multivectors. We will employ many of the concepts previously described in order to develop inverse transforms and rotors.

8.2 Frames of Reference

You have probably been on a train waiting to depart from a railway station, and through the window see another stationary train. Then suddenly you notice movement. To begin with, it is difficult to decide which train is moving, and the problem is often resolved when the entire scenario is seen with reference to some fixed object such as a tree or a building. This phenomena reminds us that motion is relative, and plays an important role in understanding transforms and frames of reference.

When a frame of reference moves – such as a train – the relationship between the seated passengers and the train remains fixed. The only thing that does change is the relationship between the train and other frames of references such as a tree or a building.

One can describe the relative motion between the train and a tree by assuming that the train remains stationary and the tree moves in an equal and opposite direction. So if the train's translation is described by $\mathbf{T}$, the tree's translation relative to the train is described by the inverse transform $\mathbf{T}^{-1}$.

Similarly, the rotation of a swivel chair can be described in two ways. The usual way is to assume that the swivel chair rotates relative to the desk where it's located. The relative motion interpretation proposes that the chair is stationary, whilst the desk rotates in an equal and opposite direction. So if the chair's rotation is described by $\mathbf{R}$, the desk's rotation relative to the chair is described by the inverse transform $\mathbf{R}^{-1}$.

J. Vince, *Rotation Transforms for Computer Graphics*,
DOI 10.1007/978-0-85729-154-7_8, © Springer-Verlag London Limited 2011

Let's assume that a tree is located in a 3D frame of reference labelled XYZ and the train's frame of reference is labelled $X'Y'Z'$. When the two frames are coincident, the tree has identical coordinates in both frames. However, if the train's frame $X'Y'Z'$ is translated by the transform $\mathbf{T}$ relative to XYZ, the tree's coordinates relative to the train have effectively moved in the opposite direction given by $\mathbf{T}^{-1}$.

Similarly, let's assume that a desk is located in a frame of reference labelled XYZ and a chair's frame of reference is labelled $X'Y'Z'$. When the two frames are coincident, the desk has identical coordinates in both frames. However, if the chair's frame $X'Y'Z'$ is rotated by the transform $\mathbf{R}$ relative to XYZ, the desk's coordinates relative to the chair have effectively rotated in the opposite direction given by $\mathbf{R}^{-1}$.

8.3 Matrix Transforms

Having established the equivalence between transforms for moving points in a fixed frame, and the inverse transforms for fixed points in a moving frame, let's explore how we construct the transforms for translated and rotated frames of reference in the plane. As in the previous chapter, we will show how matrices and multivectors offer two approaches to this problem.

In computer graphics most frame of reference transforms are expressed by a translation or a rotation, or a combination of both. We will explore these different scenarios and develop transforms for converting coordinates in the original frame of reference to coordinates in the second frame.

8.3.1 Translated Frame of Reference

Figure 8.1 shows two coincident 2D frames of reference $X'Y'$ and XY, where a point in one frame has identical coordinates in the other. Therefore, the two frames of reference are related as follows

$$\begin{bmatrix} x' \\ y' \\ 1 \end{bmatrix} = \mathbf{I} \begin{bmatrix} x \\ y \\ 1 \end{bmatrix}$$

where $\mathbf{I}$ is the identity transform

$$\mathbf{I} = \begin{bmatrix} 1 & 0 & 0 \\ 0 & 1 & 0 \\ 0 & 0 & 1 \end{bmatrix}$$

which ensures that $P' = P$.

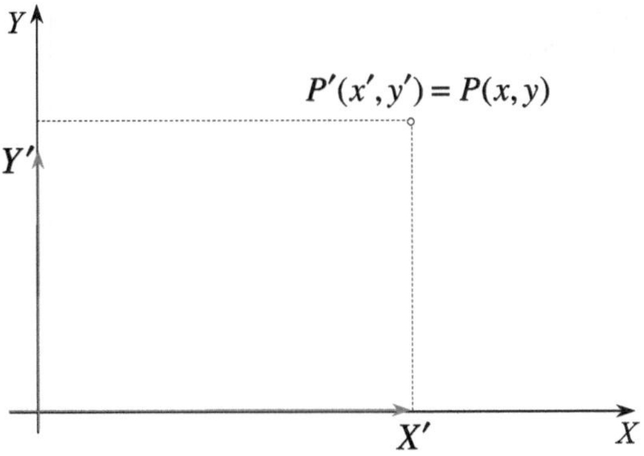

Fig. 8.1 The $X'Y'$ axial system is coincident with XY

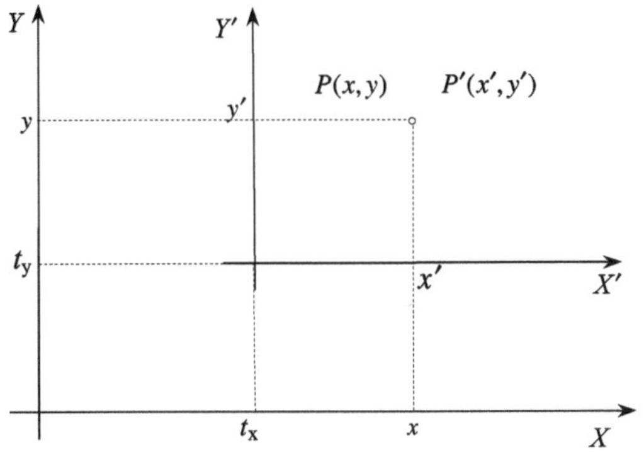

Fig. 8.2 The $X'Y'$ axial system is translated (t_x, t_y)

Figure 8.2 shows the frame $X'Y'$ translated by (t_x, t_y) which is equivalent to translating P by $\mathbf{T}_{-t_x,-t_y}$ which is also the inverse transform $\mathbf{T}^{-1}_{t_x,t_y}$. Therefore, a point $P(x, y)$ in XY has coordinates $P'(x', y')$ in $X'Y'$ given by

$$\begin{bmatrix} x' \\ y' \\ 1 \end{bmatrix} = \mathbf{T}^{-1}_{t_x,t_y} \begin{bmatrix} x \\ y \\ 1 \end{bmatrix}$$

where

$$\mathbf{T}^{-1}_{t_x,t_y} = \begin{bmatrix} 1 & 0 & -t_x \\ 0 & 1 & -t_y \\ 0 & 0 & 1 \end{bmatrix}.$$

For example, the point (t_x, t_y) in XY should have coordinates $(0, 0)$ in $X'Y'$:

$$\begin{bmatrix} 0 \\ 0 \\ 1 \end{bmatrix} = \begin{bmatrix} 1 & 0 & -t_x \\ 0 & 1 & -t_y \\ 0 & 0 & 1 \end{bmatrix} \begin{bmatrix} t_x \\ t_y \\ 1 \end{bmatrix}$$

which is confirmed.

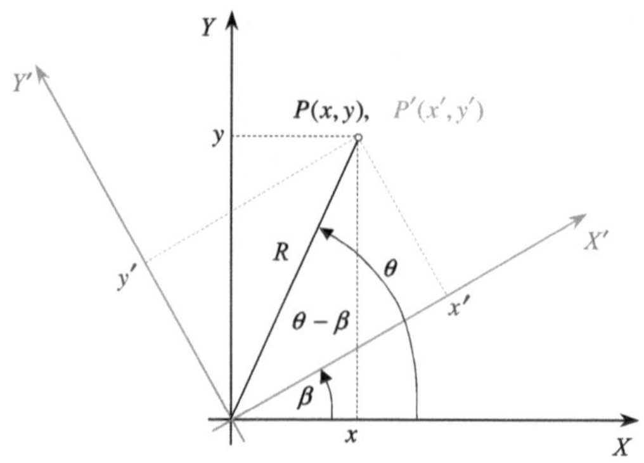

Fig. 8.3 The $X'Y'$ axial system is rotated β

8.3.2 Rotated Frame of Reference

Figure 8.3 shows the frame $X'Y'$ rotated β which is equivalent to rotating P by $-\beta$ and is effected by the transform $\mathbf{R}_\beta^{-1}$. Therefore, a point $P(x, y)$ in XY has coordinates $P'(x', y')$ in $X'Y'$ given by

$$\begin{bmatrix} x' \\ y' \\ 1 \end{bmatrix} = \mathbf{R}_\beta^{-1} \begin{bmatrix} x \\ y \\ 1 \end{bmatrix}$$

where

$$\mathbf{R}_\beta^{-1} = \begin{bmatrix} \cos\beta & \sin\beta & 0 \\ -\sin\beta & \cos\beta & 0 \\ 0 & 0 & 1 \end{bmatrix}.$$

We can also confirm this using the geometry shown in Fig. 8.3,

$$\begin{aligned} x &= R\cos\theta \\ y &= R\sin\theta \\ x' &= R\cos(\theta-\beta) \\ y' &= R\sin(\theta-\beta) \\ x' &= R\cos\theta\cos\beta + R\sin\theta\sin\beta \\ &= x\cos\beta + y\sin\beta \\ y' &= R\sin\theta\cos\beta - R\cos\theta\sin\beta \\ &= -x\sin\beta + y\cos\beta \end{aligned}$$

which as a homogeneous matrix is

$$\begin{bmatrix} x' \\ y' \\ 1 \end{bmatrix} = \begin{bmatrix} \cos\beta & \sin\beta & 0 \\ -\sin\beta & \cos\beta & 0 \\ 0 & 0 & 1 \end{bmatrix} \begin{bmatrix} x \\ y \\ 1 \end{bmatrix}$$

which is the inverse rotation transform $\mathbf{R}_\beta^{-1}$ or $\mathbf{R}_{-\beta}$. For example, the point $(1, 1)$ in XY, will have coordinates $(\sqrt{2}, 0)$ in the frame of reference rotated 45°:

$$\begin{bmatrix} \sqrt{2} \\ 0 \\ 1 \end{bmatrix} = \begin{bmatrix} \sqrt{2}/2 & \sqrt{2}/2 & 0 \\ -\sqrt{2}/2 & \sqrt{2}/2 & 0 \\ 0 & 0 & 1 \end{bmatrix} \begin{bmatrix} 1 \\ 1 \\ 1 \end{bmatrix}$$

which is confirmed.

We have previously shown that two separate rotations of a point is equivalent to a single composite rotation of a point. Similarly, it is a trivial exercise to prove that two separate rotations of a frame is equivalent to single composite rotation of a frame.

8.3.3 Rotated and Translated Frame of Reference

Having looked at translated and rotated frames of reference, let's combine the two operations and develop a single transform. This is not too difficult to follow, so long as we are careful with our definitions and diagrams.

When a point is rotated and translated we use the operation

$$P' = \mathbf{T}_{t_x,t_y}\mathbf{R}_\beta P.$$

We know that the transform for moving a frame of reference – whilst keeping a point fixed – is the inverse of that used for moving points – whilst keeping the frame fixed. Which suggests that the transform for a rotated and translated frame of reference is the inverse of $\mathbf{T}_{t_x,t_y}\mathbf{R}_\beta$ which is

$$(\mathbf{T}_{t_x,t_y}\mathbf{R}_\beta)^{-1} = \mathbf{R}_\beta^{-1}\mathbf{T}_{t_x,t_y}^{-1}$$

and makes

$$P' = \mathbf{R}_\beta^{-1}\mathbf{T}_{t_x,t_y}^{-1} P$$

or

$$P' = \mathbf{R}_{-\beta}\mathbf{T}_{-t_x,-t_y} P.$$

Substituting the matrices for $\mathbf{R}_{-\beta}$ and $\mathbf{T}_{-t_x,-t_y}$ we have

$$\begin{bmatrix} x' \\ y' \\ 1 \end{bmatrix} = \begin{bmatrix} \cos\beta & \sin\beta & 0 \\ -\sin\beta & \cos\beta & 0 \\ 0 & 0 & 1 \end{bmatrix} \begin{bmatrix} 1 & 0 & -t_x \\ 0 & 1 & -t_y \\ 0 & 0 & 1 \end{bmatrix} \begin{bmatrix} x \\ y \\ 1 \end{bmatrix}$$

which simplifies to

$$\begin{bmatrix} x' \\ y' \\ 1 \end{bmatrix} = \begin{bmatrix} \cos\beta & \sin\beta & -t_x\cos\beta - t_y\sin\beta \\ -\sin\beta & \cos\beta & t_x\sin\beta - t_y\cos\beta \\ 0 & 0 & 1 \end{bmatrix} \begin{bmatrix} x \\ y \\ 1 \end{bmatrix}. \tag{8.1}$$

Equation (8.1) is the homogeneous matrix for converting points in the XY coordinate system to the translated and rotated $X'Y'$ coordinate system.

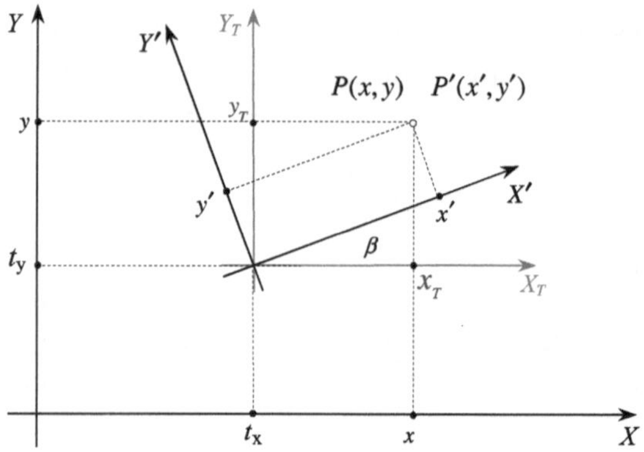

Fig. 8.4 The $X'Y'$ axial system translated (t_x, t_y) and rotated β

A quick test confirms that $\mathbf{T}_{t_x,t_y}\mathbf{R}_\beta\mathbf{R}_\beta^{-1}\mathbf{T}_{t_x,t_y}^{-1} = \mathbf{I}$, i.e.

$$\mathbf{T}_{t_x,t_y}\mathbf{R}_\beta\mathbf{R}_\beta^{-1}\mathbf{T}_{t_x,t_y}^{-1} = \mathbf{T}_{t_x,t_y}\mathbf{T}_{t_x,t_y}^{-1} = \mathbf{I}$$

or

$$\begin{bmatrix} \cos\beta & -\sin\beta & t_x \\ \sin\beta & \cos\beta & t_y \\ 0 & 0 & 1 \end{bmatrix} \begin{bmatrix} \cos\beta & \sin\beta & -t_x\cos\beta - t_y\sin\beta \\ -\sin\beta & \cos\beta & t_x\sin\beta - t_y\cos\beta \\ 0 & 0 & 1 \end{bmatrix} = \begin{bmatrix} 1 & 0 & 0 \\ 0 & 1 & 0 \\ 0 & 0 & 1 \end{bmatrix}.$$

Figure 8.4 shows how the frame of reference $X_T Y_T$ is created for the intermediate translated frame, followed by a rotation of β about the new origin, ending with the final frame of reference $X'Y'$.

Let's test (8.1) with the example shown in Fig. 8.5 where

$$\beta = 90^\circ$$
$$(t_x, t_y) = (10, 5)$$
$$(x, y) = (9, 6)$$

$$\begin{bmatrix} 1 \\ 1 \\ 1 \end{bmatrix} = \begin{bmatrix} 0 & 1 & -5 \\ -1 & 0 & 10 \\ 0 & 0 & 1 \end{bmatrix} \begin{bmatrix} 9 \\ 6 \\ 1 \end{bmatrix} \tag{8.2}$$

and we see that $(9, 6)$ in XY becomes $(1, 1)$ in $X'Y'$.

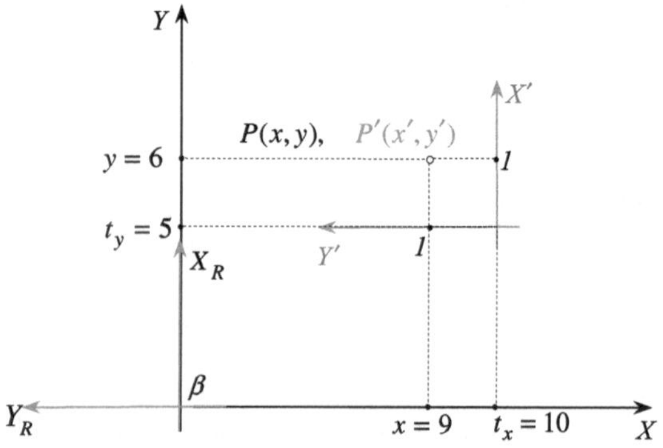

Fig. 8.5 The $X'Y'$ axial system is translated $(10, 5)$ and rotated 90°

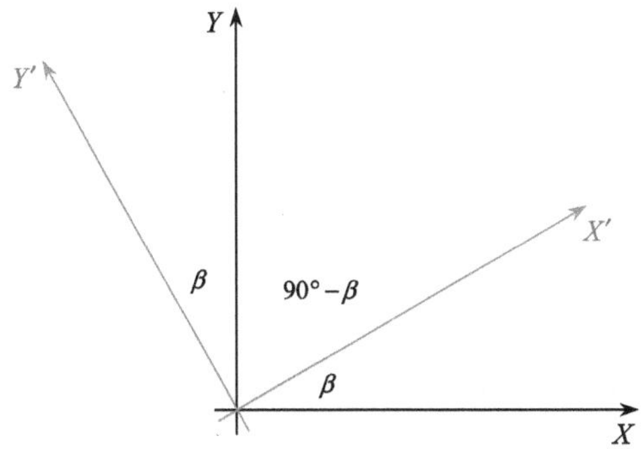

Fig. 8.6 If the X'- and Y'-axes are assumed to be unit vectors, their direction cosines form the elements of the rotation matrix

8.4 Direction Cosines

Direction cosines are the cosines of the angles between a vector and the Cartesian axes, and for a unit vector they are its components.

Figure 8.6 shows the rotated frame $X'Y'$, and by inspection the direction cosines for a vector lying on X' are $\cos\beta$ and $\cos(90° - \beta)$, which can be rewritten as $\cos\beta$ and $\sin\beta$. The direction cosines for a vector lying on Y' are $\cos(90° + \beta)$ and $\cos\beta$, which can be rewritten as $-\sin\beta$ and $\cos\beta$. But these direction cosines $\cos\beta$, $\sin\beta$, $-\sin\beta$ and $\cos\beta$ are the four elements of the inverse rotation matrix $\mathbf{R}_\beta^{-1}$:

$$\mathbf{R}_\beta^{-1} = \begin{bmatrix} \cos\beta & \sin\beta \\ -\sin\beta & \cos\beta \end{bmatrix}.$$

The top row contains the direction cosines for the X'-axis and the bottom row contains the direction cosines for the Y'-axis. This relationship also holds in 3D. Consequently, if we have access to these cosines we can construct a transform that relates rotated frames of reference.

Figure 8.7 shows an axial system $X'Y'$ rotated 45°, and the associated transform is

$$\begin{bmatrix} x' \\ y' \\ 1 \end{bmatrix} \approx \begin{bmatrix} 0.707 & 0.707 & 0 \\ -0.707 & 0.707 & 0 \\ 0 & 0 & 1 \end{bmatrix} \begin{bmatrix} x \\ y \\ 1 \end{bmatrix}.$$

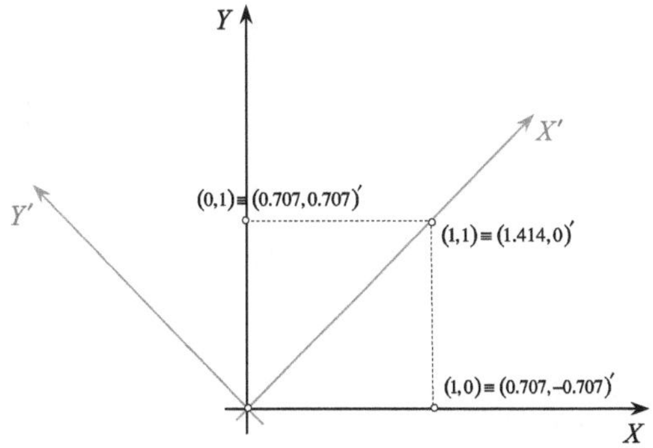

Fig. 8.7 The four vertices of the unit square shown in both frames

The four vertices on a unit square become

$$
\begin{aligned}
(0,0) &\to (0,0)\\
(1,0) &\to (0.707,-0.707)\\
(1,1) &\to (1.1414,0)\\
(0,1) &\to (0.707,0.707)
\end{aligned}
$$

which by inspection of Fig. 8.7 are correct.

8.5 Multivector Transforms

Geometric algebra is relatively new compared to other branches of mathematics and has still not found its way into mainstream computer graphics software. Nevertheless, it has been included in this chapter to demonstrate that it can be used alongside matrix transforms and quaternions.

8.5.1 Translated Frame of Reference

Figure 8.8 shows a frame of reference $X'Y'$ translated (t_x, t_y) relative to the original frame of reference XY. Therefore, if $\mathbf{p}$ points to a point $P(x, y)$ in XY, and $\mathbf{t}$ points to the origin of $X'Y'$, then $\mathbf{p}' = \mathbf{p} - \mathbf{t}$ points to the point $P'(x', y')$ relative to the $X'Y'$ frame of reference, and means that $\mathbf{p}'$ is given by

$$\mathbf{p}' = (x - t_x)\mathbf{e}_1 + (y - t_y)\mathbf{e}_2.$$

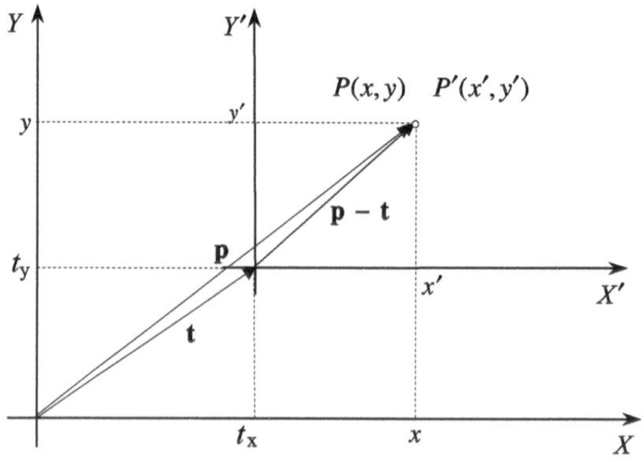

Fig. 8.8 The $X'Y'$ axial system translated (t_x, t_y)

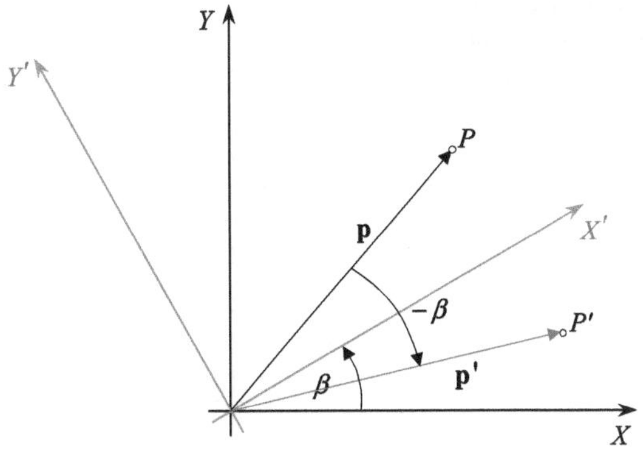

Fig. 8.9 Using a rotor **R** to rotate XY to $X'Y'$

8.5.2 Rotated Frame of Reference

We have already shown that in order to compute the coordinates of a point P in a rotated frame of reference $X'Y'$, we rotate the point by an angle in the opposite direction as shown in Fig. 8.9 to P'. Thus if the new frame of reference is rotated β, and $\mathbf{p}$ is P's position vector, then $\mathbf{p}'$ points to the new point P' and is computed as follows:

$$\mathbf{p}' = \mathbf{R}^{\dagger}_{\beta}\mathbf{p}$$

where

$$\mathbf{R}^{\dagger}_{\beta} = \cos\beta + \sin\beta\mathbf{e}_{12}.$$

Let's test this with the same example used above by rotating the frame of reference 45° and computing the coordinates of the point $(1, 1)$

$$\begin{aligned}
\mathbf{p} &= \mathbf{e}_1 + \mathbf{e}_2\\
\mathbf{R}^{\dagger}_{45^\circ} &= \cos 45^\circ + \sin 45^\circ\mathbf{e}_{12}\\
&= \frac{\sqrt{2}}{2} + \frac{\sqrt{2}}{2}\mathbf{e}_{12}\\
\mathbf{p}' &= \left(\frac{\sqrt{2}}{2} + \frac{\sqrt{2}}{2}\mathbf{e}_{12}\right)(\mathbf{e}_1 + \mathbf{e}_2)\\
&= \frac{\sqrt{2}}{2}\mathbf{e}_1 + \frac{\sqrt{2}}{2}\mathbf{e}_2 - \frac{\sqrt{2}}{2}\mathbf{e}_2 + \frac{\sqrt{2}}{2}\mathbf{e}_1\\
&= \sqrt{2}\mathbf{e}_1
\end{aligned}$$

and $P' = (\sqrt{2}, 0)$, which is correct.

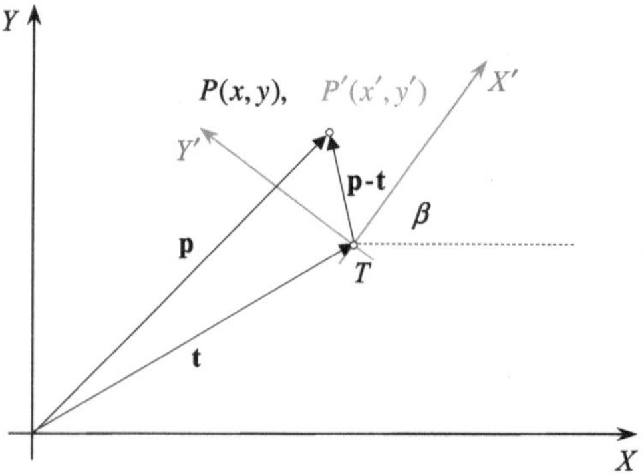

Fig. 8.10 The $X'Y'$ axial system is rotated 90° and translated (10, 5)

8.5.3 Rotated and Translated Frame of Reference

Earlier, we saw how two inverse transforms are used to compute the coordinates of a point in a rotated and translated frame of reference. We can achieve the same result using multivectors as follows.

We begin with point P and its frame of reference XY. The first step is to establish a translated frame of reference $X_T Y_T$ with position vector $\mathbf{t}$. Which means that

$$\mathbf{p}_T = \mathbf{p} - \mathbf{t}. \tag{8.3}$$

Next, as shown in Fig. 8.10, we rotate $\mathbf{p}_T$ by $-\beta$ to effectively rotate the frame of reference $X_T Y_T$ to $X'Y'$. Which means that

$$\mathbf{p}' = \mathbf{R}_\beta^\dagger \mathbf{p}_T. \tag{8.4}$$

Substituting (8.3) in (8.4) we have

$$\mathbf{p}' = \mathbf{R}_\beta^\dagger (\mathbf{p} - \mathbf{t})$$

or

$$\mathbf{p}' = (\cos\beta + \sin\beta\mathbf{e}_{12})\big((x - t_x)\mathbf{e}_1 + (y - t_y)\mathbf{e}_2\big). \tag{8.5}$$

Let's test (8.5) using the same values in the previous example where

$$\begin{aligned}
\beta &= 90^\circ \\
(t_x, t_y) &= (10, 5) \\
(x, y) &= (9, 6) \\
\mathbf{p}' &= \big(\cos 90^\circ + \sin 90^\circ \mathbf{e}_{12}\big)\big((9 - 10)\mathbf{e}_1 + (6 - 5)\mathbf{e}_2\big) \\
&= \mathbf{e}_{12}(-\mathbf{e}_1 + \mathbf{e}_2) \\
&= \mathbf{e}_1 + \mathbf{e}_2
\end{aligned}$$

which makes $P' = (1, 1)$ the same as (8.2).

Although multivectors provide an alternative way of solving vector-based problems, they still have a matrix background. For example, expanding (8.5) we have

$$\begin{aligned}
\mathbf{p}' &= (\cos\beta + \sin\beta\mathbf{e}_{12})\big((x-t_x)\mathbf{e}_1 + (y-t_y)\mathbf{e}_2\big)\\
&= (x\cos\beta - t_x\cos\beta)\mathbf{e}_1 + (y\cos\beta - t_y\cos\beta)\mathbf{e}_2\\
&\quad - (x\sin\beta - t_x\sin\beta)\mathbf{e}_2 + (y\sin\beta - t_y\sin\beta)\mathbf{e}_1\\
&= (x\cos\beta + y\sin\beta - t_x\cos\beta - t_y\sin\beta)\mathbf{e}_1\\
&\quad + (-x\sin\beta + y\cos\beta + t_x\sin\beta - t_y\cos\beta)\mathbf{e}_2
\end{aligned}$$

or in matrix form

$$\begin{bmatrix} x' \\ y' \\ 1 \end{bmatrix} = \begin{bmatrix} \cos\beta & \sin\beta & -t_x\cos\beta - t_y\sin\beta \\ -\sin\beta & \cos\beta & t_x\sin\beta - t_y\cos\beta \\ 0 & 0 & 1 \end{bmatrix} \begin{bmatrix} x \\ y \\ 1 \end{bmatrix}$$

which is identical to (8.1).

8.6 Summary

In this chapter we have discovered that if a transform such as $\mathbf{T}_{t_x,t_y}$ or $\mathbf{R}_\beta$ is used for moving points, whilst keeping the frame fixed, their inverses $\mathbf{T}_{t_x,t_y}^{-1}$ and $\mathbf{R}_\beta^{-1}$ can be used for moving the frame, whilst keeping the point fixed. It goes without saying that the converse also holds, in that we could have declared a transform for moving a frame, and its inverse could be used for moving a point.

We have also seen how geometric algebra provides an alternative approach to transforms based upon vectors, bivectors and rotors, and can undertake the same tasks.

In order to show the patterns that exist between these two mathematical approaches, all the commands are summarised.

8.6.1 Summary of Matrix Transforms

Given

$$\mathbf{T}_{t_x,t_y} = \begin{bmatrix} 1 & 0 & t_x \\ 0 & 1 & t_y \\ 0 & 0 & 1 \end{bmatrix}$$

$$\mathbf{R}_\beta = \begin{bmatrix} \cos\beta & -\sin\beta & 0 \\ \sin\beta & \cos\beta & 0 \\ 0 & 0 & 1 \end{bmatrix}.$$

Translate frame

$$\mathbf{T}_{t_x,t_y}^{-1} = \begin{bmatrix} 1 & 0 & -t_x \\ 0 & 1 & -t_y \\ 0 & 0 & 1 \end{bmatrix}.$$

Rotate frame

$$\mathbf{R}_\beta^{-1} = \begin{bmatrix} \cos\beta & \sin\beta & 0 \\ -\sin\beta & \cos\beta & 0 \\ 0 & 0 & 1 \end{bmatrix}.$$

Translate and rotate frame

$$\mathbf{R}_\beta^{-1}\mathbf{T}_{t_x,t_y}^{-1} = \begin{bmatrix} \cos\beta & \sin\beta & -t_x\cos\beta - t_y\sin\beta \\ -\sin\beta & \cos\beta & t_x\sin\beta - t_y\cos\beta \\ 0 & 0 & 1 \end{bmatrix}.$$

8.6.2 Summary of Multivector Transforms

Frame rotor

$$\mathbf{R}_\beta^\dagger = \cos\beta + \sin\beta\mathbf{e}_{12}.$$

Translate frame

$$\mathbf{p}' = \mathbf{p} - \mathbf{t} = (x - t_x)\mathbf{e}_1 + (y - t_y)\mathbf{e}_2.$$

Rotate frame

$$\mathbf{p}' = \mathbf{R}_\beta^\dagger\mathbf{p} = (\cos\beta + \sin\beta\mathbf{e}_{12})(x\mathbf{e}_1 + y\mathbf{e}_2).$$

Translate and rotate frame

$$\mathbf{p}' = \mathbf{R}_\beta^\dagger(\mathbf{p} - \mathbf{t}) = (\cos\beta + \sin\beta\mathbf{e}_{12})\big((x - t_x)\mathbf{e}_1 + (y - t_y)\mathbf{e}_2\big).$$

Chapter 9
Rotation Transforms in Space

9.1 Introduction

In this chapter we generalise the 2D transforms covered in Chap. 7 to three dimensions. In particular, we examine rotating points about the fixed three Cartesian axes and off-set, parallel axes, and about an arbitrary axis. We also explore Euler angles, and their *Achilles' heel* – gimbal lock. Matrices will be used to describe these geometric scenarios.

9.2 3D Transforms

3D transforms include the scale, translate, reflect, shear and rotate transforms, and in this chapter we need only consider the translate and rotate operations, which we will explore individually and in combination. Let's start with the translate transform.

9.2.1 Translate a Point

The matrix for a 3D point translation is basically the same as previously described but with one extra dimension $\mathbf{T}_{t_x,t_y,t_z}$. It requires the homogeneous form and is written

$$\mathbf{T}_{t_x,t_y,t_z} = \begin{bmatrix} 1 & 0 & 0 & t_x \\ 0 & 1 & 0 & t_y \\ 0 & 0 & 1 & t_z \\ 0 & 0 & 0 & 1 \end{bmatrix}$$

J. Vince, *Rotation Transforms for Computer Graphics*,
DOI 10.1007/978-0-85729-154-7_9,

where t_x, t_y, t_z are the x-, y-, z-offsets respectively. It goes without further explanation that the inverse transform $\mathbf{T}^{-1}_{t_x,t_y,t_z}$ is

$$\mathbf{T}^{-1}_{t_x,t_y,t_z} = \begin{bmatrix} 1 & 0 & 0 & -t_x \\ 0 & 1 & 0 & -t_y \\ 0 & 0 & 1 & -t_z \\ 0 & 0 & 0 & 1 \end{bmatrix}.$$

9.2.2 *Rotate a Point About the Cartesian Axes*

Although we talk about rotating points about another point in space, we require more precise information to describe this mathematically. We could, for example, associate a plane with the point of rotation and confine the rotated point to this plane, but it's much easier to visualise an axis perpendicular to this plane, about which the rotation occurs. Unfortunately, the matrix algebra for such an operation starts to become fussy, and ultimately we have seek the help of quaternions or multivectors. So let us begin this investigation by rotating a point about the three fixed Cartesian axes. Such rotations are called *Euler rotations* after the Swiss mathematician Leonhard Euler.

Recall that the transform for rotating a point about the origin in the plane is given by

$$\mathbf{R}_\beta = \begin{bmatrix} \cos\beta & -\sin\beta & 0 \\ \sin\beta & \cos\beta & 0 \\ 0 & 0 & 1 \end{bmatrix}.$$

This can be generalised into a 3D rotation $\mathbf{R}_{\beta,z}$ about the z-axis by adding a z-coordinate as follows

$$\mathbf{R}_{\beta,z} = \begin{bmatrix} \cos\beta & -\sin\beta & 0 & 0 \\ \sin\beta & \cos\beta & 0 & 0 \\ 0 & 0 & 1 & 0 \\ 0 & 0 & 0 & 1 \end{bmatrix}$$

which is illustrated in Fig. 9.1

To rotate a point about the x-axis, the x-coordinate remains constant whilst the y- and z-coordinates are changed according to the 2D rotation transform. This is expressed algebraically as

$$x' = x$$

$$y' = y\cos\beta - z\sin\beta$$

$$z' = y\sin\beta + z\cos\beta$$

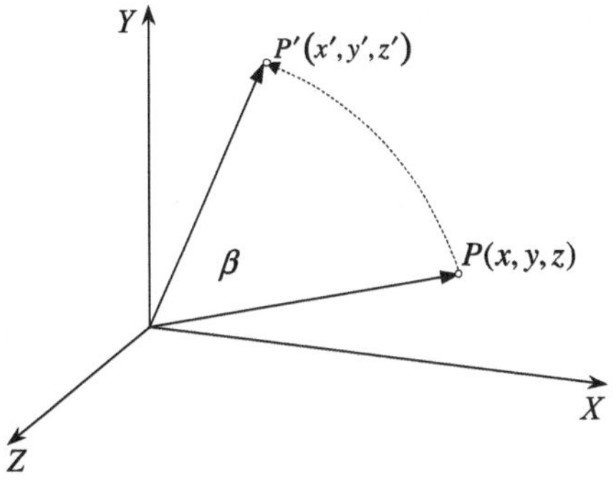

Fig. 9.1 Rotating the point P about the z-axis

or in matrix form as $\mathbf{R}_{\beta,x}$

$$\mathbf{R}_{\beta,x} = \begin{bmatrix} 1 & 0 & 0 & 0 \\ 0 & \cos\beta & -\sin\beta & 0 \\ 0 & \sin\beta & \cos\beta & 0 \\ 0 & 0 & 0 & 1 \end{bmatrix}.$$

To rotate about the y-axis, the y-coordinate remains constant whilst the x- and z-coordinates are changed. This is expressed algebraically as

$$\begin{aligned} x' &= z\sin\beta + x\cos\beta \\ y' &= y \\ z' &= z\cos\beta - x\sin\beta \end{aligned}$$

or in matrix form as $\mathbf{R}_{\beta,y}$

$$\mathbf{R}_{\beta,y} = \begin{bmatrix} \cos\beta & 0 & \sin\beta & 0 \\ 0 & 1 & 0 & 0 \\ -\sin\beta & 0 & \cos\beta & 0 \\ 0 & 0 & 0 & 1 \end{bmatrix}.$$

Note that the matrix terms don't appear to share the symmetry enjoyed by the previous two matrices. Nothing really has gone wrong, it's just the way the axes are paired together to rotate the coordinates. Now let's consider similar rotations about off-set axes parallel to the Cartesian axes.

9.2.3 Rotating About an Off-Set Axis

To begin, let's develop a transform to rotate a point about a fixed axis parallel with the z-axis, as shown in Fig. 9.2. The scenario is very reminiscent of the 2D case for rotating a point about an arbitrary point, and the general transform is given by

$$\begin{bmatrix} x' \\ y' \\ z' \\ 1 \end{bmatrix} = \mathbf{T}_{t_x,t_y,0}\mathbf{R}_{\beta,z}\mathbf{T}_{-t_x,-t_y,0}\begin{bmatrix} x \\ y \\ z \\ 1 \end{bmatrix}$$

Fig. 9.2 Rotating a point about an axis parallel with the z-axis

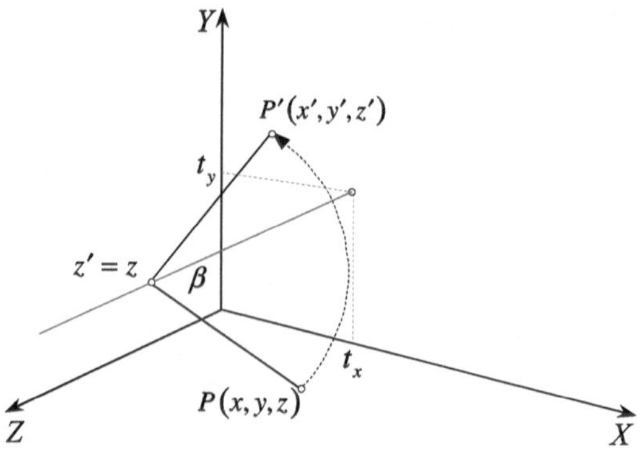

where

$\mathbf{T}_{-t_x,-t_y,0}$	creates a temporary origin
$\mathbf{R}_{\beta,z}$	rotates β about the temporary z-axis
$\mathbf{T}_{t_x,t_y,0}$	returns to the original position

and the matrix transform is

$$\mathbf{T}_{t_x,t_y,0}\mathbf{R}_{\beta,z}\mathbf{T}_{-t_x,-t_y,0} = \begin{bmatrix} \cos\beta & -\sin\beta & 0 & t_x(1-\cos\beta)+t_y\sin\beta \\ \sin\beta & \cos\beta & 0 & t_y(1-\cos\beta)-t_x\sin\beta \\ 0 & 0 & 1 & 0 \\ 0 & 0 & 0 & 1 \end{bmatrix}.$$

I hope you can see the similarity between rotating in 3D and 2D – the x- and y-coordinates are updated while the z-coordinate is held constant. We can now state the other two matrices for rotating about an off-set axis parallel with the x-axis and parallel with the y-axis:

$$\mathbf{T}_{0,t_y,t_z}\mathbf{R}_{\beta,x}\mathbf{T}_{0,-t_y,-t_z} = \begin{bmatrix} 1 & 0 & 0 & 0 \\ 0 & \cos\beta & -\sin\beta & t_y(1-\cos\beta)+t_z\sin\beta \\ 0 & \sin\beta & \cos\beta & t_z(1-\cos\beta)-t_y\sin\beta \\ 0 & 0 & 0 & 1 \end{bmatrix}$$

$$\mathbf{T}_{t_x,0,t_z}\mathbf{R}_{\beta,y}\mathbf{T}_{-t_x,0,-t_z} = \begin{bmatrix} \cos\beta & 0 & \sin\beta & t_x(1-\cos\beta)-t_z\sin\beta \\ 0 & 1 & 0 & 0 \\ -\sin\beta & 0 & \cos\beta & t_z(1-\cos\beta)+t_x\sin\beta \\ 0 & 0 & 0 & 1 \end{bmatrix}.$$

9.3 Composite Rotations

So far we have only considered single rotations about a Cartesian axis or a parallel off-set axis, but there is nothing to stop us constructing a sequence of rotations to create a composite rotation. For example, we could begin by rotating a point α about the x-axis followed by a rotation β about the y-axis, which in turn could be

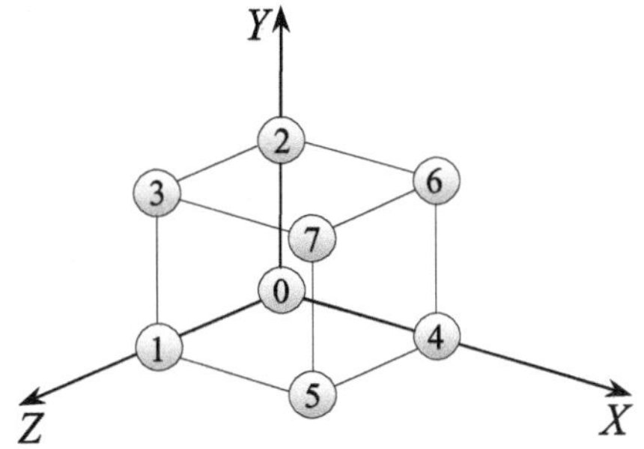

Fig. 9.3 A unit cube with vertices coded as shown in Table 9.1

Table 9.1 Vertex coordinates for the cube in Fig. 9.3

vertex	0	1	2	3	4	5	6	7
x	0	0	0	0	1	1	1	1
y	0	0	1	1	0	0	1	1
z	0	1	0	1	0	1	0	1

followed by a rotation γ about the z-axis. As mentioned above, these rotations are called Euler rotations.

One of the problems with Euler rotations is visualising exactly what is happening at each step, and predicting the orientation of an object after a composite rotation. To simplify the problem we will employ a unit cube whose vertices are numbered 0 to 7 as shown in Fig. 9.3. We will also employ the following binary coded expression that uses the Cartesian coordinates of the vertex in the vertex number:

$$vertex = 4x + 2y + z.$$

For example, vertex 0 has coordinates $(0, 0, 0)$, and vertex 7 has coordinates $(1, 1, 1)$. All the codes are shown in Table 9.1.

Let's repeat the three rotation transforms for rotating points about the x-, y- and z-axes respectively, in their non-homogeneous form and substitute c for cos and s for sin to save space:

$$\text{rotate } \alpha \text{ about the } x\text{-axis} \quad \mathbf{R}_{\alpha,x} = \begin{bmatrix} 1 & 0 & 0 \\ 0 & c_\alpha & -s_\alpha \\ 0 & s_\alpha & c_\alpha \end{bmatrix}$$

$$\text{rotate } \beta \text{ about the } y\text{-axis} \quad \mathbf{R}_{\beta,y} = \begin{bmatrix} c_\beta & 0 & s_\beta \\ 0 & 1 & 0 \\ -s_\beta & 0 & c_\beta \end{bmatrix}$$

$$\text{rotate } \gamma \text{ about the } z\text{-axis} \quad \mathbf{R}_{\gamma,z} = \begin{bmatrix} c_\gamma & -s_\gamma & 0 \\ s_\gamma & c_\gamma & 0 \\ 0 & 0 & 1 \end{bmatrix}.$$

We can create a composite, moving-point, fixed-frame rotation by placing $\mathbf{R}_{\alpha,x}$, $\mathbf{R}_{\beta,y}$ and $\mathbf{R}_{\gamma,z}$ in any sequence. As an example, let's choose the sequence $\mathbf{R}_{\gamma,z}\mathbf{R}_{\beta,y}\mathbf{R}_{\alpha,x}$

$$\mathbf{R}_{\gamma,z}\mathbf{R}_{\beta,y}\mathbf{R}_{\alpha,x} = \begin{bmatrix} c_\gamma & -s_\gamma & 0 \\ s_\gamma & c_\gamma & 0 \\ 0 & 0 & 1 \end{bmatrix} \begin{bmatrix} c_\beta & 0 & s_\beta \\ 0 & 1 & 0 \\ -s_\beta & 0 & c_\beta \end{bmatrix} \begin{bmatrix} 1 & 0 & 0 \\ 0 & c_\alpha & -s_\alpha \\ 0 & s_\alpha & c_\alpha \end{bmatrix}. \quad (9.1)$$

Multiplying the three matrices in (9.1) together we obtain

$$\begin{bmatrix} c_\gamma c_\beta & c_\gamma s_\beta s_\alpha - s_\gamma c_\alpha & c_\gamma s_\beta c_\alpha + s_\gamma s_\alpha \\ s_\gamma c_\beta & s_\gamma s_\beta s_\alpha + c_\gamma c_\alpha & s_\gamma s_\beta c_\alpha - c_\gamma s_\alpha \\ -s_\beta & c_\beta s_\alpha & c_\beta c_\alpha \end{bmatrix} \quad (9.2)$$

or using the more familiar notation:

$$\begin{bmatrix} \cos\gamma\cos\beta & \cos\gamma\sin\beta\sin\alpha - \sin\gamma\cos\alpha & \cos\gamma\sin\beta\cos\alpha + \sin\gamma\sin\alpha \\ \sin\gamma\cos\beta & \sin\gamma\sin\beta\sin\alpha + \cos\gamma\cos\alpha & \sin\gamma\sin\beta\cos\alpha - \cos\gamma\sin\alpha \\ -\sin\beta & \cos\beta\sin\alpha & \cos\beta\cos\alpha \end{bmatrix}.$$

Let's evaluate (9.2) by making $\alpha = \beta = \gamma = 90°$:

$$\begin{bmatrix} 0 & 0 & 1 \\ 0 & 1 & 0 \\ -1 & 0 & 0 \end{bmatrix}. \quad (9.3)$$

The matrix (9.3) is equivalent to rotating a point 90° about the fixed x-axis, followed by a rotation of 90° about the fixed y-axis, followed by a rotation of 90° about the fixed z-axis. This rotation sequence is illustrated in Fig. 9.4 (a)–(d).

Figure 9.4 (a) shows the starting position of the cube; (b) shows its orientation after a 90° rotation about the x-axis; (c) shows its orientation after a further rotation of 90° about the y-axis; and (d) the cube's resting position after a rotation of 90° about the z-axis.

From Fig. 9.4 (d) we see that the cube's coordinates are as shown in Table 9.2. We can confirm that these coordinates are correct by multiplying the cube's original coordinates shown in Table 9.1 by the matrix (9.3). Although it is not mathematically correct, we will show the matrix multiplying an array of coordinates as follows

$$\begin{bmatrix} 0 & 0 & 1 \\ 0 & 1 & 0 \\ -1 & 0 & 0 \end{bmatrix} \begin{bmatrix} 0 & 0 & 0 & 0 & 1 & 1 & 1 & 1 \\ 0 & 0 & 1 & 1 & 0 & 0 & 1 & 1 \\ 0 & 1 & 0 & 1 & 0 & 1 & 0 & 1 \end{bmatrix}$$
$$= \begin{bmatrix} 0 & 1 & 0 & 1 & 0 & 1 & 0 & 1 \\ 0 & 0 & 1 & 1 & 0 & 0 & 1 & 1 \\ 0 & 0 & 0 & 0 & -1 & -1 & -1 & -1 \end{bmatrix}$$

which agree with the coordinates in Table 9.2.

Naturally, any three angles can be chosen to rotate a point about the fixed axes, but it does become difficult to visualise without an interactive cgi system.

Note that the determinant of (9.3) is 1, which is as expected.

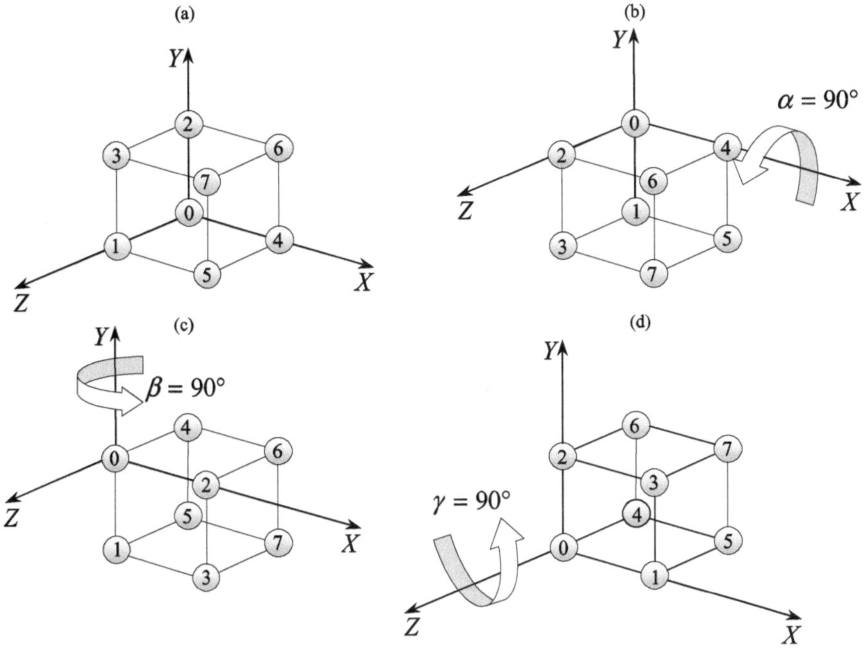

Fig. 9.4 Four views of the unit cube before and during the three rotations

Table 9.2 Vertex coordinates for the cube in Fig. 9.4 (d)

vertex	0	1	2	3	4	5	6	7
x	0	1	0	1	0	1	0	1
y	0	0	1	1	0	0	1	1
z	0	0	0	0	−1	−1	−1	−1

An observation we made with 2D rotations is that they are additive: i.e. $\mathbf{R}_\alpha$ followed by $\mathbf{R}_\beta$ is equivalent to $\mathbf{R}_{\alpha+\beta}$. But something equally important is that rotations in 2D commute:

$$\mathbf{R}_\alpha \mathbf{R}_\beta = \mathbf{R}_\beta \mathbf{R}_\alpha = \mathbf{R}_{\alpha+\beta} = \mathbf{R}_{\beta+\alpha}$$

whereas, in general, 3D rotations are non-commutative. This is seen by considering a composite rotation formed by a rotation α about the x-axis $\mathbf{R}_{\alpha,x}$, followed by a rotation β about the z-axis $\mathbf{R}_{\beta,z}$, and

$$\mathbf{R}_{\alpha,x}\mathbf{R}_{\beta,z} \neq \mathbf{R}_{\beta,z}\mathbf{R}_{\alpha,x}.$$

As an illustration, let's reverse the composite rotation computed above to $\mathbf{R}_{\alpha,x}\mathbf{R}_{\beta,y}\mathbf{R}_{\gamma,z}$:

$$\mathbf{R}_{\alpha,x}\mathbf{R}_{\beta,y}\mathbf{R}_{\gamma,z} = \begin{bmatrix} 1 & 0 & 0 \\ 0 & c_\alpha & -s_\alpha \\ 0 & s_\alpha & c_\alpha \end{bmatrix} \begin{bmatrix} c_\beta & 0 & s_\beta \\ 0 & 1 & 0 \\ -s_\beta & 0 & c_\beta \end{bmatrix} \begin{bmatrix} c_\gamma & -s_\gamma & 0 \\ s_\gamma & c_\gamma & 0 \\ 0 & 0 & 1 \end{bmatrix}. \quad (9.4)$$

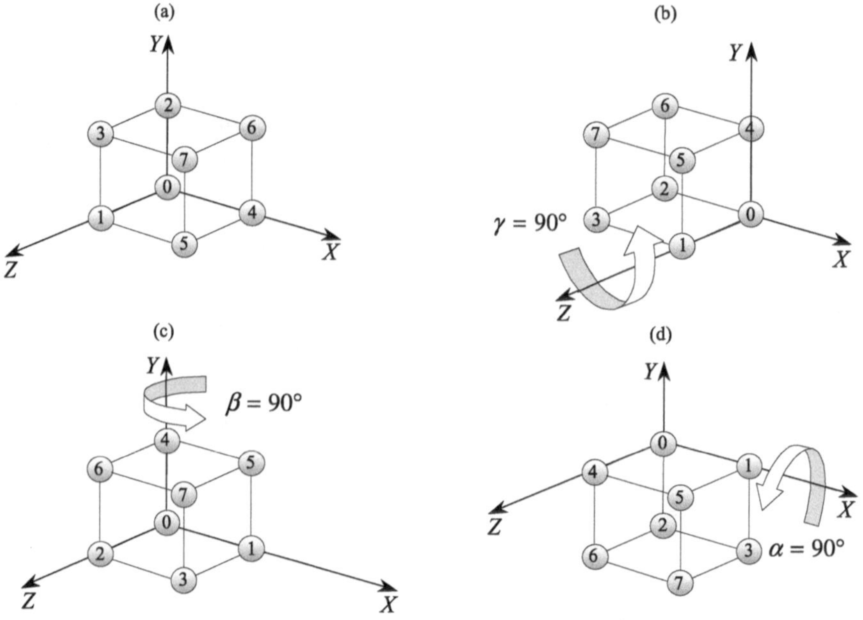

Fig. 9.5 Four views of the unit cube using the rotation sequence $\mathbf{R}_{\alpha,x}\mathbf{R}_{\beta,y}\mathbf{R}_{\gamma,z}$

Multiplying the three matrices in (9.4) together we obtain

$$\begin{bmatrix} c_\beta c_\gamma & -c_\beta s_\gamma & s_\beta \\ s_\alpha s_\beta c_\gamma + c_\alpha s_\gamma & -s_\alpha s_\beta s_\gamma + c_\alpha c_\gamma & -s_\alpha c_\beta \\ -c_\alpha s_\beta c_\gamma + s_\alpha s_\gamma & c_\alpha s_\beta s_\gamma + s_\alpha c_\gamma & c_\alpha c_\beta \end{bmatrix} \tag{9.5}$$

or using the more familiar notation:

$$\begin{bmatrix} \cos\beta\cos\gamma & -\cos\beta\sin\gamma & \sin\beta \\ \sin\alpha\sin\beta\cos\gamma + \cos\alpha\sin\gamma & -\sin\alpha\sin\beta\sin\gamma + \cos\alpha\cos\gamma & -\sin\alpha\cos\beta \\ -\cos\alpha\sin\beta\cos\gamma + \sin\alpha\sin\gamma & \cos\alpha\sin\beta\sin\gamma + \sin\alpha\cos\gamma & \cos\alpha\cos\beta \end{bmatrix}.$$

Comparing (9.3) and (9.5) it can be seen that they are completely different.

Let's evaluate (9.5) by making $\alpha = \beta = \gamma = 90°$:

$$\begin{bmatrix} 0 & 0 & 1 \\ 0 & -1 & 0 \\ 1 & 0 & 0 \end{bmatrix}. \tag{9.6}$$

The matrix (9.6) is equivalent to rotating a point 90° about the fixed z-axis, followed by a rotation of 90° about the fixed y-axis, followed by a rotation of 90° about the fixed x-axis. This rotation sequence is illustrated in Fig. 9.5 (a)–(d).

From Fig. 9.5 (d) we see that the cube's coordinates are as shown in Table 9.3. We can confirm that these coordinates are correct by multiplying the cube's original

Table 9.3 Vertex coordinates for the cube in Fig. 9.5 (d)

vertex	0	1	2	3	4	5	6	7
x	0	1	0	1	0	1	0	1
y	0	0	−1	−1	0	0	−1	−1
z	0	0	0	0	1	1	1	1

coordinates shown in Table 9.1 by the matrix (9.6). We show the matrix multiplying an array of coordinates as before:

$$\begin{bmatrix} 0 & 0 & 1 \\ 0 & -1 & 0 \\ 1 & 0 & 0 \end{bmatrix} \begin{bmatrix} 0 & 0 & 0 & 0 & 1 & 1 & 1 & 1 \\ 0 & 0 & 1 & 1 & 0 & 0 & 1 & 1 \\ 0 & 1 & 0 & 1 & 0 & 1 & 0 & 1 \end{bmatrix}$$

$$= \begin{bmatrix} 0 & 1 & 0 & 1 & 0 & 1 & 0 & 1 \\ 0 & 0 & -1 & -1 & 0 & 0 & -1 & -1 \\ 0 & 0 & 0 & 0 & 1 & 1 & 1 & 1 \end{bmatrix}$$

which agree with the coordinates in Table 9.3, and we can safely conclude that, in general, 3D rotation transforms do not commute. Inspection of Fig. 9.5 (d) shows that the unit cube has been rotated 180° about a vector $[1 \quad 0 \quad 1]^{\mathrm{T}}$.

Now let's explore the role eigenvectors and eigenvalues play in 3D rotations.

9.3.1 3D Eigenvectors

In Chap. 4 we examined the characteristic equation used to identify any eigenvectors associated with a matrix. The eigenvector $\mathbf{v}$ satisfies the relationship

$$\mathbf{A}\mathbf{v} = \lambda\mathbf{v}$$

where λ is a scaling factor.

In the context of a 3D rotation matrix, an eigenvector is a vector scaled by λ but not rotated, which implies that it is the axis of rotation. To illustrate this, let's identify the eigenvector for the composite rotation (9.3) above:

$$\mathbf{R}_{90°,z}\mathbf{R}_{90°,y}\mathbf{R}_{90°,x} = \begin{bmatrix} 0 & 0 & 1 \\ 0 & 1 & 0 \\ -1 & 0 & 0 \end{bmatrix}.$$

Figure 9.4 (a)–(d) shows the effect of this composite rotation, which is nothing more than a rotation of 90° about the y-axis. Therefore, we should be able to extract this information from the above matrix.

We begin by writing the characteristic equation for the matrix:

$$\begin{vmatrix} 0-\lambda & 0 & 1 \\ 0 & 1-\lambda & 0 \\ -1 & 0 & 0-\lambda \end{vmatrix} = 0. \tag{9.7}$$

Expanding (9.7) using the top row we have

$$\begin{aligned} -\lambda\begin{vmatrix} 1-\lambda & 0 \\ 0 & -\lambda \end{vmatrix} + 1\begin{vmatrix} 0 & 1-\lambda \\ -1 & 0 \end{vmatrix} &= 0 \\ -\lambda(-\lambda+\lambda^2)+1-\lambda &= 0 \\ \lambda^2-\lambda^3+1-\lambda &= 0 \\ -\lambda^3+\lambda^2-\lambda+1 &= 0 \\ \lambda^3-\lambda^2+\lambda &= 1. \end{aligned}$$

When working with 3×3 matrices we always end up with a cubic in λ, for which there can be three types of solution:

1. One real and two complex conjugate solutions.
2. Three real solutions including the possibility of a double solution.
3. Three distinct real solutions.

It is clear that $\lambda = 1$ is one such real root, which satisfies our requirement for an eigenvalue. We could also show that the other two roots are complex conjugates.

Substituting $\lambda = 1$ in the original equations associated with (9.7) to reveal the eigenvector, we have

$$\begin{cases} -x+0y+\ \ z=0 \\ 0x+0y+0z=0 \\ -x+0y-\ \ z=0. \end{cases}$$

It is obvious from the 1st and 3rd equations that $x = z = 0$. However, all three equations multiply the y term by zero, which implies that the associated eigenvector is of the form $[0 \quad k \quad 0]^T$, which is the y-axis, as anticipated. Now let's find the angle of rotation.

Using one of the above rotation matrices $\mathbf{R}_{\beta,y}$ and the trace operation:

$$\begin{aligned} \mathbf{R}_{\beta,y} &= \begin{bmatrix} \cos\beta & 0 & \sin\beta \\ 0 & 1 & 0 \\ -\sin\beta & 0 & \cos\beta \end{bmatrix} \\ \mathrm{Tr}(\mathbf{R}_{\beta,y}) &= 1+2\cos\beta \end{aligned}$$

therefore,

$$\beta = \arccos((\mathrm{Tr}(\mathbf{R}_{\beta,y})-1)/2).$$

To illustrate this, let $\beta = 90°$:

$$\begin{aligned} \mathbf{R}_{90°,y} &= \begin{bmatrix} 0 & 0 & 1 \\ 0 & 1 & 0 \\ -1 & 0 & 0 \end{bmatrix} \\ \mathrm{Tr}(\mathbf{R}_{90°,y}) &= 1 \end{aligned}$$

therefore,

$$\beta = \arccos\left(\left(1-1\right)/2\right) = 90°.$$

Let's choose another matrix and repeat the above:

$$\mathbf{R}_{\alpha,x} = \begin{bmatrix} 1 & 0 & 0 \\ 0 & \cos\alpha & -\sin\alpha \\ 0 & \sin\alpha & \cos\alpha \end{bmatrix}.$$

This time, let $\alpha = 45°$:

$$\mathbf{R}_{45°,x} = \begin{bmatrix} 1 & 0 & 0 \\ 0 & \sqrt{2}/2 & -\sqrt{2}/2 \\ 0 & \sqrt{2}/2 & \sqrt{2}/2 \end{bmatrix}$$

$$\text{Tr}(\mathbf{R}_{45°,x}) = 1 + \sqrt{2}$$

therefore,

$$\alpha = \arccos((1 + \sqrt{2} - 1)/2) = 45°.$$

So we now have a mechanism to extract the axis and angle of rotation from a rotation matrix. However, the algorithm for identifying the axis is far from satisfactory, and later on we will discover that there is a similar technique which is readily programable.

For completeness, let's identify the axis and angle of rotation for the matrix (9.6):

$$\mathbf{R}_{90°,x}\mathbf{R}_{90°,y}\mathbf{R}_{90°,z} = \begin{bmatrix} 0 & 0 & 1 \\ 0 & -1 & 0 \\ 1 & 0 & 0 \end{bmatrix}.$$

Once more, we begin by writing the characteristic equation for the matrix:

$$\begin{vmatrix} 0-\lambda & 0 & 1 \\ 0 & -1-\lambda & 0 \\ 1 & 0 & 0-\lambda \end{vmatrix} = 0. \tag{9.8}$$

Expanding (9.8) using the top row we have

$$\begin{aligned} -\lambda \begin{vmatrix} -1-\lambda & 0 \\ 0 & -\lambda \end{vmatrix} + 1 \begin{vmatrix} 0 & -1-\lambda \\ 1 & 0 \end{vmatrix} &= 0 \\ -\lambda(-\lambda + \lambda^2) + 1 - \lambda &= 0 \\ \lambda^2 - \lambda^3 + 1 - \lambda &= 0 \\ -\lambda^3 + \lambda^2 - \lambda + 1 &= 0 \\ \lambda^3 - \lambda^2 + \lambda &= 1. \end{aligned}$$

Again, there is a single real root: $\lambda = 1$, and substituting this in the original equations associated with (9.8) to reveal the eigenvector, we have

$$\begin{cases} -x + 0y + \ z = 0 \\ 0x - 2y + 0z = 0 \\ \ x + 0y - \ z = 0. \end{cases}$$

It is obvious from the 1st and 3rd equations that $x = z$, and from the 2nd equation that $y = 0$, which implies that the associated eigenvector is of the form $[k \quad 0 \quad k]$, which is correct.

Using the trace operation, we can write

$$\mathrm{Tr}(\mathbf{R}_{90^\circ,x}\mathbf{R}_{90^\circ,y}\mathbf{R}_{90^\circ,z}) = -1$$

therefore,

$$\beta = \arccos\left((-1-1)/2\right) = 180^\circ.$$

As promised, let's explore another way of identifying the fixed axis of rotation, which is an eigenvector. Consider the following argument where $\mathbf{A}$ is a simple rotation transform:

If $\mathbf{v}$ is a fixed axis of rotation and $\mathbf{A}$ a rotation transform, then $\mathbf{v}$ suffers no rotation:

$$\mathbf{A}\mathbf{v} = \mathbf{v} \tag{9.9}$$

similarly,

$$\mathbf{A}^{\mathrm{T}}\mathbf{v} = \mathbf{v}. \tag{9.10}$$

Subtracting (9.10) from (9.9), we have

$$\mathbf{A}\mathbf{v} - \mathbf{A}^{\mathrm{T}}\mathbf{v} = \mathbf{0} \tag{9.11}$$

$$\left(\mathbf{A} - \mathbf{A}^{\mathrm{T}}\right)\mathbf{v} = \mathbf{0} \tag{9.12}$$

where $\mathbf{0}$ is a null vector.

In Chap. 4 we defined an antisymmetric matrix $\mathbf{Q}$ as

$$\mathbf{Q} = \frac{1}{2}\left(\mathbf{A} - \mathbf{A}^{\mathrm{T}}\right) \tag{9.13}$$

therefore,

$$\left(\mathbf{A} - \mathbf{A}^{\mathrm{T}}\right) = 2\mathbf{Q}. \tag{9.14}$$

Substituting (9.14) in (9.12) we have

$$2\mathbf{Q}\mathbf{v} = \mathbf{0}$$

$$\mathbf{Q}\mathbf{v} = \mathbf{0}$$

which permits us to write

$$\begin{bmatrix} 0 & q_3 & -q_2 \\ -q_3 & 0 & q_1 \\ q_2 & -q_1 & 0 \end{bmatrix} \begin{bmatrix} v_1 \\ v_2 \\ v_3 \end{bmatrix} = \begin{bmatrix} 0 \\ 0 \\ 0 \end{bmatrix} \tag{9.15}$$

where

$$q_1 = a_{23} - a_{32}$$

$$q_2 = a_{31} - a_{13}$$

$$q_3 = a_{12} - a_{21}.$$

Expanding (9.15) we have

$$\begin{aligned}0v_1 + q_3v_2 - q_2v_3 &= 0\\ -q_3v_1 + 0v_2 + q_1v_3 &= 0\\ q_2v_1 - q_1v_2 + 0v_3 &= 0.\end{aligned}$$

Obviously, one possible solution is $v_1 = v_2 = v_3 = 0$, but we seek a solution for $\mathbf{v}$ in terms of q_1, q_2 and q_3. A standard technique is to relax one of the v terms, such as making $v_1 = 1$. Then

$$q_3v_2 - q_2v_3 = 0 \tag{9.16}$$

$$-q_3 + q_1v_3 = 0 \tag{9.17}$$

$$q_2 - q_1v_2 = 0. \tag{9.18}$$

From (9.18) we have

$$v_2 = \frac{q_2}{q_1}.$$

From (9.17) we have

$$v_3 = \frac{q_3}{q_1}$$

therefore, a solution is

$$\mathbf{v} = \begin{bmatrix} \frac{q_1}{q_1} & \frac{q_2}{q_1} & \frac{q_3}{q_1} \end{bmatrix}^{\mathrm{T}}$$

which in a non-homogeneous form is

$$\mathbf{v} = \begin{bmatrix} q_1 & q_2 & q_3 \end{bmatrix}^{\mathrm{T}}$$

or in terms of the original matrix:

$$\mathbf{v} = \begin{bmatrix} (a_{23} - a_{32}) & (a_{31} - a_{13}) & (a_{12} - a_{21}) \end{bmatrix}^{\mathrm{T}} \tag{9.19}$$

which appears to be a rather elegant solution for finding the fixed axis of revolution.

Now let's put (9.19) to the test by recomputing the axis of rotation for the pure rotations $\mathbf{R}_{\alpha,x}$, $\mathbf{R}_{\beta,y}$ and $\mathbf{R}_{\gamma,z}$ where $\alpha = \beta = \gamma = 90°$:

$$\mathbf{R}_{90°,x} = \begin{bmatrix} 1 & 0 & 0\\ 0 & 0 & -1\\ 0 & 1 & 0 \end{bmatrix}$$

using (9.19) we have

$$\mathbf{v} = \begin{bmatrix} (-1-1) & (0-0) & (0-0) \end{bmatrix} = [-2 \quad 0 \quad 0]^{\mathrm{T}}$$

which is the x-axis.

$$\mathbf{R}_{90°,y} = \begin{bmatrix} 0 & 0 & 1\\ 0 & 1 & 0\\ -1 & 0 & 0 \end{bmatrix}$$

using (9.19) we have

$$\mathbf{v} = \begin{bmatrix}(0-0) & (-1-1) & (0-0)\end{bmatrix} = [0 \quad -2 \quad 0]^{\mathrm{T}}$$

which is the y-axis.

$$\mathbf{R}_{90^\circ,z} = \begin{bmatrix} 0 & -1 & 0 \\ 1 & 0 & 0 \\ 0 & 0 & 1 \end{bmatrix}$$

using (9.19) we have

$$\mathbf{v} = \begin{bmatrix}(0-0) & (0-0) & (-1-1)\end{bmatrix} = [0 \quad 0 \quad -2]^{\mathrm{T}}$$

which is the z-axis.

However, if we attempt to extract the axis of rotation from

$$\mathbf{R}_{90^\circ,x}\mathbf{R}_{90^\circ,y}\mathbf{R}_{90^\circ,z} = \begin{bmatrix} 0 & 0 & 1 \\ 0 & -1 & 0 \\ 1 & 0 & 0 \end{bmatrix}$$

we have a problem, because $q_1 = q_2 = q_3 = 0$. This is because $\mathbf{A} = \mathbf{A}^{\mathrm{T}}$ and the technique relies upon $\mathbf{A} \neq \mathbf{A}^{\mathrm{T}}$.

So let's consider another approach based upon the fact that a rotation matrix always has a real eigenvalue $\lambda = 1$, which permits us to write

$$\begin{aligned} \mathbf{Av} &= \lambda\mathbf{v} \\ \mathbf{Av} &= \lambda\mathbf{Iv} = \mathbf{Iv} \\ (\mathbf{A} - \mathbf{I})\,\mathbf{v} &= \mathbf{0} \end{aligned}$$

therefore,

$$\begin{bmatrix} (a_{11}-1) & a_{12} & a_{13} \\ a_{21} & (a_{22}-1) & a_{23} \\ a_{31} & a_{32} & (a_{33}-1) \end{bmatrix} \begin{bmatrix} v_1 \\ v_2 \\ v_3 \end{bmatrix} = \begin{bmatrix} 0 \\ 0 \\ 0 \end{bmatrix}. \tag{9.20}$$

Expanding (9.20) we have

$$\begin{aligned} (a_{11}-1)v_1 + a_{12}v_2 + a_{13}v_3 &= 0 \\ a_{21}v_1 + (a_{22}-1)v_2 + a_{23}v_3 &= 0 \\ a_{31}v_1 + a_{32}v_2 + (a_{33}-1)v_3 &= 0. \end{aligned}$$

Once more, there exists a trivial solution where $v_1 = v_2 = v_3 = 0$, but to discover something more useful we can relax any one of the v terms which gives us three equations in two unknowns. Let's make $v_1 = 0$:

$$a_{12}v_2 + a_{13}v_3 = -(a_{11}-1) \tag{9.21}$$

$$(a_{22}-1)\,v_2 + a_{23}v_3 = -a_{21} \tag{9.22}$$

$$a_{32}v_2 + (a_{33}-1)\,v_3 = -a_{31}. \tag{9.23}$$

We are now faced with choosing a pair of equations to isolate v_2 and v_3. In fact, we have to consider all three pairings because it is possible that a future rotation matrix will contain a column with two zero elements, which could conflict with any pairing we make at this stage.

Let's begin by choosing (9.21) and (9.22). The solution employs the following strategy: Given the following matrix equation

$$\begin{bmatrix} a_1 & b_1 \\ a_2 & b_2 \end{bmatrix}\begin{bmatrix} x \\ y \end{bmatrix} = \begin{bmatrix} c_1 \\ c_2 \end{bmatrix}$$

then

$$\frac{x}{\begin{vmatrix} c_1 & b_1 \\ c_2 & b_2 \end{vmatrix}} = \frac{y}{\begin{vmatrix} a_1 & c_1 \\ a_2 & c_2 \end{vmatrix}} = \frac{1}{\begin{vmatrix} a_1 & b_1 \\ a_2 & b_2 \end{vmatrix}}.$$

Therefore, using the 1st and 2nd (9.21) and (9.22) we have

$$\frac{v_2}{\begin{vmatrix} -(a_{11}-1) & a_{13} \\ -a_{21} & a_{23} \end{vmatrix}} = \frac{v_3}{\begin{vmatrix} a_{12} & -(a_{11}-1) \\ (a_{22}-1) & -a_{21} \end{vmatrix}} = \frac{1}{\begin{vmatrix} a_{12} & a_{13} \\ (a_{22}-1) & a_{23} \end{vmatrix}}$$

$$v_1 = a_{12}a_{23} - a_{13}(a_{22}-1)$$
$$v_2 = a_{13}a_{21} - a_{23}(a_{11}-1)$$
$$v_3 = (a_{11}-1)(a_{22}-1) - a_{12}a_{21}.$$

Similarly, using the 1st and 3rd (9.21) and (9.23) we have

$$v_1 = a_{12}(a_{33}-1) - a_{13}a_{32}$$
$$v_2 = a_{13}a_{31} - (a_{11}-1)(a_{33}-1)$$
$$v_3 = a_{32}(a_{11}-1) - a_{12}a_{31}$$

and using the 2nd and 3rd (9.22) and (9.23) we have

$$v_1 = (a_{22}-1)(a_{33}-1) - a_{23}a_{32}$$
$$v_2 = a_{23}a_{31} - a_{21}(a_{33}-1)$$
$$v_3 = a_{21}a_{32} - a_{31}(a_{22}-1).$$

Now we have nine equations to cope with any eventuality. In fact, there is nothing to stop us from choosing any three that take our fancy, for example these three equations look interesting and sound:

$$v_1 = (a_{22}-1)(a_{33}-1) - a_{23}a_{32} \tag{9.24}$$
$$v_2 = (a_{33}-1)(a_{11}-1) - a_{31}a_{13} \tag{9.25}$$
$$v_3 = (a_{11}-1)(a_{22}-1) - a_{12}a_{21}. \tag{9.26}$$

Therefore, the solution for the eigenvector is $[v_1 \quad v_2 \quad v_3]^T$. Note that the sign of v_2 has been reversed to maintain symmetry.

Let's test (9.24), (9.25) and (9.26) with the transforms used above:

$$\mathbf{R}_{90^\circ,x} = \begin{bmatrix} 1 & 0 & 0 \\ 0 & 0 & -1 \\ 0 & 1 & 0 \end{bmatrix} \begin{cases} v_1 = (-1)(-1) - (-1) \times 1 = 2 \\ v_2 = (-1)(0) - 0 \times 0 = 0 \\ v_3 = (0)(-1) - 0 \times 0 = 0 \end{cases}$$

$$\mathbf{R}_{90^\circ,y} = \begin{bmatrix} 0 & 0 & 1 \\ 0 & 1 & 0 \\ -1 & 0 & 0 \end{bmatrix} \begin{cases} v_1 = (0)(-1) - 0 \times 0 = 0 \\ v_2 = (-1)(-1) - (-1) \times 1 = 2 \\ v_3 = (-1)(0) - 0 \times 0 = 0 \end{cases}$$

$$\mathbf{R}_{90^\circ,z} = \begin{bmatrix} 0 & -1 & 0 \\ 1 & 0 & 0 \\ 0 & 0 & 1 \end{bmatrix} \begin{cases} v_1 = (-1)(0) - 0 \times 0 = 0 \\ v_2 = (0)(-1) - 0 \times 0 = 0 \\ v_3 = (-1)(-1) - (-1) \times 1 = 2 \end{cases}$$

$$\mathbf{R}_{90^\circ,x}\mathbf{R}_{90^\circ,y}\mathbf{R}_{90^\circ,z} = \begin{bmatrix} 0 & 0 & 1 \\ 0 & -1 & 0 \\ 1 & 0 & 0 \end{bmatrix} \begin{cases} v_1 = (-2)(-1) - 0 \times 0 = 2 \\ v_2 = (-1)(-1) - 1 \times 1 = 0 \\ v_3 = (-1)(-2) - 0 \times (-1) = 2 \end{cases}$$

$$\mathbf{R}_{90^\circ,z}\mathbf{R}_{90^\circ,y}\mathbf{R}_{90^\circ,x} = \begin{bmatrix} 0 & 0 & 1 \\ 0 & 1 & 0 \\ -1 & 0 & 0 \end{bmatrix} \begin{cases} v_1 = (0)(-1) - 0 \times 0 = 0 \\ v_2 = (-1)(-1) - (-1) \times 1 = 2 \\ v_3 = (-1)(0) - 0 \times 0 = 0. \end{cases}$$

We can see why the resulting vectors have components of 2 by evaluating a normal rotation transform:

$$\mathbf{R}_{\alpha,x} = \begin{bmatrix} 1 & 0 & 0 \\ 0 & c_\alpha & -s_\alpha \\ 0 & s_\alpha & c_\alpha \end{bmatrix} \begin{cases} v_1 = (c_\alpha - 1)(c_\alpha - 1) - (-s_\alpha) \times (s_\alpha) = 2(1 - c_\alpha) \\ v_2 = (c_\alpha - 1)(0) - 0 \times 0 = 0 \\ v_3 = (0)(c_\alpha - 1) - 0 \times 0 = 0. \end{cases}$$

We can see that when $\alpha = 90^\circ$, $v_1 = 2$.

So far we have created three composite rotations comprising individual rotations about the x-, y- and z-axes: $\mathbf{R}_{\alpha,x}\mathbf{R}_{\beta,y}\mathbf{R}_{\gamma,z}$ and $\mathbf{R}_{\gamma,z}\mathbf{R}_{\beta,y}\mathbf{R}_{\alpha,x}$. But there is nothing stopping us from creating other combinations such as $\mathbf{R}_{\alpha,x}\mathbf{R}_{\beta,y}\mathbf{R}_{\gamma,x}$ or $\mathbf{R}_{\alpha,z}\mathbf{R}_{\beta,y}\mathbf{R}_{\gamma,z}$ that include two rotations about the same axis. In fact, there are twelve possible combinations:

$$\begin{array}{llll} \mathbf{R}_{\alpha,x}\mathbf{R}_{\beta,y}\mathbf{R}_{\gamma,x}, & \mathbf{R}_{\alpha,x}\mathbf{R}_{\beta,y}\mathbf{R}_{\gamma,z}, & \mathbf{R}_{\alpha,x}\mathbf{R}_{\beta,z}\mathbf{R}_{\gamma,x}, & \mathbf{R}_{\alpha,x}\mathbf{R}_{\beta,z}\mathbf{R}_{\gamma,y} \\ \mathbf{R}_{\alpha,y}\mathbf{R}_{\beta,x}\mathbf{R}_{\gamma,y}, & \mathbf{R}_{\alpha,y}\mathbf{R}_{\beta,x}\mathbf{R}_{\gamma,z}, & \mathbf{R}_{\alpha,y}\mathbf{R}_{\beta,z}\mathbf{R}_{\gamma,x}, & \mathbf{R}_{\alpha,y}\mathbf{R}_{\beta,z}\mathbf{R}_{\gamma,y} \\ \mathbf{R}_{\alpha,z}\mathbf{R}_{\beta,x}\mathbf{R}_{\gamma,y}, & \mathbf{R}_{\alpha,z}\mathbf{R}_{\beta,x}\mathbf{R}_{\gamma,z}, & \mathbf{R}_{\alpha,z}\mathbf{R}_{\beta,y}\mathbf{R}_{\gamma,x}, & \mathbf{R}_{\alpha,z}\mathbf{R}_{\beta,y}\mathbf{R}_{\gamma,z} \end{array}$$

which are covered in detail in Appendix A.

9.4 Gimbal Lock

There are two potential problems with all of the above composite transforms. The first is the difficulty visualising the orientation of an object subjected to several rotations; the second is that they all suffer from what is called *gimbal lock*. From a visualisation point of view, if we use the transform $\mathbf{R}_{\gamma,z}\mathbf{R}_{\beta,y}\mathbf{R}_{\alpha,x}$ to animate an object and change γ, β and α over a period of frames, it can be very difficult to predict the final movement and adjust the angles to achieve a desired effect. Gimbal

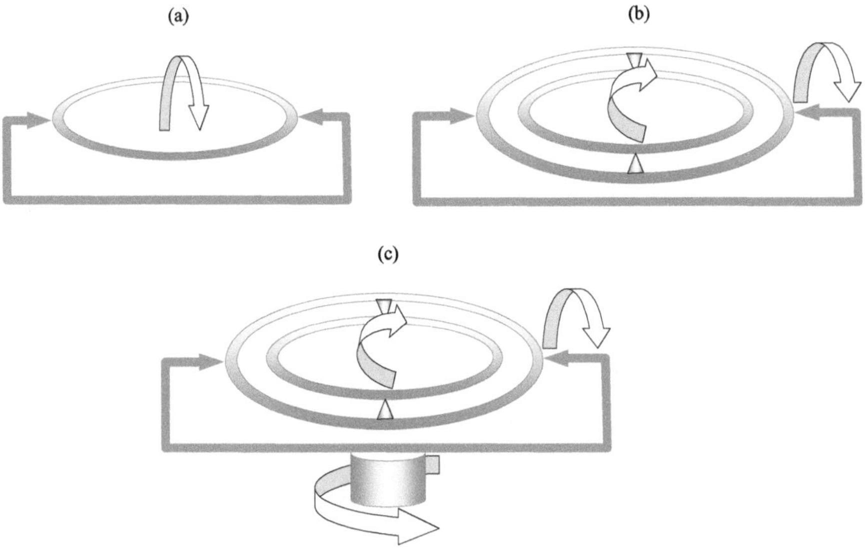

Fig. 9.6 Three types of gimbal joints

lock, on the other hand, is a weakness associated with Euler rotations when certain combinations of angles are used.

To understand this phenomenon, consider a simple gimbal which is a pivoted support that permits rotation about an axis, as shown in Fig. 9.6 (a). If two gimbals are combined, as shown in Fig. 9.6 (b), the inner cradle remains level with some reference plane as the assembly rolls and pitches. Such a combination has two degrees of rotational freedom. By adding a third gimbal so that the entire structure is free to rotate about a vertical axis, an extra degree of rotational freedom is introduced and is often used for mounting a camera on a tripod, as shown in Fig. 9.6 (c).

A mechanical gimbal joint with three degrees of freedom is represented mathematically by a composite Euler rotation transform. For example, say we choose $\mathbf{R}_{90°,y}\mathbf{R}_{90°,x}\mathbf{R}_{90°,z}$ to rotate our unit cube as shown in Fig. 9.7 (a). The cube's faces containing vertices 1, 5, 7, 3 and 0, 2, 6, 4 are first rotated about the perpendicular z-axis, as shown in Fig. 9.7 (b). The second transform rotates the cube's faces containing vertices 0, 4, 5, 1 and 2, 3, 7, 6 about the perpendicular x-axis, as shown in Fig. 9.7 (c). If we now attempt to rotate the cube about the y-axis, as shown in Fig. 9.7 (d), the cube's faces containing 0, 2, 6, 4 and 1, 5, 7, 3 are rotated again. Effectively we have lost the ability to rotate a cube about one of its axes, and such a condition is called *gimbal lock*. There is little we can do about this, apart from use another composite transform, but it, too, will have a similar restriction. For example, Appendix A shows that $\mathbf{R}_{90°,x}\mathbf{R}_{90°,z}\mathbf{R}_{90°,y}$, $\mathbf{R}_{90°,y}\mathbf{R}_{90°,z}\mathbf{R}_{90°,x}$, $\mathbf{R}_{90°,z}\mathbf{R}_{90°,x}\mathbf{R}_{90°,y}$ and $\mathbf{R}_{90°,z}\mathbf{R}_{90°,y}\mathbf{R}_{90°,x}$ all possess a similar affliction. Fortunately, there are other ways of rotating an object, which we will explore later.

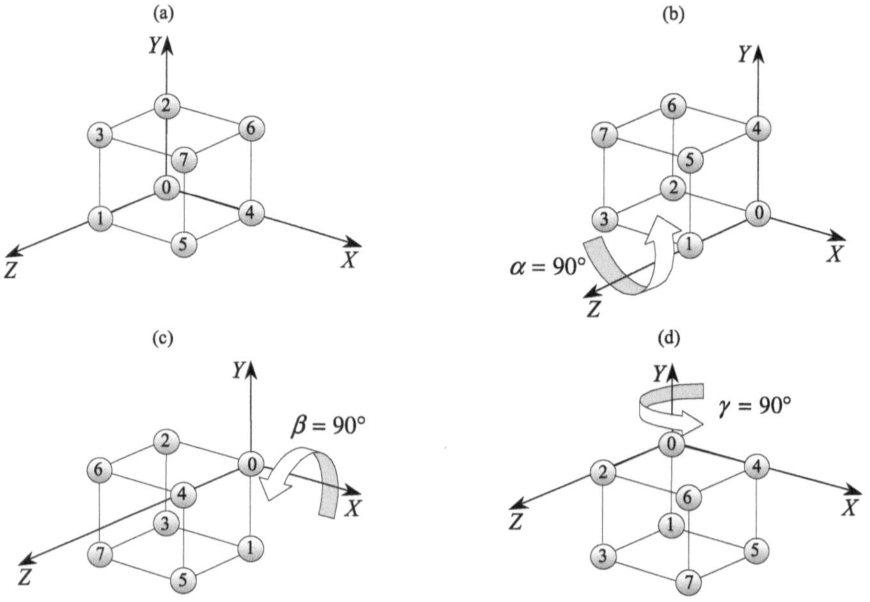

Fig. 9.7 An example of gimbal lock

9.5 Yaw, Pitch and Roll

The above Euler rotations are also known as *yaw*, *pitch* and *roll*, and great care should be taken with these angles when referring to other books and technical papers. Sometimes a left-handed system of axes is used rather than a right-handed set, and the vertical axis may be the y-axis or the z-axis, and might even point downwards. Consequently, the matrices representing the rotations can vary greatly. In this text all Cartesian coordinate systems are right-handed, and the vertical axis is always the y-axis.

The terms yaw, pitch and roll are often used in aviation and to describe the motion of ships. For example, if a ship or aeroplane is heading in a particular direction, the axis aligned with the heading is the roll axis, as shown in Fig. 9.8 (a). A perpendicular axis in the horizontal plane containing the heading axis is the pitch axis, as shown in Fig. 9.8 (b). The axis perpendicular to both these axes is the yaw axis, as shown in Fig. 9.8 (c).

Clearly, there are many ways of aligning a set of Cartesian axes with the yaw, pitch and roll axes, and consequently, it is impossible to define an absolute set of yaw, pitch and roll transforms. However, if we choose the following alignment:

- the *roll* axis is the z-axis
- the *pitch* axis is the x-axis
- the *yaw* axis is the y-axis

we have the situation as shown in Fig. 9.9, and the transforms representing these rotations are as follows:

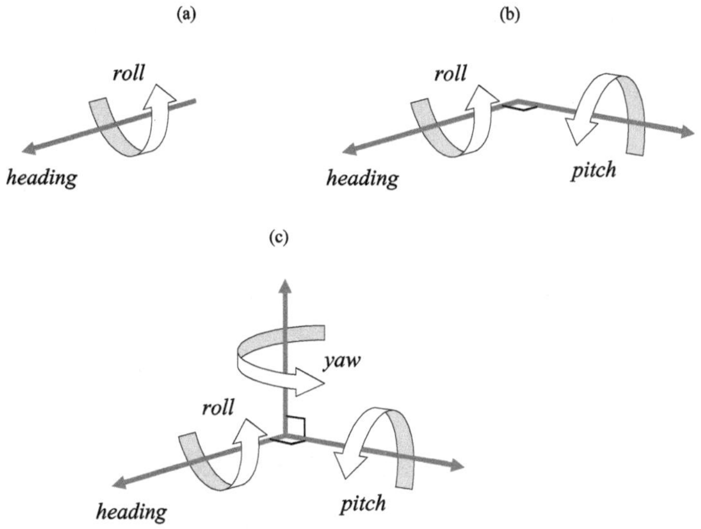

Fig. 9.8 Definitions of *yaw*, *pitch* and *roll*

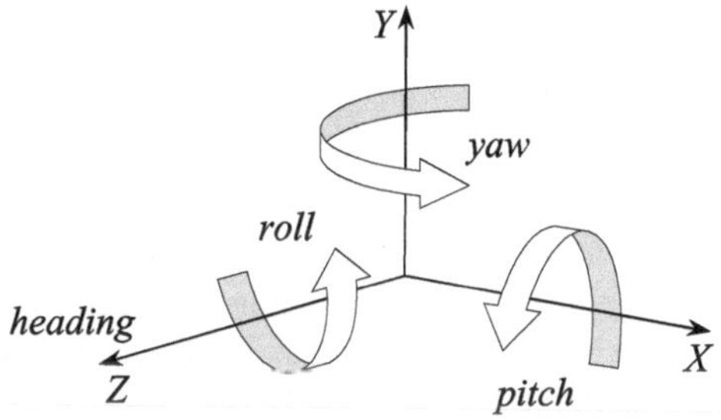

Fig. 9.9 A convention for *roll*, *pitch* and *yaw* angles

$$\mathbf{R}_{roll,z} = \begin{bmatrix} \cos roll & -\sin roll & 0 \\ \sin roll & \cos roll & 0 \\ 0 & 0 & 1 \end{bmatrix}$$

$$\mathbf{R}_{pitch,x} = \begin{bmatrix} 1 & 0 & 0 \\ 0 & \cos pitch & -\sin pitch \\ 0 & \sin pitch & \cos pitch \end{bmatrix}$$

$$\mathbf{R}_{yaw,y} = \begin{bmatrix} \cos yaw & 0 & \sin yaw \\ 0 & 1 & 0 \\ -\sin yaw & 0 & \cos yaw \end{bmatrix}.$$

A common sequence for applying these rotations is *roll*, *pitch*, *yaw*, as seen in the following transform:

$$\begin{bmatrix} x' \\ y' \\ z' \end{bmatrix} = \mathbf{R}_{yaw,y}\mathbf{R}_{pitch,x}\mathbf{R}_{roll,z}\begin{bmatrix} x \\ y \\ z \end{bmatrix}$$

and if a translation is involved,

$$\begin{bmatrix} x' \\ y' \\ z' \\ 1 \end{bmatrix} = \mathbf{T}_{t_x,t_y,t_z}\mathbf{R}_{yaw,y}\mathbf{R}_{pitch,x}\mathbf{R}_{roll,z}\begin{bmatrix} x \\ y \\ z \\ 1 \end{bmatrix}.$$

9.6 Rotating a Point About an Arbitrary Axis

Now let's examine two ways of rotating a point about an arbitrary axis. The first technique uses matrices and trigonometry and is rather laborious. The second approach employs vector analysis and is quite succinct. Fortunately, they both arrive at the same result!

9.6.1 Matrices

We begin by defining an axis using a unit vector $\hat{\mathbf{n}}$ about which a point P is rotated α to P' as shown in Fig. 9.10. And as we only have access to matrices that rotate points about the Cartesian axes, this unit vector has to be temporarily aligned with a Cartesian axis. In the following example we choose the x-axis. During the alignment process, the point P is subjected to the transforms necessary to align the unit vector with the x-axis. We then rotate P, α about the x-axis. To complete the operation, the rotated point is subjected to the transforms that return the unit vector to its original position. Although matrices provide a powerful tool for undertaking this sort of

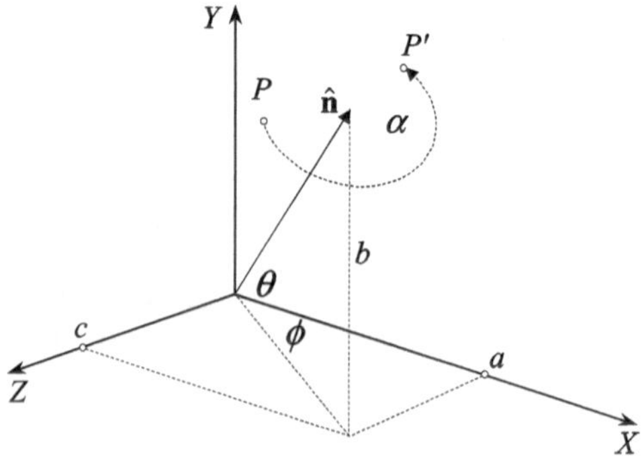

Fig. 9.10 The geometry associated with rotating a point about an arbitrary axis

work, it is nevertheless extremely tedious, but is a good exercise for improving one's algebraic skills!

Figure 9.10 shows a point $P(x, y, z)$ to be rotated through an angle α to $P'(x', y', z')$ about an axis defined by

$$\hat{\mathbf{n}} = a\mathbf{i} + b\mathbf{j} + c\mathbf{k}.$$

The transforms to achieve this operation can be expressed as follows:

$$\begin{bmatrix} x' \\ y' \\ z' \end{bmatrix} = \mathbf{R}_{-\phi,y}\mathbf{R}_{\theta,z}\mathbf{R}_{\alpha,x}\mathbf{R}_{-\theta,z}\mathbf{R}_{\phi,y}\begin{bmatrix} x \\ y \\ z \end{bmatrix}$$

which aligns the axis of rotation with the x-axis, performs the rotation of P through an angle α about the x-axis, and returns the axis of rotation back to its original position. Therefore,

$$\mathbf{R}_{\phi,y} = \begin{bmatrix} \cos\phi & 0 & \sin\phi \\ 0 & 1 & 0 \\ -\sin\phi & 0 & \cos\phi \end{bmatrix}, \qquad \mathbf{R}_{-\theta,z} = \begin{bmatrix} \cos\theta & \sin\theta & 0 \\ -\sin\theta & \cos\theta & 0 \\ 0 & 0 & 1 \end{bmatrix}$$

$$\mathbf{R}_{\alpha,x} = \begin{bmatrix} 1 & 0 & 0 \\ 0 & \cos\alpha & -\sin\alpha \\ 0 & \sin\alpha & \cos\alpha \end{bmatrix}, \qquad \mathbf{R}_{\theta,z} = \begin{bmatrix} \cos\theta & -\sin\theta & 0 \\ \sin\theta & \cos\theta & 0 \\ 0 & 0 & 1 \end{bmatrix}$$

$$\mathbf{R}_{-\phi,y} = \begin{bmatrix} \cos\phi & 0 & -\sin\phi \\ 0 & 1 & 0 \\ \sin\phi & 0 & \cos\phi \end{bmatrix}.$$

Let

$$\mathbf{R}_{-\phi,y}\mathbf{R}_{\theta,z}\mathbf{R}_{\alpha,x}\mathbf{R}_{-\theta,z}\mathbf{R}_{\phi,y} = \begin{bmatrix} a_{11} & a_{12} & a_{13} \\ a_{21} & a_{22} & a_{23} \\ a_{31} & a_{32} & a_{33} \end{bmatrix}$$

where by multiplying the matrices together we find that:

$$\begin{aligned}
a_{11} &= \cos^2\phi\cos^2\theta + \cos^2\phi\sin^2\theta\cos\alpha + \sin^2\phi\cos\alpha \\
a_{12} &= \cos\phi\cos\theta\sin\theta - \cos\phi\sin\theta\cos\theta\cos\alpha - \sin\phi\cos\theta\sin\alpha \\
a_{13} &= \cos\phi\sin\phi\cos^2\theta + \cos\phi\sin\phi\sin^2\theta\cos\alpha + \sin^2\phi\sin\theta\sin\alpha \\
&\quad + \cos^2\phi\sin\theta\sin\alpha - \cos\phi\sin\phi\cos\alpha \\
a_{21} &= \sin\theta\cos\theta\cos\phi - \cos\theta\sin\theta\cos\phi\cos\alpha + \cos\theta\sin\phi\sin\alpha \\
a_{22} &= \sin^2\theta + \cos^2\theta\cos\alpha \\
a_{23} &= \sin\theta\cos\theta\sin\phi - \cos\theta\sin\theta\sin\phi\cos\alpha - \cos\theta\cos\phi\sin\alpha \\
a_{31} &= \cos\phi\sin\phi\cos^2\theta + \cos\phi\sin\phi\sin^2\theta\cos\alpha - \cos^2\phi\sin\theta\sin\alpha \\
&\quad - \cos\phi\sin\phi\cos\alpha \\
a_{32} &= \sin\phi\cos\theta\sin\theta - \sin\phi\sin\theta\cos\theta\cos\alpha + \cos\phi\cos\theta\sin\alpha \\
a_{33} &= \sin^2\phi\cos^2\theta + \sin^2\phi\sin^2\theta\cos\alpha - \cos\phi\sin\phi\sin\theta\sin\alpha \\
&\quad + \cos\phi\sin\phi\sin\theta\sin\alpha + \cos^2\phi\cos\alpha.
\end{aligned}$$

From Fig. 9.10 we compute the sin and cos of θ and ϕ in terms of a, b and c, and then compute their equivalent $\sin^2$ and $\cos^2$ values:

$$
\begin{aligned}
\cos\theta &= \sqrt{1-b^2} &&\Rightarrow\quad \cos^2\theta = 1-b^2\\
\sin\theta &= b &&\Rightarrow\quad \sin^2\theta = b^2\\
\cos\phi &= a/\sqrt{1-b^2} &&\Rightarrow\quad \cos^2\phi = a^2/(1-b^2)\\
\sin\phi &= c/\sqrt{1-b^2} &&\Rightarrow\quad \sin^2\phi = c^2/(1-b^2).
\end{aligned}
$$

To find a_{11}:

$$
\begin{aligned}
a_{11} &= \cos^2\phi\cos^2\theta + \cos^2\phi\sin^2\theta\cos\alpha + \sin^2\phi\cos\alpha\\
&= a^2 + \frac{a^2b^2}{1-b^2}\cos\alpha + \frac{c^2}{1-b^2}\cos\alpha\\
&= a^2 + \left(\frac{c^2+a^2b^2}{1-b^2}\right)\cos\alpha
\end{aligned}
$$

but

$$
a^2+b^2+c^2=1 \quad\Rightarrow\quad c^2 = 1-a^2-b^2
$$

$$
\begin{aligned}
a_{11} &= a^2 + \left(\frac{1-a^2-b^2+a^2b^2}{1-b^2}\right)\cos\alpha\\
&= a^2 + \left(\frac{(1-a^2)(1-b^2)}{1-b^2}\right)\cos\alpha\\
&= a^2(1-\cos\alpha)+\cos\alpha.
\end{aligned}
$$

Let

$$
K = 1-\cos\alpha
$$

then

$$
a_{11} = a^2K + \cos\alpha.
$$

To find a_{12}:

$$
\begin{aligned}
a_{12} &= \cos\phi\cos\theta\sin\theta - \cos\phi\sin\theta\cos\theta\cos\alpha - \sin\phi\cos\theta\sin\alpha\\
&= \frac{a}{\sqrt{1-b^2}}\sqrt{1-b^2}b - \frac{a}{\sqrt{1-b^2}}b\sqrt{1-b^2}\cos\alpha - \frac{c}{\sqrt{1-b^2}}\sqrt{1-b^2}\sin\alpha\\
&= ab - ab\cos\alpha - c\sin\alpha\\
&= ab(1-\cos\alpha) - c\sin\alpha\\
a_{12} &= abK - c\sin\alpha.
\end{aligned}
$$

To find a_{13}:

$$\begin{aligned}
a_{13} &= \cos\phi\sin\phi\cos^2\theta + \cos\phi\sin\phi\sin^2\theta\cos\alpha + \sin^2\phi\sin\theta\sin\alpha \\
&\quad + \cos^2\phi\sin\theta\sin\alpha - \cos\phi\sin\phi\cos\alpha \\
&= \cos\phi\sin\phi\cos^2\theta + \cos\phi\sin\phi\sin^2\theta\cos\alpha + \sin\theta\sin\alpha - \cos\phi\sin\phi\cos\alpha \\
&= \frac{a}{\sqrt{1-b^2}}\frac{c}{\sqrt{1-b^2}}\left(1-b^2\right) + \frac{a}{\sqrt{1-b^2}}\frac{c}{\sqrt{1-b^2}}b^2\cos\alpha + b\sin\alpha \\
&\quad - \frac{a}{\sqrt{1-b^2}}\frac{c}{\sqrt{1-b^2}}\cos\alpha \\
&= ac + ac\frac{b^2}{(1-b^2)}\cos\alpha + b\sin\alpha - \frac{ac}{(1-b^2)}\cos\alpha \\
&= ac + ac\frac{(b^2-1)}{(1-b^2)}\cos\alpha + b\sin\alpha \\
&= ac(1-\cos\alpha) + b\sin\alpha \\
a_{13} &= acK + b\sin\alpha.
\end{aligned}$$

Using similar algebraic methods, we discover that:

$$\begin{aligned}
a_{21} &= abK + c\sin\alpha \\
a_{22} &= b^2K + \cos\alpha \\
a_{23} &= bcK - a\sin\alpha \\
a_{31} &= acK - b\sin\alpha \\
a_{32} &= bcK + a\sin\alpha \\
a_{33} &= c^2K + \cos\alpha
\end{aligned}$$

and our original matrix transform becomes:

$$\begin{bmatrix} x'_p \\ y'_p \\ z'_p \end{bmatrix} = \begin{bmatrix} a^2K + \cos\alpha & abK - c\sin\alpha & acK + b\sin\alpha \\ abK + c\sin\alpha & b^2K + \cos\alpha & bcK - a\sin\alpha \\ acK - b\sin\alpha & bcK + a\sin\alpha & c^2K + \cos\alpha \end{bmatrix} \begin{bmatrix} x_p \\ y_p \\ z_p \end{bmatrix}$$

where

$$K = 1 - \cos\alpha.$$

9.6.2 Vectors

Now let's solve the same problem using vectors. Figure 9.11 shows a view of the geometry associated with the task at hand. For clarification, Fig. 9.12 shows a cross-section and a plan view of the geometry.

The axis of rotation is given by the unit vector:

$$\hat{\mathbf{n}} = a\mathbf{i} + b\mathbf{j} + c\mathbf{k}.$$

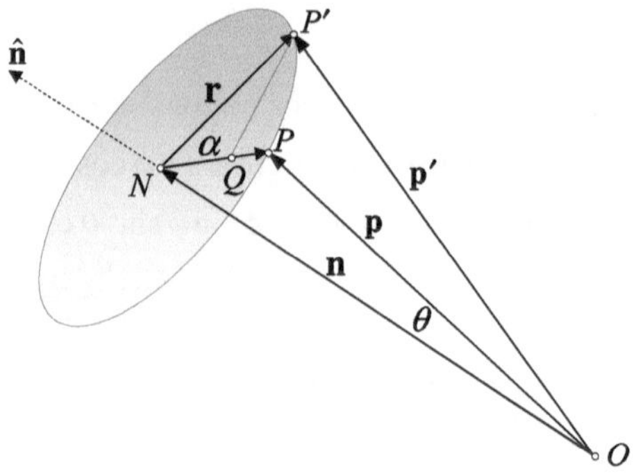

Fig. 9.11 A view of the geometry associated with rotating a point about an arbitrary axis

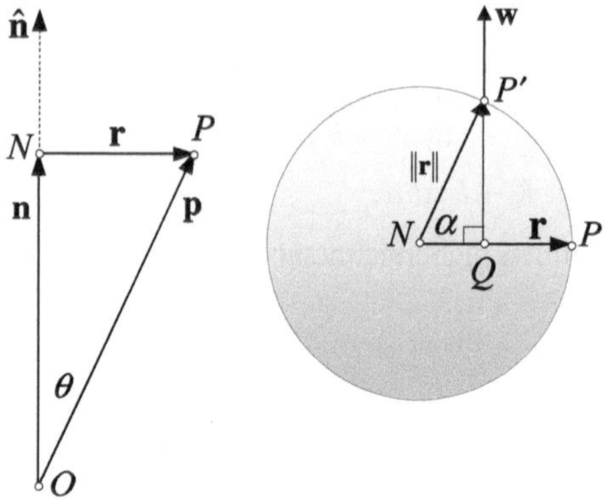

Fig. 9.12 A cross-section and plan view of the geometry associated with rotating a point about an arbitrary axis

- $P(x_p, y_p z_p)$ is the point to be rotated by angle α to $P'(x'_p, y'_p, z'_p)$.
- O is the origin, whilst $\mathbf{p}$ and $\mathbf{p}'$ are position vectors for P and P' respectively.

From Fig. 9.11 and Fig. 9.12:

$$\mathbf{p}' = \overrightarrow{ON} + \overrightarrow{NQ} + \overrightarrow{QP'}.$$

To find $\overrightarrow{ON}$:

$$|\mathbf{n}| = |\mathbf{p}| \cos\theta = \hat{\mathbf{n}} \cdot \mathbf{p}$$

therefore,

$$\overrightarrow{ON} = \mathbf{n} = \hat{\mathbf{n}}(\hat{\mathbf{n}} \cdot \mathbf{p}).$$

To find $\overrightarrow{NQ}$:

$$\overrightarrow{NQ} = \frac{NQ}{NP}\mathbf{r} = \frac{NQ}{NP'}\mathbf{r} = \cos\alpha\, \mathbf{r}$$

but

$$\mathbf{p} = \mathbf{n} + \mathbf{r} = \hat{\mathbf{n}}(\hat{\mathbf{n}} \cdot \mathbf{p}) + \mathbf{r}$$

therefore,

$$\mathbf{r} = \mathbf{p} - \hat{\mathbf{n}}(\hat{\mathbf{n}} \cdot \mathbf{p})$$

and

$$\overrightarrow{NQ} = \big(\mathbf{p} - \hat{\mathbf{n}}(\hat{\mathbf{n}} \cdot \mathbf{p})\big)\cos\alpha.$$

To find $\overrightarrow{QP'}$:
Let

$$\hat{\mathbf{n}} \times \mathbf{p} = \mathbf{w}$$

where

$$|\mathbf{w}| = |\hat{\mathbf{n}}| \cdot |\mathbf{p}|\sin\theta = |\mathbf{p}|\sin\theta$$

but

$$|\mathbf{r}| = |\mathbf{p}|\sin\theta$$

therefore,

$$|\mathbf{w}| = |\mathbf{r}|.$$

Now

$$\frac{QP'}{NP'} = \frac{QP'}{|\mathbf{r}|} = \frac{QP'}{|\mathbf{w}|} = \sin\alpha$$

therefore,

$$\overrightarrow{QP'} = \mathbf{w}\sin\alpha = \hat{\mathbf{n}} \times \mathbf{p}\sin\alpha$$

then

$$\mathbf{p}' = \hat{\mathbf{n}}(\hat{\mathbf{n}} \cdot \mathbf{p}) + \big(\mathbf{p} - \hat{\mathbf{n}}(\hat{\mathbf{n}} \cdot \mathbf{p})\big)\cos\alpha + \hat{\mathbf{n}} \times \mathbf{p}\sin\alpha$$

and

$$\mathbf{p}' = \mathbf{p}\cos\alpha + \hat{\mathbf{n}}(\hat{\mathbf{n}} \cdot \mathbf{p})(1 - \cos\alpha) + \hat{\mathbf{n}} \times \mathbf{p}\sin\alpha.$$

This is known as the Rodrigues rotation formula, as it was developed by the French mathematician, Olinde Rodrigues (1795–1851), who had also invented the ideas behind quaternions before Hamilton. This has been documented by Simon Altmann in the Mathematics Magazine under the title "*Hamilton, Rodrigues and the quaternion scandal*" [7].

If we let

$$K = 1 - \cos\alpha$$

then

$$\begin{aligned}
\mathbf{p}' &= \mathbf{p}\cos\alpha + \hat{\mathbf{n}}(\hat{\mathbf{n}} \cdot \mathbf{p})K + \hat{\mathbf{n}} \times \mathbf{p}\sin\alpha \\
&= (x_p\mathbf{i} + y_p\mathbf{j} + z_p\mathbf{k})\cos\alpha + (a\mathbf{i} + b\mathbf{j} + c\mathbf{k})(ax_p + by_p + cz_p)K \\
&\quad + \big((bz_p - cy_p)\mathbf{i} + (cx_p - az_p)\mathbf{j} + (ay_p - bx_p)\mathbf{k}\big)\sin\alpha \\
&= \big(x_p\cos\alpha + a(ax_p + by_p + cz_p)K + (bz_p - cy_p)\sin\alpha\big)\mathbf{i} \\
&\quad + \big(y_p\cos\alpha + b(ax_p + by_p + cz_p)K + (cx_p - az_p)\sin\alpha\big)\mathbf{j}
\end{aligned}$$

$$
\begin{aligned}
&+ \big(z_p \cos\alpha + c(ax_p + by_p + cz_p)K + (ay_p - bx_p)\sin\alpha\big)\mathbf{k}\\
&= \big(x_p(a^2K + \cos\alpha) + y_p(abK - c\sin\alpha) + z_p(acK + b\sin\alpha)\big)\mathbf{i}\\
&+ \big(x_p(abK + c\sin\alpha) + y_p\big(b^2K + \cos\alpha\big) + z_p(bcK - a\sin\alpha)\big)\mathbf{j}\\
&+ \big(x_p(acK - b\sin\alpha) + y_p(bcK + a\sin\alpha) + z_p\big(c^2K + \cos\alpha\big)\big)\mathbf{k}
\end{aligned}
$$

and the transform is:

$$
\begin{bmatrix} x'_p \\ y'_p \\ z'_p \end{bmatrix} = \begin{bmatrix} a^2K + \cos\alpha & abK - c\sin\alpha & acK + b\sin\alpha \\ abK + c\sin\alpha & b^2K + \cos\alpha & bcK - a\sin\alpha \\ acK - b\sin\alpha & bcK + a\sin\alpha & c^2K + \cos\alpha \end{bmatrix} \begin{bmatrix} x_p \\ y_p \\ z_p \end{bmatrix}
$$

which is identical to the transform derived using matrices.

Now let's test the transform with a simple example that can be easily verified. If we rotate the point $P(10, 0, 0)$, 180° about an axis defined by $\mathbf{n} = \mathbf{i} + \mathbf{j}$, it should end up at $P'(0, 10, 0)$.

Therefore

$$
\alpha = 180^\circ, \quad \cos\alpha = -1, \quad \sin\alpha = 0, \quad K = 2
$$

$$
a = \frac{\sqrt{2}}{2}, \quad b = \frac{\sqrt{2}}{2}, \quad c = 0
$$

and

$$
\begin{bmatrix} 0 \\ 10 \\ 0 \end{bmatrix} = \begin{bmatrix} 0 & 1 & 0 \\ 1 & 0 & 0 \\ 0 & 0 & 0 \end{bmatrix} \begin{bmatrix} 10 \\ 0 \\ 0 \end{bmatrix}
$$

which is correct.

9.7 Summary

In this chapter we have seen how the 2×2 matrix for rotating a point in the plane is developed for rotating points in space. In its simplest form, the rotations are restricted to one of the three Cartesian axes, but by employing homogeneous coordinates, the translation transform can be used to rotate points about an off-set axis parallel with one of the Cartesian axes.

Composite Euler rotations are created by combining the matrices representing the individual rotations about three successive axes, for which there are twelve combinations. Unfortunately, one of the problems with such transforms is that they suffer from gimbal lock, where one degree of freedom is lost under certain angle combinations. Another problem, is that it is difficult to predict how a point moves in space when animated by a composite transform, although they are widely used for positioning objects in world space.

We have also seen how to compute the eigenvector associated with a rotation transform, and how it represents the axis about which rotation occurs, and the eigenvalue represents the angle of rotation.

Finally, matrices and vectors were used to develop a transform for rotating a point about an arbitrary axis.

9.7.1 Summary of Matrix Transforms

Translate a point

$$\mathbf{T}_{t_x,t_y,t_z} = \begin{bmatrix} 1 & 0 & 0 & t_x \\ 0 & 1 & 0 & t_y \\ 0 & 0 & 1 & t_z \\ 0 & 0 & 0 & 1 \end{bmatrix}.$$

Rotate a point about the x-, y-, z-axes

$$\mathbf{R}_{\beta,x} = \begin{bmatrix} 1 & 0 & 0 & 0 \\ 0 & \cos\beta & -\sin\beta & 0 \\ 0 & \sin\beta & \cos\beta & 0 \\ 0 & 0 & 0 & 1 \end{bmatrix}$$

$$\mathbf{R}_{\beta,y} = \begin{bmatrix} \cos\beta & 0 & \sin\beta & 0 \\ 0 & 1 & 0 & 0 \\ -\sin\beta & 0 & \cos\beta & 0 \\ 0 & 0 & 0 & 1 \end{bmatrix}$$

$$\mathbf{R}_{\beta,z} = \begin{bmatrix} \cos\beta & -\sin\beta & 0 & 0 \\ \sin\beta & \cos\beta & 0 & 0 \\ 0 & 0 & 1 & 0 \\ 0 & 0 & 0 & 1 \end{bmatrix}.$$

Rotate a point about off-set x-, y-, z-axes

$$\mathbf{T}_{0,t_y,t_z}\mathbf{R}_{\beta,x}\mathbf{T}_{0,-t_y,-t_z} = \begin{bmatrix} 1 & 0 & 0 & 0 \\ 0 & \cos\beta & -\sin\beta & t_y(1-\cos\beta)+t_z\sin\beta \\ 0 & \sin\beta & \cos\beta & t_z(1-\cos\beta)-t_y\sin\beta \\ 0 & 0 & 0 & 1 \end{bmatrix}$$

$$\mathbf{T}_{t_x,0,t_z}\mathbf{R}_{\beta,y}\mathbf{T}_{-t_x,0,-t_z} = \begin{bmatrix} \cos\beta & 0 & \sin\beta & t_x(1-\cos\beta)-t_z\sin\beta \\ 0 & 1 & 0 & 0 \\ -\sin\beta & 0 & \cos\beta & t_z(1-\cos\beta)+t_x\sin\beta \\ 0 & 0 & 0 & 1 \end{bmatrix}$$

$$\mathbf{T}_{t_x,t_y,0}\mathbf{R}_{\beta,z}\mathbf{T}_{-t_x,-t_y,0} = \begin{bmatrix} \cos\beta & -\sin\beta & 0 & t_x(1-\cos\beta)+t_y\sin\beta \\ \sin\beta & \cos\beta & 0 & t_y(1-\cos\beta)-t_x\sin\beta \\ 0 & 0 & 1 & 0 \\ 0 & 0 & 0 & 1 \end{bmatrix}.$$

Rotate a point about an arbitrary axis

$$\mathbf{p}' = \mathbf{p}\cos\alpha + \hat{\mathbf{n}}(\hat{\mathbf{n}}\cdot\mathbf{p})(1-\cos\alpha) + \hat{\mathbf{n}}\times\mathbf{p}\sin\alpha$$

$$\mathbf{R}_{\alpha,\hat{\mathbf{n}}} = \begin{bmatrix} a^2K+\cos\alpha & abK-c\sin\alpha & acK+b\sin\alpha \\ abK+c\sin\alpha & b^2K+\cos\alpha & bcK-a\sin\alpha \\ acK-b\sin\alpha & bcK+a\sin\alpha & c^2K+\cos\alpha \end{bmatrix}$$

$$K = 1-\cos\alpha$$

$$\hat{\mathbf{n}} = a\mathbf{i}+b\mathbf{j}+c\mathbf{k}.$$

Extracting the angle and axis of rotation from a transform

$$\mathbf{R}_{\alpha,\mathbf{v}} = \begin{bmatrix} a_{11} & a_{12} & a_{13} \\ a_{21} & a_{22} & a_{23} \\ a_{31} & a_{32} & a_{33} \end{bmatrix}$$

$$\alpha = \arccos((\mathrm{Tr}(\mathbf{R}_{\alpha,\mathbf{v}}) - 1)/2)$$

$$\mathbf{v} = v_1\mathbf{i} + v_2\mathbf{j} + v_3\mathbf{k}$$

$$v_1 = (a_{22} - 1)(a_{33} - 1) - a_{23}a_{32}$$

$$v_2 = (a_{33} - 1)(a_{11} - 1) - a_{31}a_{13}$$

$$v_3 = (a_{11} - 1)(a_{22} - 1) - a_{12}a_{21}.$$

Chapter 10
Frames of Reference in Space

10.1 Introduction

In Chap. 8 we discovered how to compute the coordinates of a point in a frame of reference in the plane. In this chapter we study the same problem but in 3D space. Again, we employ many of the concepts previously described in order to develop the transforms for translating and rotating frames in space.

The relativity of motion, previously discussed, implies that we cannot absolutely claim that one frame of reference is stationary whilst another is in motion – it is simply a question of interpretation and convenience. Fortunately, a matrix transform can be used to support moving points and moving frames, which means that the transform used for rotating a point in a fixed frame of reference, can be used for rotating the frame of reference in the opposite direction, whilst keeping the point fixed.

In a 3D space context, this implies that the rotation transform $\mathbf{R}_{\alpha,x}$, which rotates a point α about the fixed x-axis, can be used to rotate the frame of reference $-\alpha$ about the x-axis, whilst the point remains fixed. Similarly, the rotation transform $\mathbf{R}_{-\alpha,x}$, which rotates a point $-\alpha$ about the fixed x-axis, can also be used to rotate the frame of reference α about the x-axis, whilst the point remains fixed.

We employ a simple notation to distinguish transforms that rotate points from those that rotate frames. For example, $\mathbf{R}_{\alpha,x}$ rotates a point α about the x-axis, whilst $\mathbf{R}^{-1}_{\alpha,x}$ rotates a frame α about the x-axis. Similarly, $\mathbf{R}_{-\alpha,x}$ rotates a point $-\alpha$ about the x-axis, whilst $\mathbf{R}^{-1}_{-\alpha,x}$ rotates a frame $-\alpha$ about the x-axis. Also, $\mathbf{T}_{t_x,t_y,t_z}$ translates a point (t_x, t_y, t_z), whilst $\mathbf{T}^{-1}_{t_x,t_y,t_z}$ translates a frame (t_x, t_y, t_z). Similarly, $\mathbf{T}_{-t_x,-t_y,-t_z}$ translates a point $(-t_x, -t_y, -t_z)$, whilst $\mathbf{T}^{-1}_{-t_x,-t_y,-t_z}$ translates a frame $(-t_x, -t_y, -t_z)$.

10.2 Frames of Reference

We have already discussed frames of reference in Chap. 8, and even though the frames were 2D, the same ideas can be generalised to 3D space without having to introduce any new concepts – we simply add an extra z-coordinate.

J. Vince, *Rotation Transforms for Computer Graphics*,
DOI 10.1007/978-0-85729-154-7_10,

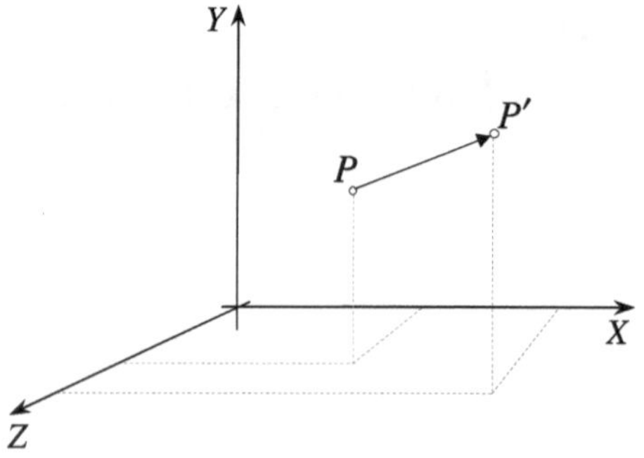

Fig. 10.1 The point P is translated by (t_x, t_y, t_z) with a fixed frame

10.3 Matrix Transforms

In general, if points in the frame $X'Y'Z'$ are related to points in the frame XYZ by the transform $\mathbf{A}$ using

$$P = \mathbf{A}P',$$

then a point P in XYZ has coordinates in $X'Y'Z'$ using

$$P' = \mathbf{A}^{-1}P.$$

In computer graphics most frame of reference transforms involve a translation or a rotation, or a combination of both. We now explore these different scenarios and develop transforms for converting coordinates in the original frame of reference to another frame.

10.3.1 Translated Frames of Reference

Figure 10.1 shows a point P translated by (t_x, t_y, t_z) to P' using the transform $\mathbf{T}_{t_x,t_y,t_z}$

$$\mathbf{T}_{t_x,t_y,t_z} = \begin{bmatrix} 1 & 0 & 0 & t_x \\ 0 & 1 & 0 & t_y \\ 0 & 0 & 1 & t_z \\ 0 & 0 & 0 & 1 \end{bmatrix}$$

where the translated point P' is given by

$$P' = \mathbf{T}_{t_x,t_y,t_z}P.$$

Thus the coordinates of P are updated relative to the fixed frame of reference XYZ.

However, there is a second interpretation for $\mathbf{T}_{t_x,t_y,t_z}$, where P remains fixed and the frame of reference XYZ is translated by $(-t_x, -t_y, -t_z)$, as shown in Fig. 10.2. Consequently, the point $P(x, y, z)$ in XYZ has coordinates $P'(x', y', z')$ in $X'Y'Z'$ given by

$$P' = \mathbf{T}^{-1}_{-t_x,-t_y,-t_z}P.$$

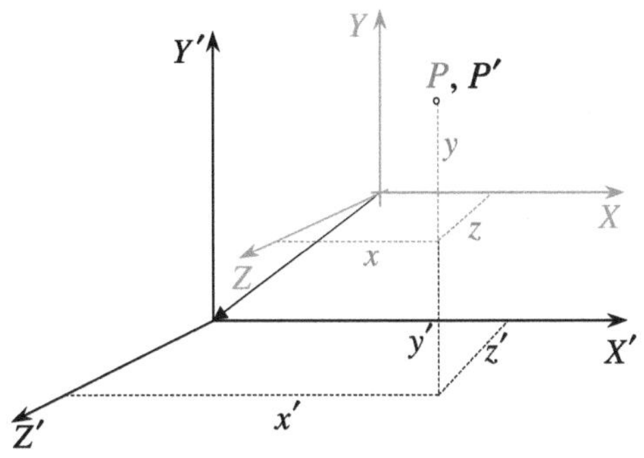

Fig. 10.2 The point P remains fixed whilst the frame XYZ is translated $(-t_x, -t_y, -t_z)$

For example, the origin of XYZ becomes (t_x, t_y, t_z) in $X'Y'Z'$:

$$\begin{bmatrix} t_x \\ t_y \\ t_z \\ 1 \end{bmatrix} = \begin{bmatrix} 1 & 0 & 0 & t_x \\ 0 & 1 & 0 & t_y \\ 0 & 0 & 1 & t_z \\ 0 & 0 & 0 & 1 \end{bmatrix} \begin{bmatrix} 0 \\ 0 \\ 0 \\ 1 \end{bmatrix}.$$

Therefore, the transform for translating a frame by (t_x, t_y, t_z) is

$$\mathbf{T}^{-1}_{t_x,t_y,t_z} = \begin{bmatrix} 1 & 0 & 0 & -t_x \\ 0 & 1 & 0 & -t_y \\ 0 & 0 & 1 & -t_z \\ 0 & 0 & 0 & 1 \end{bmatrix}.$$

Now we consider rotated frames of reference in space.

10.3.2 Rotated Frames of Reference About Cartesian Axes

A 2D frame of reference can only be rotated within its plane, whereas a 3D frame can be rotated about any axis, whether it be a Cartesian axis or some arbitrary axis. Let's look at how $\mathbf{R}_{\alpha,x}$ behaves when rotating frames about the x-axis, and apply the results to the other axes.

We know that we can rotate a point, α about the x-axis using

$$\mathbf{R}_{\alpha,x} = \begin{bmatrix} 1 & 0 & 0 \\ 0 & \cos\alpha & -\sin\alpha \\ 0 & \sin\alpha & \cos\alpha \end{bmatrix}.$$

However, $\mathbf{R}_{\alpha,x}$ can also be used to rotate a frame $-\alpha$ about the x-axis. Similarly, $\mathbf{R}^{-1}_{\alpha,x}$ rotates a frame α about the same axis. Therefore, in general, we can use the same technique for all three Cartesian axes.

Figure 10.3 (a) and (b) show our unit cube rotated by $-90°$ about the x-axis, whilst (c) and (d) show the frame rotated $90°$ about the same axis, with the cube fixed.

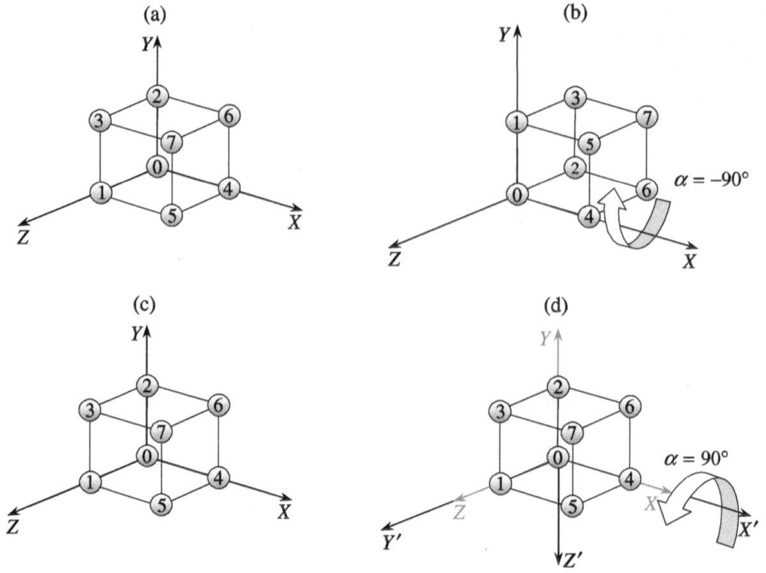

Fig. 10.3 (**a**) and (**b**) The cube is rotated −90°. (**c**) and (**d**) The XYZ frame is rotated 90°

The transform for $\mathbf{R}^{-1}_{90^\circ,x}$ is

$$\mathbf{R}^{-1}_{90^\circ,x} = \mathbf{R}_{-90^\circ,x} = \begin{bmatrix} 1 & 0 & 0 \\ 0 & 0 & 1 \\ 0 & -1 & 0 \end{bmatrix}$$

which when used on the cube's coordinates create

$$\begin{bmatrix} 1 & 0 & 0 \\ 0 & 0 & 1 \\ 0 & -1 & 0 \end{bmatrix} \begin{bmatrix} 0 & 0 & 0 & 0 & 1 & 1 & 1 & 1 \\ 0 & 0 & 1 & 1 & 0 & 0 & 1 & 1 \\ 0 & 1 & 0 & 1 & 0 & 1 & 0 & 1 \end{bmatrix}$$

$$= \begin{bmatrix} 0 & 0 & 0 & 0 & 1 & 1 & 1 & 1 \\ 0 & 1 & 0 & 1 & 0 & 1 & 0 & 1 \\ 0 & 0 & -1 & -1 & 0 & 0 & -1 & -1 \end{bmatrix}$$

which are confirmed by Fig. 10.3 (d).

In summary, the transforms for rotating frames α about the x-, y- and z-axes are:

$$\mathbf{R}^{-1}_{\alpha,x} = \begin{bmatrix} 1 & 0 & 0 \\ 0 & \cos\alpha & \sin\alpha \\ 0 & -\sin\alpha & \cos\alpha \end{bmatrix}$$

$$\mathbf{R}^{-1}_{\alpha,y} = \begin{bmatrix} \cos\alpha & 0 & -\sin\alpha \\ 0 & 1 & 0 \\ \sin\alpha & 0 & \cos\alpha \end{bmatrix}$$

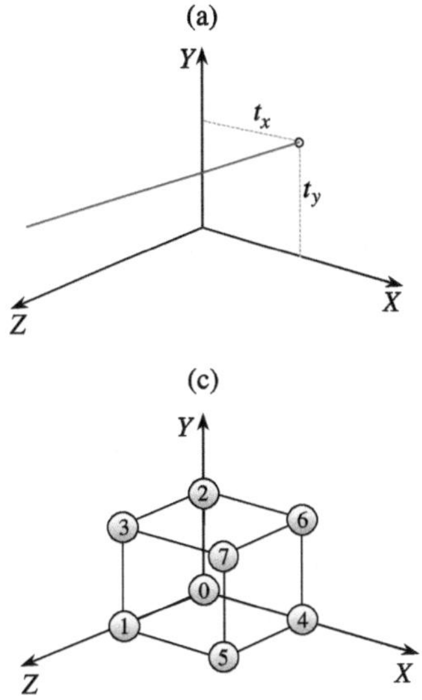

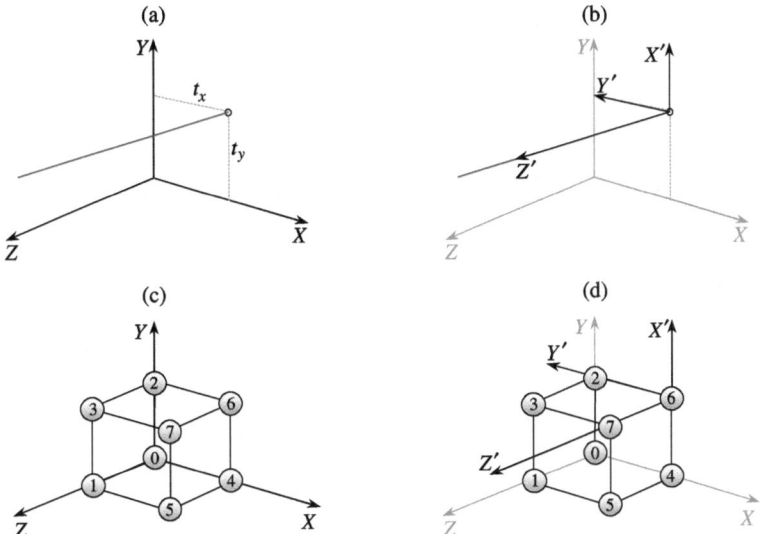

Fig. 10.4 (**a**) and (**b**) The off-set axis and rotated frame. (**c**) and (**d**) The unit cube and rotated frame

$$\mathbf{R}_{\alpha,z}^{-1} = \begin{bmatrix} \cos\alpha & \sin\alpha & 0 \\ -\sin\alpha & \cos\alpha & 0 \\ 0 & 0 & 1 \end{bmatrix}.$$

10.3.3 Rotated Frames About Off-Set Axes

In Chap. 9 we developed three transforms for rotating a point about an off-set axis parallel with one of the three Cartesian axes. Let's develop three complementary transforms for rotating a frame about the same off-set axes.

To ensure that we compute the correct transform, we must be very clear in our own minds what we are attempting to do. The objective is to identify an off-set axis parallel with the z-axis, for example, in the current XYZ frame of reference, about which a frame is rotated. The first step, then, is to translate the frame, and then rotate it.

Let's assume that the axis passes through the point $(t_x, t_y, 0)$, as shown in Fig. 10.4 (a). Therefore, given the following definitions for $\mathbf{T}_{t_x,t_y,0}^{-1}$ and $\mathbf{R}_{\alpha,z}^{-1}$

$$\mathbf{T}_{t_x,t_y,0}^{-1} = \begin{bmatrix} 1 & 0 & 0 & -t_x \\ 0 & 1 & 0 & -t_y \\ 0 & 0 & 1 & 0 \\ 0 & 0 & 0 & 1 \end{bmatrix}$$

$$\mathbf{R}_{\alpha,z}^{-1} = \begin{bmatrix} \cos\alpha & \sin\alpha & 0 & 0 \\ -\sin\alpha & \cos\alpha & 0 & 0 \\ 0 & 0 & 1 & 0 \\ 0 & 0 & 0 & 1 \end{bmatrix}$$

$$\mathbf{R}_{\alpha,z}^{-1}\mathbf{T}_{t_x,t_y,0}^{-1} = \begin{bmatrix} \cos\alpha & \sin\alpha & 0 & -t_x\cos\alpha - t_y\sin\alpha \\ -\sin\alpha & \cos\alpha & 0 & -t_y\cos\alpha + t_x\sin\alpha \\ 0 & 0 & 1 & 0 \\ 0 & 0 & 0 & 1 \end{bmatrix}.$$

Let's test this transform by making $\alpha = 90°$, and $t_x = t_y = 1$, as shown in Fig. 10.4 (b):

$$\mathbf{R}_{90°,z}^{-1}\mathbf{T}_{1,1,0}^{-1} = \begin{bmatrix} 0 & 1 & 0 & -1 \\ -1 & 0 & 0 & 1 \\ 0 & 0 & 1 & 0 \\ 0 & 0 & 0 & 1 \end{bmatrix}$$

which, if used on the unit cube shown in Fig. 10.4 (c), produces

$$\begin{bmatrix} 0 & 1 & 0 & -1 \\ -1 & 0 & 0 & 1 \\ 0 & 0 & 1 & 0 \\ 0 & 0 & 0 & 1 \end{bmatrix} \begin{bmatrix} 0 & 0 & 0 & 0 & 1 & 1 & 1 & 1 \\ 0 & 0 & 1 & 1 & 0 & 0 & 1 & 1 \\ 0 & 1 & 0 & 1 & 0 & 1 & 0 & 1 \\ 1 & 1 & 1 & 1 & 1 & 1 & 1 & 1 \end{bmatrix}$$

$$= \begin{bmatrix} -1 & -1 & 0 & 0 & -1 & -1 & 0 & 0 \\ 1 & 1 & 1 & 1 & 0 & 0 & 0 & 0 \\ 0 & 1 & 0 & 1 & 0 & 1 & 0 & 1 \\ 1 & 1 & 1 & 1 & 1 & 1 & 1 & 1 \end{bmatrix}$$

as confirmed by Fig. 10.4 (d).

Let's explore what happens if we swap the rotation and translation transforms to $\mathbf{T}_{t_x,t_y,0}^{-1}\mathbf{R}_{\alpha,z}^{-1}$. This now implies that the frame of reference is rotated α about the z-axis, and then translated $(t_x, t_y, 0)$ in the rotated frame's space, which is not what we had planned. Here are the three transforms for rotating a frame of reference about an off-set axis:

$$\mathbf{R}_{\alpha,x}^{-1}\mathbf{T}_{0,t_y,t_z}^{-1} = \begin{bmatrix} 1 & 0 & 0 & 0 \\ 0 & \cos\alpha & \sin\alpha & -t_y\cos\alpha - t_z\sin\alpha \\ 0 & -\sin\alpha & \cos\alpha & -t_z\cos\alpha + t_y\sin\alpha \\ 0 & 0 & 0 & 1 \end{bmatrix}$$

$$\mathbf{R}_{\alpha,y}^{-1}\mathbf{T}_{t_x,0,t_z}^{-1} = \begin{bmatrix} \cos\alpha & 0 & -\sin\alpha & -t_x\cos\alpha + t_z\sin\alpha \\ 0 & 1 & 0 & 0 \\ \sin\alpha & 0 & \cos\alpha & -t_z\cos\alpha - t_x\sin\alpha \\ 0 & 0 & 0 & 1 \end{bmatrix}$$

$$\mathbf{R}_{\alpha,z}^{-1}\mathbf{T}_{t_x,t_y,0}^{-1} = \begin{bmatrix} \cos\alpha & \sin\alpha & 0 & -t_x\cos\alpha - t_y\sin\alpha \\ -\sin\alpha & \cos\alpha & 0 & -t_y\cos\alpha + t_x\sin\alpha \\ 0 & 0 & 1 & 0 \\ 0 & 0 & 0 & 1 \end{bmatrix}.$$

10.4 Composite Rotations

In Chap. 9 we went into some detail describing how point rotation transforms can be combined into composite rotations about the three Cartesian axes. There are twelve possible combinations that are listed in Appendix A. As any rotation transform can be used to rotate a point in one direction, or a frame of reference in the opposite direction, the previously computed composite transforms for rotating points, can be used for rotating frames in the opposite direction.

For example, we previously computed $\mathbf{R}_{\gamma,z}\mathbf{R}_{\beta,y}\mathbf{R}_{\alpha,x}$:

$$\mathbf{R}_{\gamma,z}\mathbf{R}_{\beta,y}\mathbf{R}_{\alpha,x} = \begin{bmatrix} c_\gamma c_\beta & c_\gamma s_\beta s_\alpha - s_\gamma c_\alpha & c_\gamma s_\beta c_\alpha + s_\gamma s_\alpha \\ s_\gamma c_\beta & s_\gamma s_\beta s_\alpha + c_\gamma c_\alpha & s_\gamma s_\beta c_\alpha - c_\gamma s_\alpha \\ -s_\beta & c_\beta s_\alpha & c_\beta c_\alpha \end{bmatrix}$$

which rotates a point about a fixed frame of reference. But it can also be used to rotate a frame of reference in the opposite directions:

$$\mathbf{R}^{-1}_{-\gamma,z}\mathbf{R}^{-1}_{-\beta,y}\mathbf{R}^{-1}_{-\alpha,x} = \begin{bmatrix} c_\gamma c_\beta & c_\gamma s_\beta s_\alpha - s_\gamma c_\alpha & c_\gamma s_\beta c_\alpha + s_\gamma s_\alpha \\ s_\gamma c_\beta & s_\gamma s_\beta s_\alpha + c_\gamma c_\alpha & s_\gamma s_\beta c_\alpha - c_\gamma s_\alpha \\ -s_\beta & c_\beta s_\alpha & c_\beta c_\alpha \end{bmatrix}.$$

In order to compute $\mathbf{R}^{-1}_{\gamma,z}\mathbf{R}^{-1}_{\beta,y}\mathbf{R}^{-1}_{\alpha,x}$ we only have to reverse the sign of the sine terms in the transform for $\mathbf{R}_{\gamma,z}\mathbf{R}_{\beta,y}\mathbf{R}_{\alpha,x}$:

$$\mathbf{R}^{-1}_{\gamma,z}\mathbf{R}^{-1}_{\beta,y}\mathbf{R}^{-1}_{\alpha,x} = \begin{bmatrix} c_\gamma c_\beta & c_\gamma s_\beta s_\alpha + s_\gamma c_\alpha & -c_\gamma s_\beta c_\alpha + s_\gamma s_\alpha \\ -s_\gamma c_\beta & -s_\gamma s_\beta s_\alpha + c_\gamma c_\alpha & s_\gamma s_\beta c_\alpha + c_\gamma s_\alpha \\ s_\beta & -c_\beta s_\alpha & c_\beta c_\alpha \end{bmatrix}. \quad (10.1)$$

Let's test (10.1) by making $\alpha = \beta = \gamma = 90°$:

$$\mathbf{R}^{-1}_{90°,z}\mathbf{R}^{-1}_{90°,y}\mathbf{R}^{-1}_{90°,x} = \begin{bmatrix} 0 & 0 & 1 \\ 0 & -1 & 0 \\ 1 & 0 & 0 \end{bmatrix}.$$

Figure 10.5 (a) shows the initial scenario, Fig. 10.5 (b) shows the frame rotated 90° about the local x-axis, Fig. 10.5 (c) shows the frame rotated 90° about the local y-axis, and Fig. 10.5 (d) shows the frame rotated 90° about the local z-axis. If we subject the coordinates of the unit cube to this composite transform we have

$$\begin{bmatrix} 0 & 0 & 1 \\ 0 & -1 & 0 \\ 1 & 0 & 0 \end{bmatrix} \begin{bmatrix} 0 & 0 & 0 & 0 & 1 & 1 & 1 & 1 \\ 0 & 0 & 1 & 1 & 0 & 0 & 1 & 1 \\ 0 & 1 & 0 & 1 & 0 & 1 & 0 & 1 \end{bmatrix}$$
$$= \begin{bmatrix} 0 & 1 & 0 & 1 & 0 & 1 & 0 & 1 \\ 0 & 0 & -1 & -1 & 0 & 0 & -1 & -1 \\ 0 & 0 & 0 & 0 & 1 & 1 & 1 & 1 \end{bmatrix}$$

which are confirmed by Fig. 10.5 (d).

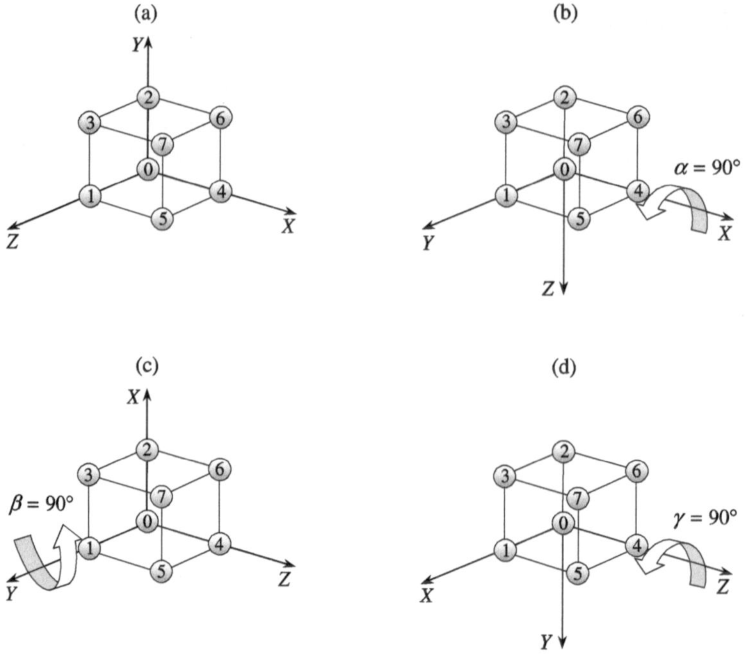

Fig. 10.5 (**a**) The original frame. (**b**) Rotated frame about the x-axis. (**c**) Rotated frame about the y-axis. (**d**) Rotated frame about the z-axis

Note that this specific transform also suffers from gimbal lock, where it is impossible to rotate the cube through an axis passing through vertices 0 and 2. Whereas, the cube is rotated twice about the axis passing through vertices 0 and 4. Effectively, the axial system has been rotated 180° about the vector $[1\ 0\ 1]^{\mathrm{T}}$, which could be confirmed by calculating the eigenvalue and eigenvector.

10.5 Rotated and Translated Frames of Reference

One of the most important frame changing transforms in cgi enables us to view an object from any location in space. The transform uses a rotation, which is often a composite transform, for example $\mathbf{R}^{-1}_{\gamma,z}\mathbf{R}^{-1}_{\beta,y}\mathbf{R}^{-1}_{\alpha,x}$, and a translation $\mathbf{T}^{-1}_{t_x,t_y,t_z}$. And as the position of the translated frame is normally specified in the original frame, we begin by translating the frame followed by the rotation:

$$\begin{bmatrix} c_\gamma c_\beta & c_\gamma s_\beta s_\alpha + s_\gamma c_\alpha & -c_\gamma s_\beta c_\alpha + s_\gamma s_\alpha & 0 \\ -s_\gamma c_\beta & -s_\gamma s_\beta s_\alpha + c_\gamma c_\alpha & s_\gamma s_\beta c_\alpha + c_\gamma s_\alpha & 0 \\ s_\beta & -c_\beta s_\alpha & c_\beta c_\alpha & 0 \\ 0 & 0 & 0 & 1 \end{bmatrix} \begin{bmatrix} 1 & 0 & 0 & -t_x \\ 0 & 1 & 0 & -t_y \\ 0 & 0 & 1 & -t_z \\ 0 & 0 & 0 & 1 \end{bmatrix}. \tag{10.2}$$

It is not worth multiplying these matrices together as it creates too many terms. However, we can test it with a simple example.

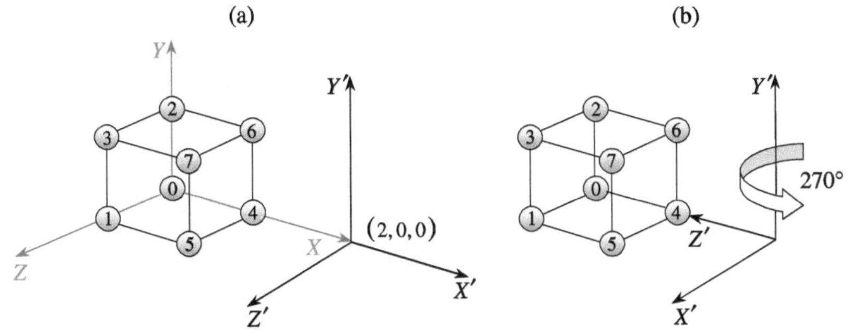

Fig. 10.6 (**a**) The translated frame. (**b**) The rotated frame about the y-axis

Using the unit cube, let's position the new frame 2 units along the initial x-axis, and then rotate the frame 270° about its local y-axis so that its z-axis is looking back towards the original origin. Figure 10.6 (a) shows the translated frame, and Fig. 10.6 (b) shows the rotated frame. Thus $t_x = 2$, $t_y = t_z = 0$, $\alpha = 0°$, $\beta = 270°$ and $\gamma = 0°$. Substituting these values in (10.2) we have

$$\mathbf{R}^{-1}_{0°,z}\mathbf{R}^{-1}_{270°,y}\mathbf{R}^{-1}_{0°,x}\mathbf{T}^{-1}_{2,0,0} = \begin{bmatrix} 0 & 0 & 1 & 0 \\ 0 & 1 & 0 & 0 \\ -1 & 0 & 0 & 0 \\ 0 & 0 & 0 & 1 \end{bmatrix} \begin{bmatrix} 1 & 0 & 0 & -2 \\ 0 & 1 & 0 & 0 \\ 0 & 0 & 1 & 0 \\ 0 & 0 & 0 & 1 \end{bmatrix} \tag{10.3}$$

$$= \begin{bmatrix} 0 & 0 & 1 & 0 \\ 0 & 1 & 0 & 0 \\ -1 & 0 & 0 & 2 \\ 0 & 0 & 0 & 1 \end{bmatrix}. \tag{10.4}$$

Using (10.4) to process the coordinates of the unit cube we have

$$\begin{bmatrix} 0 & 0 & 1 & 0 \\ 0 & 1 & 0 & 0 \\ -1 & 0 & 0 & 2 \\ 0 & 0 & 0 & 1 \end{bmatrix} \begin{bmatrix} 0 & 0 & 0 & 0 & 1 & 1 & 1 & 1 \\ 0 & 0 & 1 & 1 & 0 & 0 & 1 & 1 \\ 0 & 1 & 0 & 1 & 0 & 1 & 0 & 1 \\ 1 & 1 & 1 & 1 & 1 & 1 & 1 & 1 \end{bmatrix}$$

$$= \begin{bmatrix} 0 & 1 & 0 & 1 & 0 & 1 & 0 & 1 \\ 0 & 0 & 1 & 1 & 0 & 0 & 1 & 1 \\ 2 & 2 & 2 & 2 & 1 & 1 & 1 & 1 \\ 1 & 1 & 1 & 1 & 1 & 1 & 1 & 1 \end{bmatrix}$$

which are confirmed by Fig. 10.6 (b).

To obtain a perspective view of the cube we simply divide its transformed x- and y-coordinates by the associated z-coordinate:

$$\begin{bmatrix} 0 & 0.5 & 0 & 0.5 & 0 & 1 & 0 & 1 \\ 0 & 0 & 0.5 & 0.5 & 0 & 0 & 1 & 1 \\ 2 & 2 & 2 & 2 & 1 & 1 & 1 & 1 \\ 1 & 1 & 1 & 1 & 1 & 1 & 1 & 1 \end{bmatrix}$$

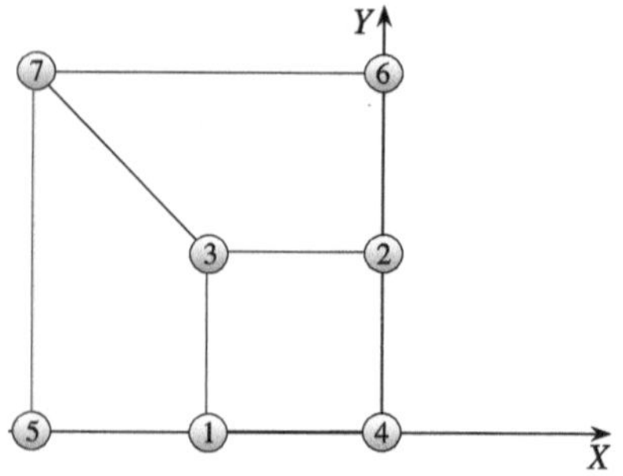

Fig. 10.7 Perspective view of the unit cube

and as the x-axis of the display screen is in the opposite direction to that of the frame, we have to switch the sign of the x-coordinates:

$$\begin{bmatrix} 0 & -0.5 & 0 & -0.5 & 0 & -1 & 0 & -1 \\ 0 & 0 & 0.5 & 0.5 & 0 & 0 & 1 & 1 \\ 2 & 2 & 2 & 2 & 1 & 1 & 1 & 1 \\ 1 & 1 & 1 & 1 & 1 & 1 & 1 & 1 \end{bmatrix}.$$

The x- and y-coordinates are used in Fig. 10.7 to show the perspective view seen from this translated, rotated frame.

10.6 Rotated Frames of Reference About Arbitrary Axes

In Chap. 9 we developed the following transform to rotate a point α about an arbitrary axis $\hat{\mathbf{n}}$:

$$\mathbf{R}_{\alpha,\hat{\mathbf{n}}} = \begin{bmatrix} a^2K + \cos\alpha & abK - c\sin\alpha & acK + b\sin\alpha \\ abK + c\sin\alpha & b^2K + \cos\alpha & bcK - a\sin\alpha \\ acK - b\sin\alpha & bcK + a\sin\alpha & c^2K + \cos\alpha \end{bmatrix}$$

$$K = 1 - \cos\alpha$$

$$\hat{\mathbf{n}} = a\mathbf{i} + b\mathbf{j} + c\mathbf{k}.$$

Therefore, there is nothing to stop us using the same transform to rotate a frame $-\alpha$ about $\hat{\mathbf{n}}$, or its inverse $\mathbf{R}^{-1}_{\alpha,\hat{\mathbf{n}}}$ to rotate a frame α about $\hat{\mathbf{n}}$. To compute the latter, we simply transpose the matrix, or change the sign of α which implies reversing the sign of the sine terms:

$$\mathbf{R}^{-1}_{\alpha,\hat{\mathbf{n}}} = \begin{bmatrix} a^2K + \cos\alpha & abK + c\sin\alpha & acK - b\sin\alpha \\ abK - c\sin\alpha & b^2K + \cos\alpha & bcK + a\sin\alpha \\ acK + b\sin\alpha & bcK - a\sin\alpha & c^2K + \cos\alpha \end{bmatrix}. \tag{10.5}$$

Let's test (10.5) using the previous example where we rotated a frame 270° about the y-axis, which makes $\alpha = 270°$, $\hat{\mathbf{n}} = \mathbf{j}$ and $K = 1$:

$$\mathbf{R}^{-1}_{270°,\mathbf{j}} = \begin{bmatrix} 0 & 0 & 1 \\ 0 & 1 & 0 \\ -1 & 0 & 0 \end{bmatrix}$$

which agrees perfectly. Naturally, this can be combined with a translation transform using

$$\mathbf{R}_{\alpha,\hat{\mathbf{n}}}^{-1}\mathbf{T}_{t_x,t_y,t_z}^{-1} = \begin{bmatrix} a^2K+\cos\alpha & abK+c\sin\alpha & acK-b\sin\alpha & 0 \\ abK-c\sin\alpha & b^2K+\cos\alpha & bcK+a\sin\alpha & 0 \\ acK+b\sin\alpha & bcK-a\sin\alpha & c^2K+\cos\alpha & 0 \\ 0 & 0 & 0 & 1 \end{bmatrix} \times \begin{bmatrix} 1 & 0 & 0 & -t_x \\ 0 & 1 & 0 & -t_y \\ 0 & 0 & 1 & -t_z \\ 0 & 0 & 0 & 1 \end{bmatrix}.$$

10.7 Summary

Hopefully, this chapter has covered most of the scenarios involving rotated and translated frames of reference in 3D space. Although composite rotation transforms offer a powerful mechanism for creating complex rotations, they are difficult to visualise and suffer from gimbal lock. Perhaps, the most useful transform is for rotating a frame about an arbitrary axis. For completeness, the important transforms are summarised below.

10.7.1 Summary of Transforms

Translating a frame

$$\mathbf{T}_{t_x,t_y,t_z}^{-1} = \begin{bmatrix} 1 & 0 & 0 & -t_x \\ 0 & 1 & 0 & -t_y \\ 0 & 0 & 1 & -t_z \\ 0 & 0 & 0 & 1 \end{bmatrix}.$$

Rotating a frame about a Cartesian axis

$$\mathbf{R}_{\alpha,x}^{-1} = \begin{bmatrix} 1 & 0 & 0 \\ 0 & \cos\alpha & \sin\alpha \\ 0 & -\sin\alpha & \cos\alpha \end{bmatrix}$$

$$\mathbf{R}_{\alpha,y}^{-1} = \begin{bmatrix} \cos\alpha & 0 & -\sin\alpha \\ 0 & 1 & 0 \\ \sin\alpha & 0 & \cos\alpha \end{bmatrix}$$

$$\mathbf{R}_{\alpha,z}^{-1} = \begin{bmatrix} \cos\alpha & \sin\alpha & 0 \\ -\sin\alpha & \cos\alpha & 0 \\ 0 & 0 & 1 \end{bmatrix}.$$

Rotating a frame about an off-set axis

$$\mathbf{R}_{\alpha,x}^{-1}\mathbf{T}_{0,t_y,t_z}^{-1} = \begin{bmatrix} 1 & 0 & 0 & 0 \\ 0 & \cos\alpha & \sin\alpha & -t_y\cos\alpha - t_z\sin\alpha \\ 0 & -\sin\alpha & \cos\alpha & -t_z\cos\alpha + t_y\sin\alpha \\ 0 & 0 & 0 & 1 \end{bmatrix}$$

$$\mathbf{R}_{\alpha,y}^{-1}\mathbf{T}_{t_x,0,t_z}^{-1} = \begin{bmatrix} \cos\alpha & 0 & -\sin\alpha & -t_x\cos\alpha + t_z\sin\alpha \\ 0 & 1 & 0 & 0 \\ \sin\alpha & 0 & \cos\alpha & -t_z\cos\alpha - t_x\sin\alpha \\ 0 & 0 & 0 & 1 \end{bmatrix}$$

$$\mathbf{R}_{\alpha,z}^{-1}\mathbf{T}_{t_x,t_y,0}^{-1} = \begin{bmatrix} \cos\alpha & \sin\alpha & 0 & -t_x\cos\alpha - t_y\sin\alpha \\ -\sin\alpha & \cos\alpha & 0 & -t_y\cos\alpha + t_x\sin\alpha \\ 0 & 0 & 1 & 0 \\ 0 & 0 & 0 & 1 \end{bmatrix}.$$

Rotating a frame using a composite transform

$$\mathbf{R}_{\gamma,z}^{-1}\mathbf{R}_{\beta,y}^{-1}\mathbf{R}_{\alpha,x}^{-1} = \begin{bmatrix} c_\gamma c_\beta & c_\gamma s_\beta s_\alpha + s_\gamma c_\alpha & -c_\gamma s_\beta c_\alpha + s_\gamma s_\alpha \\ -s_\gamma c_\beta & -s_\gamma s_\beta s_\alpha + c_\gamma c_\alpha & s_\gamma s_\beta c_\alpha + c_\gamma s_\alpha \\ s_\beta & -c_\beta s_\alpha & c_\beta c_\alpha \end{bmatrix}.$$

Rotating and translating a frame

$$\mathbf{R}_{\gamma,z}^{-1}\mathbf{R}_{\beta,y}^{-1}\mathbf{R}_{\alpha,x}^{-1}\mathbf{T}_{t_x,t_y,t_z}^{-1} = \begin{bmatrix} c_\gamma c_\beta & c_\gamma s_\beta s_\alpha + s_\gamma c_\alpha & -c_\gamma s_\beta c_\alpha + s_\gamma s_\alpha & 0 \\ -s_\gamma c_\beta & -s_\gamma s_\beta s_\alpha + c_\gamma c_\alpha & s_\gamma s_\beta c_\alpha + c_\gamma s_\alpha & 0 \\ s_\beta & -c_\beta s_\alpha & c_\beta c_\alpha & 0 \\ 0 & 0 & 0 & 1 \end{bmatrix}$$

$$\times \begin{bmatrix} 1 & 0 & 0 & -t_x \\ 0 & 1 & 0 & -t_y \\ 0 & 0 & 1 & -t_z \\ 0 & 0 & 0 & 1 \end{bmatrix}.$$

Rotating a frame about an arbitrary axis

$$\mathbf{R}_{\alpha,\hat{\mathbf{n}}}^{-1} = \begin{bmatrix} a^2K + \cos\alpha & abK + c\sin\alpha & acK - b\sin\alpha \\ abK - c\sin\alpha & b^2K + \cos\alpha & bcK + a\sin\alpha \\ acK + b\sin\alpha & bcK - a\sin\alpha & c^2K + \cos\alpha \end{bmatrix}$$

$$K = 1 - \cos\alpha$$

$$\hat{\mathbf{n}} = a\mathbf{i} + b\mathbf{j} + c\mathbf{k}.$$

Chapter 11
Quaternion Transforms in Space

11.1 Introduction

Quaternions were introduced in Chap. 5 as a mathematical object that combines a scalar with a vector, in the same way a complex number combines a scalar with an imaginary quantity. Quaternions, like complex numbers, possess rotational qualities, but work in four dimensions rather than on the complex plane.

Hamilton invented quaternions in October 1843, and by December of the same year his friend, John T. Graves, had invented octonions. Arthur Cayley had also been intrigued by Hamilton's quaternions, and independently discovered octonions in 1845.

There are four such *composition algebras*: real $\mathbb{R}$, complex $\mathbb{C}$, quaternion $\mathbb{H}$, and octonion $\mathbb{O}$ that obey an n-square identity used to compute their magnitudes. Adolf Hurwitz (1859–1919) proved that the product of the sum of n squares by the sum of n squares is the sum of n squares only when n is equal to 1, 2, 4 and 8, which are represented by reals, complex, quaternions and octonions. No other system is possible, which shows how important quaternions are within the realm of mathematics. Appendix C provides further information on this topic.

In this chapter we investigate how quaternions are used to rotate 3D vectors about an arbitrary axis.

11.2 Definition

A quaternion $\mathbf{q}$ is the union of a scalar and a vector:

$$\mathbf{q} = s + \mathbf{v}$$

where s is a scalar and $\mathbf{v}$ is a 3D vector. If we express the vector $\mathbf{v}$ in terms of its components, we have

$$\mathbf{q} = s + x\mathbf{i} + y\mathbf{j} + z\mathbf{k}$$

where s, x, y and z are all real numbers.

J. Vince, *Rotation Transforms for Computer Graphics*,
DOI 10.1007/978-0-85729-154-7_11,

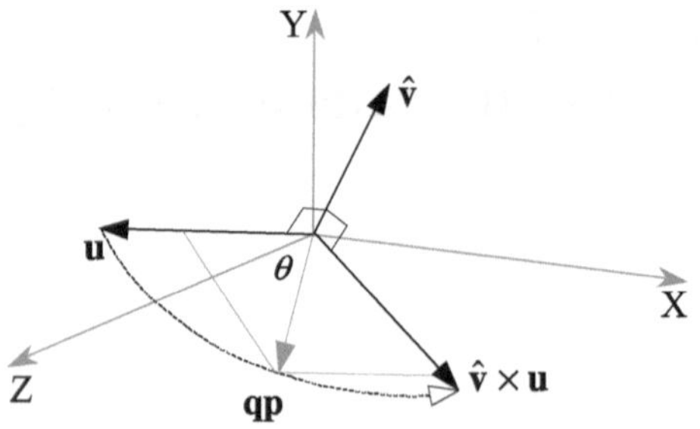

Fig. 11.1 Three orthogonal vectors $\mathbf{u}$, $\hat{\mathbf{v}}$ and $\hat{\mathbf{v}} \times \mathbf{u}$

Hamilton had hoped that a quaternion could be used like a complex rotor, where we saw in Chap. 2 that

$$\mathbf{R}_\theta = \cos\theta + i\sin\theta$$

rotates a complex number by θ. Could the product of a quaternion $\mathbf{q}$ and a vector $\mathbf{u}$ rotate $\mathbf{u}$ about the quaternion's vector? Well yes, but only in a restricted sense. To understand this, consider the scenario where we form the product of a unit quaternion $\mathbf{q}$ and a pure quaternion $\mathbf{p}$. The unit quaternion $\mathbf{q}$ is defined as

$$\mathbf{q} = s + \lambda\hat{\mathbf{v}}$$

where

$$s^2 + \lambda^2 = 1$$

and the pure quaternion $\mathbf{p}$ encodes the vector $\mathbf{u}$ with a zero scalar term

$$\mathbf{p} = 0 + \mathbf{u}.$$

In Chap. 5 we computed product $\mathbf{qp}$, which in this context is

$$\begin{aligned}\mathbf{qp} &= (s + \lambda\hat{\mathbf{v}})(0 + \mathbf{u})\\ &= -\lambda\hat{\mathbf{v}} \cdot \mathbf{u} + s\mathbf{u} + \lambda\hat{\mathbf{v}} \times \mathbf{u}.\end{aligned}$$

However, if $\hat{\mathbf{v}}$ is orthogonal to $\mathbf{u}$, the dot product term $-\lambda\hat{\mathbf{v}} \cdot \mathbf{u}$ vanishes, and we are left with

$$\mathbf{qp} = s\mathbf{u} + \lambda\hat{\mathbf{v}} \times \mathbf{u}.$$

Figure 11.1 illustrates this scenario, where $\mathbf{u}$ is perpendicular to $\hat{\mathbf{v}}$, and $\hat{\mathbf{v}} \times \mathbf{u}$ is perpendicular to the plane containing $\mathbf{u}$ and $\hat{\mathbf{v}}$.

Now because $\hat{\mathbf{v}}$ is a unit vector, the length of $\hat{\mathbf{v}} \times \mathbf{u}$ is $|\mathbf{u}|$, which means that we have two orthogonal vectors, i.e. $\mathbf{u}$ and $\hat{\mathbf{v}} \times \mathbf{u}$, with the same length. Therefore, to rotate $\mathbf{u}$ about $\hat{\mathbf{v}}$, all that we have to do is to substitute $\cos\theta$ for s and $\sin\theta$ for λ:

$$\mathbf{qp} = \cos\theta\mathbf{u} + \sin\theta\hat{\mathbf{v}} \times \mathbf{u}.$$

For example, if we create a quaternion whose vector is aligned with the z-axis as shown in Fig. 11.2 with

$$\mathbf{q} = \cos\theta + \sin\theta\mathbf{k}$$

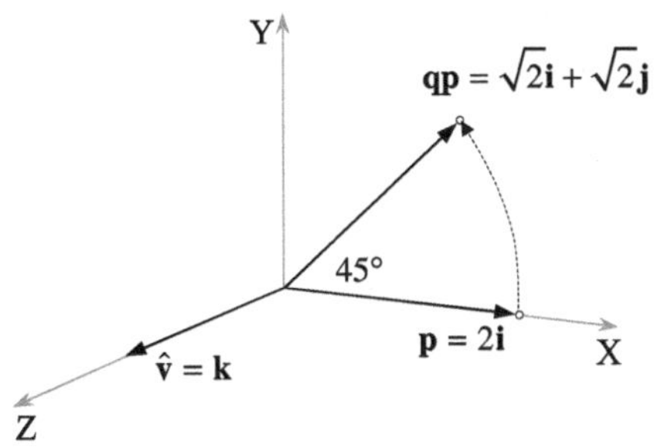

Fig. 11.2 The vector $2\mathbf{i}$ is rotated 45° by the quaternion $\mathbf{q} = \cos\theta + \sin\theta\mathbf{k}$

and a pure quaternion to represent the vector $2\mathbf{i}$

$$\mathbf{p} = 0 + 2\mathbf{i}$$

then

$$\mathbf{qp} = 2\cos\theta\mathbf{i} + \sin\theta\mathbf{k} \times 2\mathbf{i}.$$

With $\theta = 45°$ then

$$\begin{aligned}\mathbf{qp} &= 2\frac{\sqrt{2}}{2}\mathbf{i} + \frac{\sqrt{2}}{2}\mathbf{k} \times 2\mathbf{i}\\ &= \sqrt{2}\mathbf{i} + \sqrt{2}\mathbf{j}\end{aligned}$$

which is a pure quaternion, i.e. a vector, and $\mathbf{p}$ has been rotated 45°.

Let's see what happens when $\theta = 180°$:

$$\mathbf{qp} = 2\cos 180°\mathbf{i} + \sin 180°\mathbf{k} \times 2\mathbf{i} = -2\mathbf{i}$$

which is also a pure quaternion, and $\mathbf{p}$ has been rotated 180°. Note that the vector has not been scaled during the rotation. This is because we are using a unit quaternion.

Now let's see what happens when we reduce the angle between $\hat{\mathbf{v}}$ and $\mathbf{p}$. Let's reduce the angle to 45° and retain the quaternion's magnitude at unity, as shown in Fig. 11.3. Therefore,

$$\begin{aligned}\hat{\mathbf{v}} &= \frac{1}{\sqrt{2}}\mathbf{i} + \frac{1}{\sqrt{2}}\mathbf{k}\\ \mathbf{q} &= \cos\theta + \sin\theta\left(\frac{1}{\sqrt{2}}\mathbf{i} + \frac{1}{\sqrt{2}}\mathbf{k}\right)\\ \mathbf{p} &= 0 + 2\mathbf{i}.\end{aligned}$$

This time we must include the dot product term:

$$\mathbf{qp} = -\sin\theta\hat{\mathbf{v}} \cdot \mathbf{u} + \cos\theta\mathbf{u} + \sin\theta\hat{\mathbf{v}} \times \mathbf{u}.$$

We let $\theta = 45°$

$$\begin{aligned}\mathbf{qp} &= -\frac{\sqrt{2}}{2}\left(\frac{1}{\sqrt{2}}\mathbf{i} + \frac{1}{\sqrt{2}}\mathbf{k}\right) \cdot (2\mathbf{i}) + \frac{\sqrt{2}}{2}2\mathbf{i} + \frac{\sqrt{2}}{2}\left(\frac{1}{\sqrt{2}}\mathbf{i} + \frac{1}{\sqrt{2}}\mathbf{k}\right) \times 2\mathbf{i}\\ &= -1 + \sqrt{2}\mathbf{i} + \mathbf{j}\end{aligned}$$

Fig. 11.3 Rotating the vector $2\mathbf{i}$ by the quaternion $\mathbf{q} = \cos\theta + \sin\theta(\frac{1}{\sqrt{2}}\mathbf{i} + \frac{1}{\sqrt{2}}\mathbf{k})$

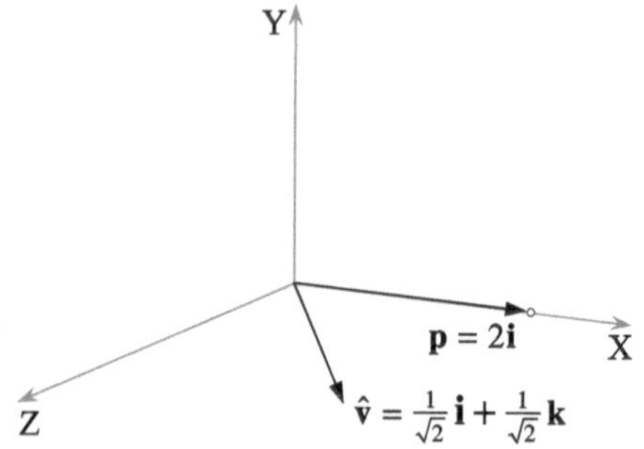

which, unfortunately, is not a pure quaternion. It has not been rotated 45° and the vector's magnitude is reduced to $\sqrt{3}$! Multiplying the vector by a non-orthogonal quaternion seems to have converted some of the vector information into the quaternion's scalar component.

Not to worry. Could it be that an inverse quaternion reverses the operation? Let's see what happens if we multiply this result, i.e. $\mathbf{qp}$, by $\mathbf{q}^{-1}$.

Given

$$\mathbf{q} = \cos\theta + \sin\theta\left(\frac{1}{\sqrt{2}}\mathbf{i} + \frac{1}{\sqrt{2}}\mathbf{k}\right)$$

then

$$\begin{aligned}\mathbf{q}^{-1} &= \cos\theta - \sin\theta\left(\frac{1}{\sqrt{2}}\mathbf{i} + \frac{1}{\sqrt{2}}\mathbf{k}\right)\\ &= \frac{\sqrt{2}}{2} - \frac{\sqrt{2}}{2}\left(\frac{1}{\sqrt{2}}\mathbf{i} + \frac{1}{\sqrt{2}}\mathbf{k}\right)\\ &= \frac{1}{2}(\sqrt{2} - \mathbf{i} - \mathbf{k}).\end{aligned}$$

Therefore,

$$\begin{aligned}\mathbf{qpq}^{-1} &= \frac{1}{2}(-1 + \sqrt{2}\mathbf{i} + \mathbf{j})(\sqrt{2} - \mathbf{i} - \mathbf{k})\\ &= \frac{1}{2}\big(-\sqrt{2} - (\sqrt{2}\mathbf{i} + \mathbf{j})\cdot(-\mathbf{i} - \mathbf{k}) + (\mathbf{i} + \mathbf{k}) + \sqrt{2}(\sqrt{2}\mathbf{i} + \mathbf{j}) - \mathbf{i} + \sqrt{2}\mathbf{j} + \mathbf{k}\big)\\ &= \frac{1}{2}(-\sqrt{2} + \sqrt{2} + \mathbf{i} + \mathbf{k} + 2\mathbf{i} + \sqrt{2}\mathbf{j} - \mathbf{i} + \sqrt{2}\mathbf{j} + \mathbf{k})\\ &= \mathbf{i} + \sqrt{2}\mathbf{j} + \mathbf{k}\end{aligned}$$

which *is* a pure quaternion. Furthermore, its magnitude is 2, but what is strange, the vector has been rotated 90° rather than 45° as shown in Fig. 11.4.

If this 'sandwiching' of the vector by $\mathbf{q}$ and $\mathbf{q}^{-1}$ is correct, it implies that increasing θ to 90° should rotate $\mathbf{p} = 2\mathbf{i}$ by 180° to $2\mathbf{k}$. Let's try this.

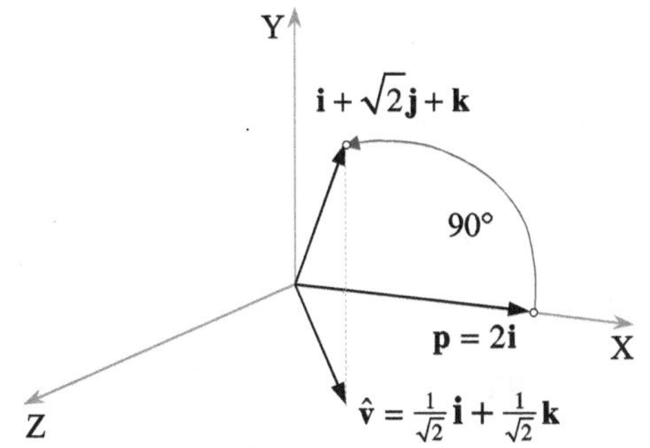

Fig. 11.4 The vector $2\mathbf{i}$ is rotated 90° to $\mathbf{i}+\sqrt{2}\mathbf{j}+\mathbf{k}$

Let $\theta = 90°$, therefore,

$$\begin{aligned}\mathbf{qp} &= \left(0+1\left(\frac{1}{\sqrt{2}}\mathbf{i}+\frac{1}{\sqrt{2}}\mathbf{k}\right)\right)(0+2\mathbf{i})\\ &= \frac{2}{\sqrt{2}}(-1+\mathbf{j}).\end{aligned}$$

Next, we post-multiply by $\mathbf{q}^{-1}$

$$\begin{aligned}\mathbf{qpq}^{-1} &= \frac{2}{\sqrt{2}}(-1+\mathbf{j})\left(0-\frac{1}{\sqrt{2}}\mathbf{i}-\frac{1}{\sqrt{2}}\mathbf{k}\right)\\ &= \frac{2}{\sqrt{2}}\left(\frac{1}{\sqrt{2}}\mathbf{i}+\frac{1}{\sqrt{2}}\mathbf{k}-\frac{1}{\sqrt{2}}\mathbf{i}+\frac{1}{\sqrt{2}}\mathbf{k}\right)\\ &= \mathbf{i}+\mathbf{k}-\mathbf{i}+\mathbf{k}\\ &= 2\mathbf{k}\end{aligned}$$

which confirms our prediction. Now let's show how this double angle arises.

We begin by defining a unit quaternion $\mathbf{q}$:

$$\mathbf{q} = s+\lambda\hat{\mathbf{v}}$$

and we will eventually assign values to s and λ. The vector $\mathbf{u}$ to be rotated is a pure quaternion:

$$\mathbf{p} = 0+\mathbf{u}.$$

The inverse quaternion $\mathbf{q}^{-1}$ is

$$\mathbf{q}^{-1} = s-\lambda\hat{\mathbf{v}}$$

therefore, the triple $\mathbf{qpq}^{-1}$ is

$$\begin{aligned}\mathbf{qpq}^{-1} &= (s+\lambda\hat{\mathbf{v}})(0+\mathbf{u})(s-\lambda\hat{\mathbf{v}})\\ &= (-\lambda\hat{\mathbf{v}}\cdot\mathbf{u}+s\mathbf{u}+\lambda\hat{\mathbf{v}}\times\mathbf{u})(s-\lambda\hat{\mathbf{v}})\\ &= -\lambda s\hat{\mathbf{v}}\cdot\mathbf{u}+\lambda s\mathbf{u}\cdot\hat{\mathbf{v}}+\lambda^2(\hat{\mathbf{v}}\times\mathbf{u})\cdot\hat{\mathbf{v}}\\ &\quad+\lambda^2(\hat{\mathbf{v}}\cdot\mathbf{u})\hat{\mathbf{v}}+s^2\mathbf{u}+\lambda s\hat{\mathbf{v}}\times\mathbf{u}\\ &\quad-\lambda s\mathbf{u}\times\hat{\mathbf{v}}-\lambda^2(\hat{\mathbf{v}}\times\mathbf{u})\times\hat{\mathbf{v}}\\ &= \lambda^2(\hat{\mathbf{v}}\times\mathbf{u})\cdot\hat{\mathbf{v}}+\lambda^2(\hat{\mathbf{v}}\cdot\mathbf{u})\hat{\mathbf{v}}+s^2\mathbf{u}+2\lambda s\hat{\mathbf{v}}\times\mathbf{u}-\lambda^2(\hat{\mathbf{v}}\times\mathbf{u})\times\hat{\mathbf{v}}.\end{aligned}$$

Note that

$$(\hat{\mathbf{v}} \times \mathbf{u}) \cdot \hat{\mathbf{v}} = 0$$

and

$$(\hat{\mathbf{v}} \times \mathbf{u}) \times \hat{\mathbf{v}} = (\hat{\mathbf{v}} \cdot \hat{\mathbf{v}})\mathbf{u} - (\mathbf{u} \cdot \hat{\mathbf{v}})\hat{\mathbf{v}} = \mathbf{u} - (\mathbf{u} \cdot \hat{\mathbf{v}})\hat{\mathbf{v}}.$$

Therefore,

$$\begin{aligned}\mathbf{qpq}^{-1} &= \lambda^2(\hat{\mathbf{v}} \cdot \mathbf{u})\hat{\mathbf{v}} + s^2\mathbf{u} + 2\lambda s\hat{\mathbf{v}} \times \mathbf{u} - \lambda^2\mathbf{u} + \lambda^2(\mathbf{u} \cdot \hat{\mathbf{v}})\hat{\mathbf{v}} \\ &= 2\lambda^2(\hat{\mathbf{v}} \cdot \mathbf{u})\hat{\mathbf{v}} + (s^2 - \lambda^2)\mathbf{u} + 2\lambda s\hat{\mathbf{v}} \times \mathbf{u}.\end{aligned}$$

Obviously, this is a pure quaternion as there is no scalar component. However, it is not obvious where the angle doubling comes from. But, look what happens when we make $s = \cos\theta$ and $\lambda = \sin\theta$:

$$\begin{aligned}\mathbf{qpq}^{-1} &= 2\sin^2\theta(\hat{\mathbf{v}} \cdot \mathbf{u})\hat{\mathbf{v}} + (\cos^2\theta - \sin^2\theta)\mathbf{u} + 2\sin\theta\cos\theta\hat{\mathbf{v}} \times \mathbf{u} \\ &= (1 - \cos 2\theta)(\hat{\mathbf{v}} \cdot \mathbf{u})\hat{\mathbf{v}} + \cos 2\theta\mathbf{u} + \sin 2\theta\hat{\mathbf{v}} \times \mathbf{u}.\end{aligned}$$

The double angle trigonometric terms emerge! Now, if we want this triple to actually rotate the vector by θ, then we must build this in from the outset by halving θ in $\mathbf{q}$:

$$\mathbf{q} = \cos(\theta/2) + \sin(\theta/2)\hat{\mathbf{v}}$$

which makes

$$\mathbf{qpq}^{-1} = (1 - \cos\theta)(\hat{\mathbf{v}} \cdot \mathbf{u})\hat{\mathbf{v}} + \cos\theta\mathbf{u} + \sin\theta\hat{\mathbf{v}} \times \mathbf{u}. \tag{11.1}$$

Equation (11.1) is the same equation we came across in Chap. 9 discovered by Rodrigues a few years before Hamilton, hence the scandal!

Let's test (11.1) using the previous example where we rotated a vector $\mathbf{u} = 2\mathbf{i}$, 90° about the quaternion's vector $\hat{\mathbf{v}} = \frac{1}{\sqrt{2}}\mathbf{i} + \frac{1}{\sqrt{2}}\mathbf{k}$:

$$\begin{aligned}\mathbf{qpq}^{-1} &= \frac{2}{\sqrt{2}}\left(\frac{1}{\sqrt{2}}\mathbf{i} + \frac{1}{\sqrt{2}}\mathbf{k}\right) + \sqrt{2}\mathbf{j} \\ &= \mathbf{i} + \sqrt{2}\mathbf{j} + \mathbf{k}\end{aligned}$$

which agrees with the previous result.

Thus, when a quaternion takes on the form

$$\mathbf{q} = \cos(\theta/2) + \sin(\theta/2)\hat{\mathbf{v}}$$

it rotates a vector $\mathbf{p}$, anticlockwise θ using the triple

$$\mathbf{qpq}^{-1}.$$

It can be shown that this triple always preserves the magnitude of the rotated vector.

You may be wondering what happens if the triple is reversed to $\mathbf{q}^{-1}\mathbf{pq}$? A guess would suggest that the rotation sequence is reversed, but let's see what an algebraic solution predicts:

$$
\begin{aligned}
\mathbf{q}^{-1}\mathbf{p}\mathbf{q} &= (s-\lambda\hat{\mathbf{v}})(0+\mathbf{u})(s+\lambda\hat{\mathbf{v}})\\
&= (\lambda\hat{\mathbf{v}}\cdot\mathbf{u}+s\mathbf{u}-\lambda\hat{\mathbf{v}}\times\mathbf{u})(s+\lambda\hat{\mathbf{v}})\\
&= \lambda s\hat{\mathbf{v}}\cdot\mathbf{u}-\lambda s\mathbf{u}\cdot\hat{\mathbf{v}}+\lambda^2\hat{\mathbf{v}}\times\mathbf{u}\cdot\hat{\mathbf{v}}+\lambda^2\hat{\mathbf{v}}\cdot\mathbf{u}\hat{\mathbf{v}}\\
&\quad+s^2\mathbf{u}-\lambda s\hat{\mathbf{v}}\times\mathbf{u}+\lambda s\mathbf{u}\times\hat{\mathbf{v}}-\lambda^2\hat{\mathbf{v}}\times\mathbf{u}\times\hat{\mathbf{v}}\\
&= \lambda^2(\hat{\mathbf{v}}\times\mathbf{u})\cdot\hat{\mathbf{v}}+\lambda^2(\hat{\mathbf{v}}\cdot\mathbf{u})\hat{\mathbf{v}}+s^2\mathbf{u}-2\lambda s\hat{\mathbf{v}}\times\mathbf{u}-\lambda^2(\hat{\mathbf{v}}\times\mathbf{u})\times\hat{\mathbf{v}}.
\end{aligned}
$$

Once again

$$(\hat{\mathbf{v}}\times\mathbf{u})\cdot\hat{\mathbf{v}}=0$$

and

$$(\hat{\mathbf{v}}\times\mathbf{u})\times\hat{\mathbf{v}}=\mathbf{u}-(\mathbf{u}\cdot\hat{\mathbf{v}})\hat{\mathbf{v}}.$$

Therefore,

$$
\begin{aligned}
\mathbf{q}^{-1}\mathbf{p}\mathbf{q} &= \lambda^2(\hat{\mathbf{v}}\cdot\mathbf{u})\hat{\mathbf{v}}+s^2\mathbf{u}-2\lambda s\hat{\mathbf{v}}\times\mathbf{u}-\lambda^2\mathbf{u}+\lambda^2(\mathbf{u}\cdot\hat{\mathbf{v}})\hat{\mathbf{v}}\\
&= 2\lambda^2(\hat{\mathbf{v}}\cdot\mathbf{u})\hat{\mathbf{v}}+\left(s^2-\lambda^2\right)\mathbf{u}-2\lambda s\hat{\mathbf{v}}\times\mathbf{u}.
\end{aligned}
$$

Again, let's make $s=\cos\theta$ and $\lambda=\sin\theta$:

$$\mathbf{q}^{-1}\mathbf{p}\mathbf{q}=(1-2\cos\theta)(\hat{\mathbf{v}}\cdot\mathbf{u})\hat{\mathbf{v}}+\cos 2\theta\mathbf{u}-\sin 2\theta\hat{\mathbf{v}}\times\mathbf{u}$$

and the only thing that has changed is the sign of the cross-product term, which reverses the direction of its vector. However, we must remember to compensate for the angle-doubling by halving θ:

$$\mathbf{q}^{-1}\mathbf{p}\mathbf{q}=(1-\cos\theta)(\hat{\mathbf{v}}\cdot\mathbf{u})\hat{\mathbf{v}}+\cos\theta\mathbf{u}-\sin\theta\hat{\mathbf{v}}\times\mathbf{u}. \tag{11.2}$$

Let's see what happens when we employ (11.2) to rotate $\mathbf{u}=2\mathbf{i}$, 90° about the quaternion's vector $\hat{\mathbf{v}}=\frac{1}{\sqrt{2}}\mathbf{i}+\frac{1}{\sqrt{2}}\mathbf{k}$:

$$
\begin{aligned}
\mathbf{q}^{-1}\mathbf{p}\mathbf{q} &= \frac{2}{\sqrt{2}}\left(\frac{1}{\sqrt{2}}\mathbf{i}+\frac{1}{\sqrt{2}}\mathbf{k}\right)-\sqrt{2}\mathbf{j}\\
&= \mathbf{i}-\sqrt{2}\mathbf{j}+\mathbf{k}
\end{aligned}
$$

which has rotated $\mathbf{u}$ clockwise 90° about the quaternion's vector. Therefore, the rotor $\mathbf{q}\mathbf{p}\mathbf{q}^{-1}$ rotates a vector anticlockwise, and $\mathbf{q}^{-1}\mathbf{p}\mathbf{q}$ rotates a vector clockwise:

$$
\begin{aligned}
\mathbf{q}\mathbf{p}\mathbf{q}^{-1} &= (1-\cos\theta)(\hat{\mathbf{v}}\cdot\mathbf{u})\hat{\mathbf{v}}+\cos\theta\mathbf{u}+\sin\theta\hat{\mathbf{v}}\times\mathbf{u}\\
\mathbf{q}^{-1}\mathbf{p}\mathbf{q} &= (1-\cos\theta)(\hat{\mathbf{v}}\cdot\mathbf{u})\hat{\mathbf{v}}+\cos\theta\mathbf{u}-\sin\theta\hat{\mathbf{v}}\times\mathbf{u}.
\end{aligned}
$$

However, we must remember that the rotor interprets θ as 2θ.

Let's compute another example. Consider the point $P(0,1,1)$ in Fig. 11.5 which is to be rotated 90° about the y-axis. We can see that the rotated point P' has the coordinates $(1,1,0)$ which we will confirm algebraically. The point P is represented by the pure quaternion

$$\mathbf{p}=0+\mathbf{u}.$$

The axis of rotation is $\hat{\mathbf{v}}=\mathbf{j}$, and the vector to be rotated is $\mathbf{u}=\mathbf{j}+\mathbf{k}$. Therefore,

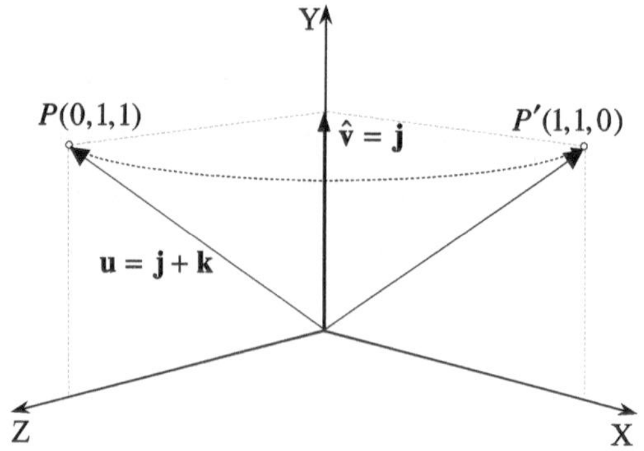

Fig. 11.5 The point $P(0,1,1)$ is rotated 90° to $P'(1,1,0)$ about the y-axis

$$
\begin{aligned}
\mathbf{qpq}^{-1} &= (1-\cos\theta)(\hat{\mathbf{v}}\cdot\mathbf{u})\hat{\mathbf{v}} + \cos\theta\mathbf{u} + \sin\theta\hat{\mathbf{v}}\times\mathbf{u}\\
&= \mathbf{j}\cdot(\mathbf{j}+\mathbf{k})\mathbf{j} + \mathbf{j}\times(\mathbf{j}+\mathbf{k})\\
&= \mathbf{i}+\mathbf{j}
\end{aligned}
$$

and confirms that P is indeed rotated to $(1,1,0)$.

Before moving onto the next section it is worth clarifying the interpretation of the two triples described above. As with the rotation transforms previously covered, we have used $\mathbf{R}_{\theta,\mathbf{v}}$ to describe a point rotation, θ about the vector $\mathbf{v}$, and $\mathbf{R}^{-1}_{\theta,\mathbf{v}}$ to describe a frame rotation θ about the vector $\mathbf{v}$. Therefore, in keeping with this convention, we will use $\mathbf{qpq}^{-1}$ to describe a point rotation of the point represented by $\mathbf{p}$ about the quaternion's vector. And $\mathbf{q}^{-1}\mathbf{pq}$ to describe a frame rotation about the same vector.

11.3 Quaternions in Matrix Form

Having discovered a vector equation to represent the triple $\mathbf{qpq}^{-1}$, let's continue and convert it into a matrix. We will explore two methods: the first is a simple vectorial method which translates the vector equation representing $\mathbf{qpq}^{-1}$ directly into a matrix form; the second method uses matrix algebra to develop a rather cunning solution. Let's start with the vectorial approach, for which it is convenient to describe the unit quaternion as

$$
\begin{aligned}
\mathbf{q} &= s+\mathbf{v}\\
&= s + x\mathbf{i} + y\mathbf{j} + z\mathbf{k}
\end{aligned}
$$

and the pure quaternion as

$$\mathbf{p} = 0 + \mathbf{u}$$

which means that the triple becomes

$$\mathbf{qpq}^{-1} = 2(\mathbf{v}\cdot\mathbf{u})\mathbf{v} + \left(s^2 - |\mathbf{v}|^2\right)\mathbf{u} + 2s\mathbf{v}\times\mathbf{u}.$$

And as we are working with unit quaternions to prevent scaling

$$s^2 + |\mathbf{v}|^2 = 1$$

therefore,

$$s^2 - |\mathbf{v}|^2 = 2s^2 - 1$$

and

$$\mathbf{qpq}^{-1} = 2(\mathbf{v}\cdot\mathbf{u})\mathbf{v} + \left(2s^2 - 1\right)\mathbf{u} + 2s\mathbf{v}\times\mathbf{u}.$$

We can now represent the three terms $2(\mathbf{v}\cdot\mathbf{u})\mathbf{v}$, $(2s^2 - 1)\mathbf{u}$ and $2s\mathbf{v}\times\mathbf{u}$ as three individual matrices, which can be summed together:

$$\begin{aligned}
2(\mathbf{v}\cdot\mathbf{u})\mathbf{v} &= 2(xx_u + yy_u + zz_u)(x\mathbf{i} + y\mathbf{j} + z\mathbf{k})\\
&= \begin{bmatrix} 2x^2 & 2xy & 2xz \\ 2xy & 2y^2 & 2yz \\ 2xz & 2yz & 2z^2 \end{bmatrix}\begin{bmatrix} x_u \\ y_u \\ z_u \end{bmatrix}\\
\left(2s^2 - 1\right)\mathbf{u} &= \left(2s^2 - 1\right)x_u\mathbf{i} + \left(2s^2 - 1\right)y_u\mathbf{j} + \left(2s^2 - 1\right)z_u\mathbf{k}\\
&= \begin{bmatrix} 2s^2 - 1 & 0 & 0 \\ 0 & 2s^2 - 1 & 0 \\ 0 & 0 & 2s^2 - 1 \end{bmatrix}\begin{bmatrix} x_u \\ y_u \\ z_u \end{bmatrix}\\
2s\mathbf{v}\times\mathbf{u} &= 2s\big((yz_u - zy_u)\mathbf{i} + (zx_u - xz_u)\mathbf{j} + (xy_u - yx_u)\mathbf{k}\big)\\
&= \begin{bmatrix} 0 & -2sz & 2sy \\ 2sz & 0 & -2sx \\ -2sy & 2sx & 0 \end{bmatrix}\begin{bmatrix} x_u \\ y_u \\ z_u \end{bmatrix}.
\end{aligned}$$

Adding these matrices together produces

$$\mathbf{qpq}^{-1} = \begin{bmatrix} 2(s^2 + x^2) - 1 & 2(xy - sz) & 2(xz + sy) \\ 2(xy + sz) & 2(s^2 + y^2) - 1 & 2(yz - sx) \\ 2(xz - sy) & 2(yz + sx) & 2(s^2 + z^2) - 1 \end{bmatrix}\begin{bmatrix} x_u \\ y_u \\ z_u \end{bmatrix} \tag{11.3}$$

or

$$\mathbf{qpq}^{-1} = \begin{bmatrix} 1 - 2(y^2 + z^2) & 2(xy - sz) & 2(xz + sy) \\ 2(xy + sz) & 1 - 2(x^2 + z^2) & 2(yz - sx) \\ 2(xz - sy) & 2(yz + sx) & 1 - 2(x^2 + y^2) \end{bmatrix}\begin{bmatrix} x_u \\ y_u \\ z_u \end{bmatrix}. \tag{11.4}$$

To compute the equivalent matrix for $\mathbf{q}^{-1}\mathbf{pq}$ all that we have to do is reverse the sign of $2s\mathbf{v}\times\mathbf{u}$:

$$\mathbf{q}^{-1}\mathbf{pq} = \begin{bmatrix} 2(s^2 + x^2) - 1 & 2(xy + sz) & 2(xz - sy) \\ 2(xy - sz) & 2(s^2 + y^2) - 1 & 2(yz + sx) \\ 2(xz + sy) & 2(yz - sx) & 2(s^2 + z^2) - 1 \end{bmatrix}\begin{bmatrix} x_u \\ y_u \\ z_u \end{bmatrix} \tag{11.5}$$

or

$$\mathbf{q}^{-1}\mathbf{p}\mathbf{q} = \begin{bmatrix} 1-2(y^2+z^2) & 2(xy+sz) & 2(xz-sy) \\ 2(xy-sz) & 1-2(x^2+z^2) & 2(yz+sx) \\ 2(xz+sy) & 2(yz-sx) & 1-2(x^2+y^2) \end{bmatrix} \begin{bmatrix} x_u \\ y_u \\ z_u \end{bmatrix} \tag{11.6}$$

which is the transpose of (11.3) for $\mathbf{q}\mathbf{p}\mathbf{q}^{-1}$.

11.3.1 Quaternion Products and Matrices

The second way to derive (11.3) depends upon representing a quaternion product in matrix form. For example, given

$$\begin{aligned} \mathbf{q}_1 &= s_1 + x_1\mathbf{i} + y_1\mathbf{j} + z_1\mathbf{k} \\ \mathbf{q}_2 &= s_2 + x_2\mathbf{i} + y_2\mathbf{j} + z_2\mathbf{k} \end{aligned}$$

their product is

$$\begin{aligned} \mathbf{q}_1\mathbf{q}_2 &= (s_1 + x_1\mathbf{i} + y_1\mathbf{j} + z_1\mathbf{k})(s_2 + x_2\mathbf{i} + y_2\mathbf{j} + z_2\mathbf{k}) \\ &= s_1s_2 - x_1x_2 - y_1y_2 - z_1z_2 \\ &\quad + s_1(x_2\mathbf{i} + y_2\mathbf{j} + z_2\mathbf{k}) \\ &\quad + s_2(x_1\mathbf{i} + y_1\mathbf{j} + z_1\mathbf{k}) \\ &\quad + (y_1z_2 - y_2z_1)\mathbf{i} + (x_2z_1 - x_1z_2)\mathbf{j} + (x_1y_2 - x_2y_1)\mathbf{k} \\ &= s_1s_2 - x_1x_2 - y_1y_2 - z_1z_2 \\ &\quad + (s_1x_2 + s_2x_1 + y_1z_2 - y_2z_1)\mathbf{i} \\ &\quad + (s_1y_2 + s_2y_1 + x_2z_1 - x_1z_2)\mathbf{j} \\ &\quad + (s_1z_2 + s_2z_1 + x_1y_2 - x_2y_1)\mathbf{k} \end{aligned}$$

$$\mathbf{q}_1\mathbf{q}_2 = \begin{bmatrix} s_1 & -x_1 & -y_1 & -z_1 \\ x_1 & s_1 & -z_1 & y_1 \\ y_1 & z_1 & s_1 & -x_1 \\ z_1 & -y_1 & x_1 & s_1 \end{bmatrix} \begin{bmatrix} s_2 \\ x_2 \\ y_2 \\ z_2 \end{bmatrix}.$$

At this stage we have quaternion $\mathbf{q}_1$ represented by a matrix, and $\mathbf{q}_2$ represented by a column vector. Now let's reverse the scenario without altering the result by making $\mathbf{q}_2$ the matrix and $\mathbf{q}_1$ the column vector:

$$\mathbf{q}_1\mathbf{q}_2 = \begin{bmatrix} s_2 & -x_2 & -y_2 & -z_2 \\ x_2 & s_2 & z_2 & -y_2 \\ y_2 & -z_2 & s_2 & x_2 \\ z_2 & y_2 & -x_2 & s_2 \end{bmatrix} \begin{bmatrix} s_1 \\ x_1 \\ y_1 \\ z_1 \end{bmatrix}.$$

So now we have two ways of computing $\mathbf{q}_1\mathbf{q}_2$ and we need a way of distinguishing between the two matrices. Let's call the matrix that preserves the left-to-right quaternion sequence L and the matrix that reverses the sequence to right-to-left, R:

$$\mathbf{q}_1\mathbf{q}_2 = L(\mathbf{q}_1)\mathbf{q}_2 = \begin{bmatrix} s_1 & -x_1 & -y_1 & -z_1 \\ x_1 & s_1 & -z_1 & y_1 \\ y_1 & z_1 & s_1 & -x_1 \\ z_1 & -y_1 & x_1 & s_1 \end{bmatrix} \begin{bmatrix} s_2 \\ x_2 \\ y_2 \\ z_2 \end{bmatrix}$$

$$\mathbf{q}_1\mathbf{q}_2 = R(\mathbf{q}_2)\mathbf{q}_1 = \begin{bmatrix} s_2 & -x_2 & -y_2 & -z_2 \\ x_2 & s_2 & z_2 & -y_2 \\ y_2 & -z_2 & s_2 & x_2 \\ z_2 & y_2 & -x_2 & s_2 \end{bmatrix} \begin{bmatrix} s_1 \\ x_1 \\ y_1 \\ z_1 \end{bmatrix}.$$

Remember that $L(\mathbf{q}_1)\mathbf{q}_2 = R(\mathbf{q}_2)\mathbf{q}_1$, as this is central to understanding the next stage. Furthermore, don't be surprised if you don't understand the logic of the argument in the first reading. It took the author many hours of anguish trying to decipher the original algorithm, and the explanation has been expanded to ensure that you do not suffer the same experience!

First, let's employ the matrices L and R to rearrange the quaternion triple product **acb** to **abc**: i.e. move **c** from the middle to the right-hand side.

We start with the quaternion triple product **acb** and divide it into two parts, **ac** and **b**. We can do this because quaternion algebra is associative:

$$\mathbf{acb} = (\mathbf{ac})\mathbf{b}.$$

We have already demonstrated above that the product **ac** can be replaced by $L(\mathbf{a})\mathbf{c}$:

$$\mathbf{acb} = L(\mathbf{a})\mathbf{cb}.$$

We now have another two parts: $L(\mathbf{a})\mathbf{c}$ and **b** which can be reversed using R without disturbing the result:

$$\mathbf{acb} = L(\mathbf{a})\mathbf{cb} = R(\mathbf{b})L(\mathbf{a})\mathbf{c}$$

which has achieved our objective to move **c** to the right-hand side.

Now let's repeat the same process to rearrange the triple product $\mathbf{qpq}^{-1}$. The objective is to remove **p** from the middle of **q** and $\mathbf{q}^{-1}$ and move it to the right hand side. The reason for doing this is to bring together **q** and $\mathbf{q}^{-1}$ in the form of two matrices, which can be multiplied together into a single matrix.

We start with the quaternion triple product $\mathbf{qpq}^{-1}$ and divide it into two parts, **qp** and $\mathbf{q}^{-1}$:

$$\mathbf{qpq}^{-1} = (\mathbf{qp})\mathbf{q}^{-1}.$$

The product **qp** can be replaced by $L(\mathbf{q})\mathbf{p}$:

$$\mathbf{qpq}^{-1} = L(\mathbf{q})\mathbf{pq}^{-1}.$$

We now have another two parts: $L(\mathbf{q})\mathbf{p}$ and $\mathbf{q}^{-1}$ which can be reversed using R without disturbing the result:

$$\mathbf{qpq}^{-1} = L(\mathbf{q})\mathbf{pq}^{-1} = R\big(\mathbf{q}^{-1}\big)L(\mathbf{q})\mathbf{p}$$

which has achieved our objective to move **p** to the right-hand side.

The next step is to compute $L(\mathbf{q})$ and $R(\mathbf{q}^{-1})$ using $\mathbf{q} = s + x\mathbf{i} + y\mathbf{j} + z\mathbf{k}$. $L(\mathbf{q})$ is easy as it is the same as $L(\mathbf{q}_1)$ without any subscripts:

$$L(\mathbf{q}) = \begin{bmatrix} s & -x & -y & -z \\ x & s & -z & y \\ y & z & s & -x \\ z & -y & x & s \end{bmatrix}.$$

$R(\mathbf{q}^{-1})$ is also easy, but requires converting $\mathbf{q}_2$ in the original definition into $\mathbf{q}_2^{-1}$ which is effected by reversing the signs of the vector components:

$$R(\mathbf{q}^{-1}) = \begin{bmatrix} s & x & y & z \\ -x & s & -z & y \\ -y & z & s & -x \\ -z & -y & x & s \end{bmatrix}.$$

So now we can write

$$\begin{aligned}
\mathbf{qpq}^{-1} &= R(\mathbf{q}^{-1})L(\mathbf{q})\mathbf{p} \\
&= \begin{bmatrix} s & x & y & z \\ -x & s & -z & y \\ -y & z & s & -x \\ -z & -y & x & s \end{bmatrix} \begin{bmatrix} s & -x & -y & -z \\ x & s & -z & y \\ y & z & s & -x \\ z & -y & x & s \end{bmatrix} \begin{bmatrix} 0 \\ x_u \\ y_u \\ z_u \end{bmatrix} \\
&= \begin{bmatrix} 1 & 0 & 0 & 0 \\ 0 & 1-2(y^2+z^2) & 2(xy-sz) & 2(xz+sy) \\ 0 & 2(xy+sz) & 1-2(x^2+z^2) & 2(yz-sx) \\ 0 & 2(xz-sy) & 2(yz+sx) & 1-2(x^2+y^2) \end{bmatrix} \begin{bmatrix} 0 \\ x_u \\ y_u \\ z_u \end{bmatrix}.
\end{aligned}$$

If we remove the first row and column and treat $\mathbf{p}$ as a vector, rather than a quaternion, we have

$$= \begin{bmatrix} 1-2(y^2+z^2) & 2(xy-sz) & 2(xz+sy) \\ 2(xy+sz) & 1-2(x^2+z^2) & 2(yz-sx) \\ 2(xz-sy) & 2(yz+sx) & 1-2(x^2+y^2) \end{bmatrix} \begin{bmatrix} x_u \\ y_u \\ z_u \end{bmatrix}$$

which is identical to (11.4)!

11.3.2 Geometric Verification

Let's illustrate the action of (11.3) by rotating the point (0, 1, 1), 90° about the y-axis, as shown in Fig. 11.6. The quaternion must take the form

$$\mathbf{q} = \cos(\theta/2) + \sin(\theta/2)\hat{\mathbf{v}}$$

which means that $\theta = 90°$ and $\hat{\mathbf{v}} = \mathbf{j}$, therefore,

$$\mathbf{q} = \cos 45° + \sin 45°\hat{\mathbf{j}}.$$

Consequently,

$$s = \frac{\sqrt{2}}{2}, \quad x = 0, \quad y = \frac{\sqrt{2}}{2}, \quad z = 0.$$

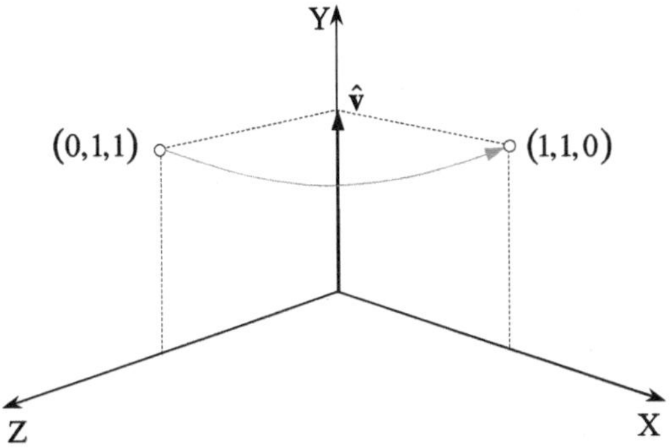

Fig. 11.6 The point $P(0,1,1)$ is rotated 90° to $P'(1,1,0)$ about the y-axis

Substituting these values in (11.3) gives

$$\begin{bmatrix} 1 \\ 1 \\ 0 \end{bmatrix} = \begin{bmatrix} 0 & 0 & 1 \\ 0 & 1 & 0 \\ -1 & 0 & 0 \end{bmatrix} \begin{bmatrix} 0 \\ 1 \\ 1 \end{bmatrix}$$

where $(0,1,1)$ is rotated to $(1,1,0)$, which is correct.

So now we have a transform that rotates a point about an arbitrary axis intersecting the origin without the problems of gimbal lock associated with Euler transforms.

Before moving on, let's evaluate one more example. Let's perform a 180° rotation about a vector $\mathbf{v} = \mathbf{i} + \mathbf{k}$ passing through the origin. To begin with, we will deliberately forget to convert the vector into a unit vector, just to see what happens to the final matrix. The quaternion should take the form

$$\mathbf{q} = \cos(\theta/2) + \sin(\theta/2)\hat{\mathbf{v}}$$

but we will use $\mathbf{v}$ as specified. Therefore, with $\theta = 180°$

$$s = 0, \quad x = 1, \quad y = 0, \quad z = 1.$$

Substituting these values in (11.3) gives

$$\begin{bmatrix} 1 & 0 & 2 \\ 0 & -1 & 0 \\ 2 & 0 & 1 \end{bmatrix}$$

which looks nothing like a rotation matrix, and reminds us how important it is to have a unit vector to represent the axis. Let's repeat these calculations normalising the vector to $\hat{\mathbf{v}} = \mathbf{i}/\sqrt{2} + \mathbf{k}/\sqrt{2}$:

$$s = 0, \quad x = \frac{1}{\sqrt{2}}, \quad y = 0, \quad z = \frac{1}{\sqrt{2}}.$$

Substituting these values in (11.3) gives

$$\begin{bmatrix} 0 & 0 & 1 \\ 0 & -1 & 0 \\ 1 & 0 & 0 \end{bmatrix}$$

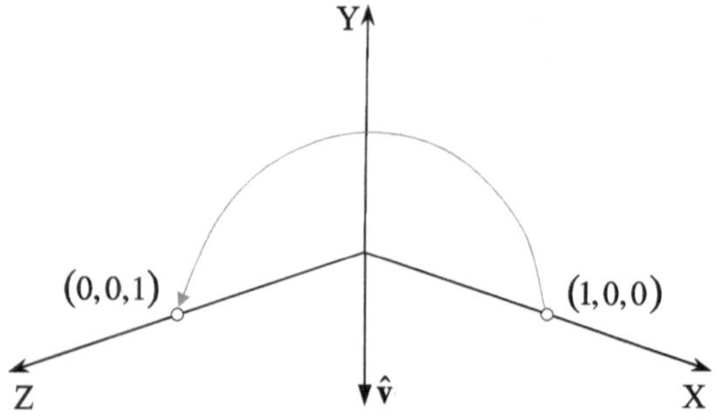

Fig. 11.7 The point $(1, 0, 0)$ is rotated $180°$ about the vector $\mathbf{v}$ to $(0, 0, 1)$

which not only looks like a rotation matrix, but has a determinant of 1 and rotates the point $(1, 0, 0)$ to $(0, 0, 1)$ as shown in Fig. 11.7,

$$\begin{bmatrix} 0 \\ 0 \\ 1 \end{bmatrix} = \begin{bmatrix} 0 & 0 & 1 \\ 0 & -1 & 0 \\ 1 & 0 & 0 \end{bmatrix} \begin{bmatrix} 1 \\ 0 \\ 0 \end{bmatrix}.$$

11.4 Multiple Rotations

Say a vector or frame of reference is subjected to two rotations specified by $\mathbf{q}_1$ followed by $\mathbf{q}_2$. There is a temptation to convert both quaternions to their respective matrix and multiply the matrices together. However, this not the most efficient way of combining the rotations. It is best to accumulate the rotations as quaternions and then convert to matrix notation, if required.

To illustrate this, consider the vector $\mathbf{p}$ subjected to the first quaternion $\mathbf{q}_1$:

$$\mathbf{q}_1\mathbf{p}\mathbf{q}_1^{-1}$$

followed by a second quaternion $\mathbf{q}_2$

$$\mathbf{q}_2\left(\mathbf{q}_1\mathbf{p}\mathbf{q}_1^{-1}\right)\mathbf{q}_2^{-1}$$

which can be expressed as

$$(\mathbf{q}_2\mathbf{q}_1)\mathbf{p}(\mathbf{q}_2\mathbf{q}_1)^{-1}.$$

Extra quaternions can be added accordingly. Let's illustrate this with an example.

To keep things simple, the first quaternion $\mathbf{q}_1$ rotates $30°$ about the y-axis:

$$\mathbf{q}_1 = \cos 15° + \sin 15°\mathbf{j}.$$

The second quaternion $\mathbf{q}_2$ rotates $60°$ also about the y-axis:

$$\mathbf{q}_1 = \cos 30° + \sin 30°\mathbf{j}.$$

Together, the two quaternions rotate $90°$ about the y-axis. To accumulate these rotations, we must multiply them together:

$$\begin{aligned}\mathbf{q}_1\mathbf{q}_2 &= (\cos 15^\circ + \sin 15^\circ \mathbf{j})(\cos 30^\circ + \sin 30^\circ \mathbf{j})\\ &= \cos 15^\circ \cos 30^\circ - \sin 15^\circ \sin 30^\circ + \cos 15^\circ \sin 30^\circ \mathbf{j} + \cos 30^\circ \sin 15^\circ \mathbf{j}\\ &= \frac{\sqrt{2}}{2} + \frac{\sqrt{2}}{2}\mathbf{j}\end{aligned}$$

which is a quaternion that rotates 90° about the y-axis. Using the matrix (11.4) we have

$$\begin{bmatrix} 0 & 0 & 1\\ 0 & 1 & 0\\ -1 & 0 & 0\end{bmatrix}$$

which rotates points about the y-axis by 90°.

11.5 Eigenvalue and Eigenvector

Although there is no doubt that (11.3) is a rotation matrix, we can secure further evidence by calculating its eigenvalue and eigenvector. The eigenvalue should be θ, where

$$\mathrm{Tr}(\mathbf{qpq}^{-1}) = 1 + 2\cos\theta.$$

The trace of (11.3) is

$$\begin{aligned}\mathrm{Tr}(\mathbf{qpq}^{-1}) &= 2(s^2 + x^2) - 1 + 2(s^2 + y^2) - 1 + 2(s^2 + z^2) - 1\\ &= 4s^2 + 2(s^2 + x^2 + y^2 + z^2) - 3\\ &= 4s^2 - 1\\ &= 4\cos^2(\theta/2) - 1\\ &= 4\cos\theta + 4\sin^2(\theta/2) - 1\\ &= 4\cos\theta + 2 - 2\cos\theta - 1\\ &= 1 + 2\cos\theta\end{aligned}$$

and

$$\cos\theta = \frac{1}{2}\big(\mathrm{Tr}(\mathbf{qpq}^{-1}) - 1\big).$$

To compute the eigenvector of (11.3) we use the three equations derived in Chap. 9:

$$\begin{aligned}v_1 &= (a_{22} - 1)(a_{33} - 1) - a_{23}a_{32}\\ v_2 &= (a_{33} - 1)(a_{11} - 1) - a_{31}a_{13}\\ v_3 &= (a_{11} - 1)(a_{22} - 1) - a_{12}a_{21}.\end{aligned}$$

Therefore,

$$\begin{aligned}
v_1 &= \big(2(s^2+y^2)-2\big)\big(2(s^2+z^2)-2\big)-2(yz-sx)2(yz+sx)\\
&=4(s^2+y^2-1)(s^2+z^2-1)-4(y^2z^2-s^2x^2)\\
&=4\big((x^2+z^2)(x^2+y^2)-y^2z^2+s^2x^2\big)\\
&=4(x^4+x^2y^2+x^2z^2+z^2y^2-y^2z^2+s^2x^2)\\
&=4x^2(s^2+x^2+y^2+z^2)\\
&=4x^2.
\end{aligned}$$

Similarly, $v_2=4y^2$ and $v_3=4z^2$, which confirms that the eigenvector has components associated with the quaternion's vector. The square terms should be no surprise, as the triple $\mathbf{qpq}^{-1}$ includes the product of two quaternions.

11.6 Rotating About an Off-Set Axis

Now that we have a matrix to represent a quaternion rotor, we can employ it to resolve problems such as rotating a point about an off-set axis using the same techniques associated with normal rotation transforms. For example, in Chap. 9 we used the following notation

$$\begin{bmatrix} x' \\ y' \\ z' \\ 1 \end{bmatrix} = \mathbf{T}_{t_x,0,t_z}\mathbf{R}_{\beta,y}\mathbf{T}_{-t_x,0,-t_z}\begin{bmatrix} x \\ y \\ z \\ 1 \end{bmatrix}$$

to rotate a point about a fixed axis parallel with the y-axis. Therefore, by substituting $\mathbf{qpq}^{-1}$ for $\mathbf{R}_{\beta,y}$ we have

$$\begin{bmatrix} x' \\ y' \\ z' \\ 1 \end{bmatrix} = \mathbf{T}_{t_x,0,t_z}\mathbf{qpq}^{-1}\mathbf{T}_{-t_x,0,-t_z}\begin{bmatrix} x \\ y \\ z \\ 1 \end{bmatrix}.$$

Let's test this by rotating our unit cube 90° about the vertical axis intersecting vertices 4 and 6 as shown in Fig. 11.8 (a) and (b).

The quaternion to achieve this is

$$\mathbf{q}=\cos 45^\circ+\sin 45^\circ\mathbf{j}$$

with the pure quaternion

$$\mathbf{p}=0+\mathbf{u}$$

and using (11.3) this creates the homogeneous matrix

$$\mathbf{qpq}^{-1}=\begin{bmatrix} 0 & 0 & 1 & 0 \\ 0 & 1 & 0 & 0 \\ -1 & 0 & 0 & 0 \\ 0 & 0 & 0 & 1 \end{bmatrix}\begin{bmatrix} x_u \\ y_u \\ z_u \\ 1 \end{bmatrix}.$$

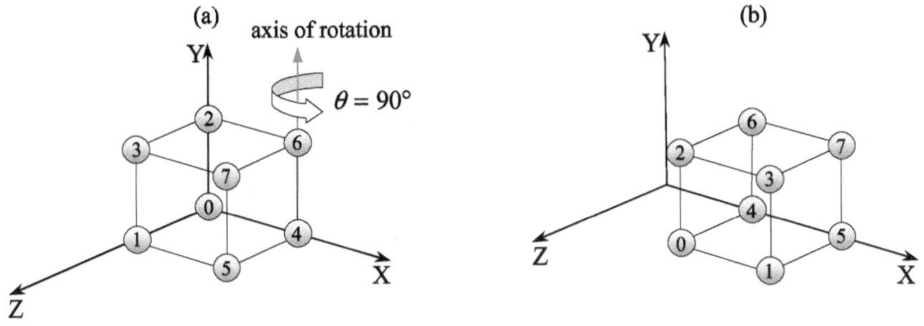

Fig. 11.8 The cube is rotated 90° about the axis intersecting vertices 4 and 6

The other two matrices are

$$\mathbf{T}_{-t_x,0,0} = \begin{bmatrix} 1 & 0 & 0 & -1 \\ 0 & 1 & 0 & 0 \\ 0 & 0 & 1 & 0 \\ 0 & 0 & 0 & 1 \end{bmatrix}$$

$$\mathbf{T}_{t_x,0,0} = \begin{bmatrix} 1 & 0 & 0 & 1 \\ 0 & 1 & 0 & 0 \\ 0 & 0 & 1 & 0 \\ 0 & 0 & 0 & 1 \end{bmatrix}.$$

Multiplying these matrices together creates

$$\begin{bmatrix} 0 & 0 & 1 & 1 \\ 0 & 1 & 0 & 0 \\ -1 & 0 & 0 & 1 \\ 0 & 0 & 0 & 1 \end{bmatrix}$$

which when applied to the cube's coordinates produces

$$\begin{bmatrix} 0 & 0 & 1 & 1 \\ 0 & 1 & 0 & 0 \\ -1 & 0 & 0 & 1 \\ 0 & 0 & 0 & 1 \end{bmatrix} \begin{bmatrix} 0 & 0 & 0 & 0 & 1 & 1 & 1 & 1 \\ 0 & 0 & 1 & 1 & 0 & 0 & 1 & 1 \\ 0 & 1 & 0 & 1 & 0 & 1 & 0 & 1 \\ 1 & 1 & 1 & 1 & 1 & 1 & 1 & 1 \end{bmatrix}$$

$$= \begin{bmatrix} 1 & 2 & 1 & 2 & 1 & 2 & 1 & 2 \\ 0 & 0 & 1 & 1 & 0 & 0 & 1 & 1 \\ 1 & 1 & 1 & 1 & 0 & 0 & 0 & 0 \\ 1 & 1 & 1 & 1 & 1 & 1 & 1 & 1 \end{bmatrix}.$$

These coordinates are confirmed by Fig. 11.8 (a) and (b).

11.7 Frames of Reference

Chapter 10 explored various techniques for changing the coordinates of objects in different frames of reference. Now that we have covered quaternions, and especially

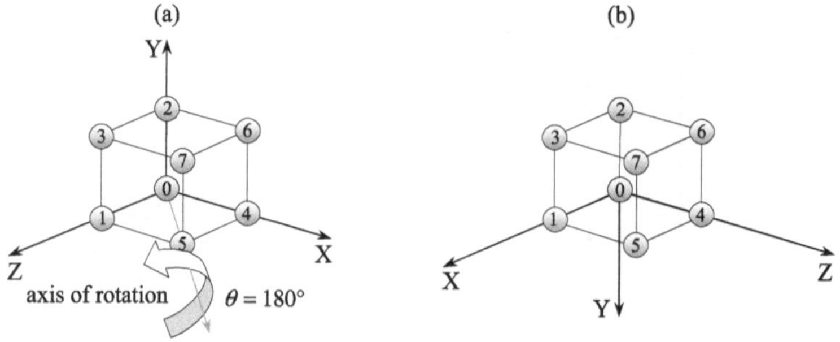

Fig. 11.9 The frame is rotated 180° about the vector $[\mathbf{i}+\mathbf{k}]$

the matrix representing the triple $\mathbf{q}^{-1}\mathbf{p}\mathbf{q}$, (11.5) we can show how quaternions can be added to these techniques.

The triple $\mathbf{q}\mathbf{p}\mathbf{q}^{-1}$ is used for rotating points about the vector associated with the quaternion $\mathbf{q}$, whereas the triple $\mathbf{q}^{-1}\mathbf{p}\mathbf{q}$ is used for rotating points about the same vector, but in the opposite direction. But we have already reasoned that this reverse rotation is equivalent to a change of frame of reference. To demonstrate this, consider the problem of rotating the frame of reference 180° about $\mathbf{i}+\mathbf{k}$ as shown in Fig. 11.9 (a) and (b). The unit quaternion for such a rotation is

$$\begin{aligned}\mathbf{q} &= \cos 90^\circ + \sin 90^\circ\left(\frac{1}{\sqrt{2}}\mathbf{i} + \frac{1}{\sqrt{2}}\mathbf{k}\right)\\ &= 0 + \frac{\sqrt{2}}{2}\mathbf{i} + \frac{\sqrt{2}}{2}\mathbf{k}.\end{aligned}$$

Consequently,

$$s = 0, \quad x = \frac{\sqrt{2}}{2}, \quad y = 0, \quad z = \frac{\sqrt{2}}{2}.$$

Substituting these values in (11.5) we obtain

$$\mathbf{q}^{-1}\mathbf{p}\mathbf{q} = \begin{bmatrix} 0 & 0 & 1\\ 0 & -1 & 0\\ 1 & 0 & 0\end{bmatrix}\begin{bmatrix} x_u\\ y_u\\ z_u\end{bmatrix}$$

which if used to process the coordinates of our unit cube produces

$$\begin{aligned}&\begin{bmatrix} 0 & 0 & 1\\ 0 & -1 & 0\\ 1 & 0 & 0\end{bmatrix}\begin{bmatrix} 0 & 0 & 0 & 0 & 1 & 1 & 1 & 1\\ 0 & 0 & 1 & 1 & 0 & 0 & 1 & 1\\ 0 & 1 & 0 & 1 & 0 & 1 & 0 & 1\end{bmatrix}\\ &= \begin{bmatrix} 0 & 1 & 0 & 1 & 0 & 1 & 0 & 1\\ 0 & 0 & -1 & -1 & 0 & 0 & -1 & -1\\ 0 & 0 & 0 & 0 & 1 & 1 & 1 & 1\end{bmatrix}.\end{aligned}$$

This scenario is shown in Fig. 11.9 (a) and (b).

Fig. 11.10 The point (0, 1, 1) is rotated 90° about the vector **v** to (1, 1, 0)

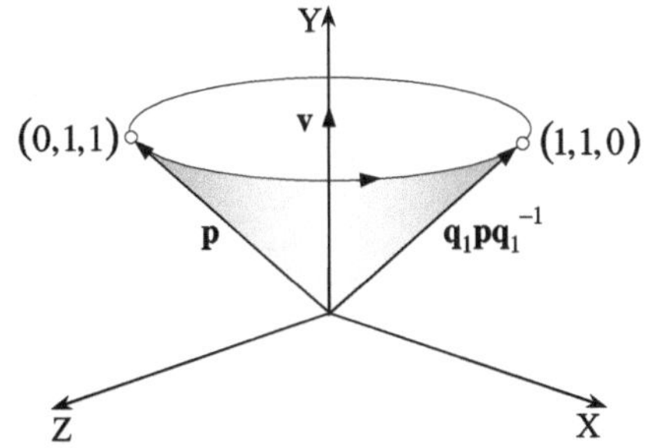

11.8 Interpolating Quaternions

Like vectors, quaternions can also be interpolated to compute an in-between quaternion. However, whereas two interpolated vectors results in a third vector that is readily visualised, two interpolated quaternions results in a third quaternion that acts as a rotor, and is not immediately visualised.

We have already seen that the spherical interpolant for vectors is

$$\mathbf{v} = \frac{\sin(1-t)\theta}{\sin\theta}\mathbf{v}_1 + \frac{\sin t\theta}{\sin\theta}\mathbf{v}_2$$

and requires no modification for quaternions:

$$\mathbf{q} = \frac{\sin(1-t)\theta}{\sin\theta}\mathbf{q}_1 + \frac{\sin t\theta}{\sin\theta}\mathbf{q}_2. \tag{11.7}$$

So, given

$$\mathbf{q}_1 = s_1 + x_1\mathbf{i} + y_1\mathbf{j} + z_1\mathbf{k}$$
$$\mathbf{q}_2 = s_2 + x_2\mathbf{i} + y_2\mathbf{j} + z_2\mathbf{k}$$

θ is obtained by taking the 4D dot product of $\mathbf{q}_1$ and $\mathbf{q}_2$:

$$\cos\theta = \frac{\mathbf{q}_1 \cdot \mathbf{q}_2}{|\mathbf{q}_1||\mathbf{q}_2|}$$
$$\cos\theta = \frac{s_1s_2 + x_1x_2 + y_1y_2 + z_1z_2}{|\mathbf{q}_1||\mathbf{q}_2|}$$

and if we are working with unit quaternions, then

$$\cos\theta = s_1s_2 + x_1x_2 + y_1y_2 + z_1z_2. \tag{11.8}$$

Let's use (11.7) in a scenario with two simple quaternions.

Figure 11.10 shows one such scenario where the point (0, 1, 1) is rotated 90° about **v**, the axis of $\mathbf{q}_1$. Figure 11.11 shows another scenario where the same point (0, 1, 1) is rotated 90° about **v**, the axis of $\mathbf{q}_2$. The quaternions are

$$\mathbf{q}_1 = \cos 45° + \sin 45°\mathbf{j} = \frac{\sqrt{2}}{2} + \frac{\sqrt{2}}{2}\mathbf{j}$$
$$\mathbf{q}_2 = \cos 45° + \sin 45°\mathbf{i} = \frac{\sqrt{2}}{2} + \frac{\sqrt{2}}{2}\mathbf{i}.$$

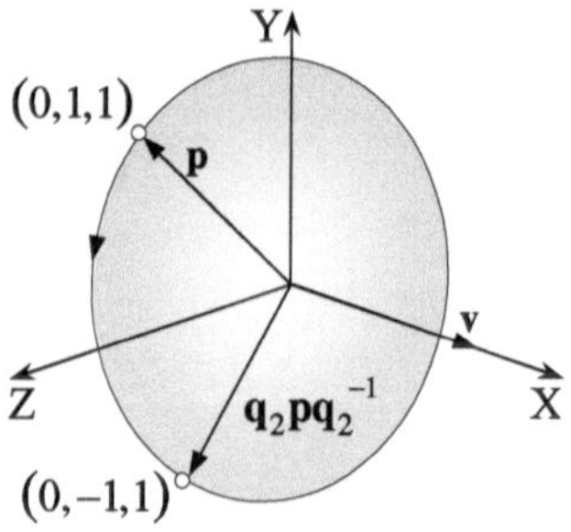

Fig. 11.11 The point (0, 1, 1) is rotated 90° about the vector **v** to (0, −1, 1)

Therefore, using (11.8)

$$\cos\theta = \frac{\sqrt{2}}{2}\frac{\sqrt{2}}{2} = 0.5$$
$$\theta = 60°.$$

Before proceeding, let's compute the two matrices for the two quaternion triples.

For $\mathbf{q}_1$

$$s = \frac{\sqrt{2}}{2}, \quad x = 0, \quad y = \frac{\sqrt{2}}{2}, \quad z = 0$$

which when substituted in (11.3) gives

$$\mathbf{q}_1\mathbf{p}\mathbf{q}_1^{-1} = \begin{bmatrix} 0 & 0 & 1 \\ 0 & 1 & 0 \\ -1 & 0 & 0 \end{bmatrix} \begin{bmatrix} x_u \\ y_u \\ z_u \end{bmatrix}.$$

Substituting the coordinates (0, 1, 1) gives

$$\begin{bmatrix} 1 \\ 1 \\ 0 \end{bmatrix} = \begin{bmatrix} 0 & 0 & 1 \\ 0 & 1 & 0 \\ -1 & 0 & 0 \end{bmatrix} \begin{bmatrix} 0 \\ 1 \\ 1 \end{bmatrix}$$

which is correct.

For $\mathbf{q}_2$

$$s = \frac{\sqrt{2}}{2}, \quad x = \frac{\sqrt{2}}{2}, \quad y = 0, \quad z = 0$$

which when substituted in (11.3) gives

$$\mathbf{q}_2\mathbf{p}\mathbf{q}_2^{-1} = \begin{bmatrix} 1 & 0 & 0 \\ 0 & 0 & -1 \\ 0 & 1 & 0 \end{bmatrix} \begin{bmatrix} x_u \\ y_u \\ z_u \end{bmatrix}.$$

Substituting the coordinates (0, 1, 1) gives

$$\begin{bmatrix} 0 \\ -1 \\ 1 \end{bmatrix} = \begin{bmatrix} 1 & 0 & 0 \\ 0 & 0 & -1 \\ 0 & 1 & 0 \end{bmatrix} \begin{bmatrix} 0 \\ 1 \\ 1 \end{bmatrix}$$

which is also correct.

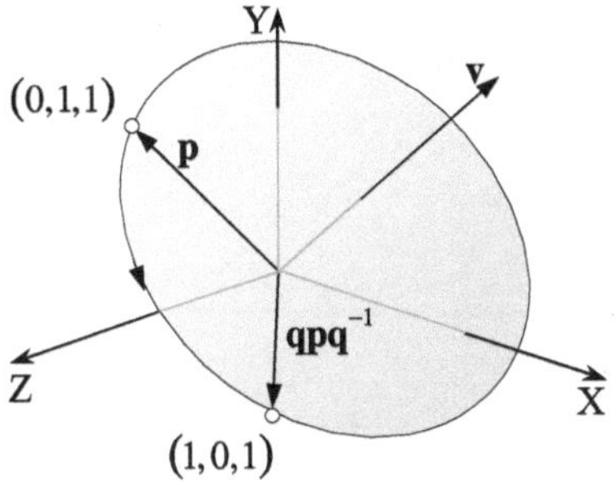

Fig. 11.12 The point (0, 1, 1) is rotated 90° about the vector **v** to (1, 0, 1)

Using (11.7) with $t = 0.5$ should compute a mid-way position for an interpolated quaternion, with its vector at 45° between the x- and y-axes, as shown in Fig. 11.12. We already know that $\theta = 60°$, therefore $\sin\theta = \sqrt{3}/2$, and using (11.7)

$$\begin{aligned}\mathbf{q} &= \frac{\sin\frac{1}{2}60°}{\sin 60°}\left(\frac{\sqrt{2}}{2}+\frac{\sqrt{2}}{2}\mathbf{j}\right)+\frac{\sin\frac{1}{2}60°}{\sin 60°}\left(\frac{\sqrt{2}}{2}+\frac{\sqrt{2}}{2}\mathbf{i}\right)\\ &= \frac{1}{\sqrt{3}}\left(\frac{\sqrt{2}}{2}+\frac{\sqrt{2}}{2}\mathbf{j}\right)+\frac{1}{\sqrt{3}}\left(\frac{\sqrt{2}}{2}+\frac{\sqrt{2}}{2}\mathbf{i}\right)\\ &= \frac{\sqrt{2}}{2\sqrt{3}}+\frac{\sqrt{2}}{2\sqrt{3}}\mathbf{j}+\frac{\sqrt{2}}{2\sqrt{3}}+\frac{\sqrt{2}}{2\sqrt{3}}\mathbf{i}\\ &= \frac{\sqrt{2}}{\sqrt{3}}+\frac{1}{\sqrt{6}}\mathbf{i}+\frac{1}{\sqrt{6}}\mathbf{j}\end{aligned}$$

therefore,

$$s = \frac{\sqrt{2}}{\sqrt{3}}, \quad x = \frac{1}{\sqrt{6}}, \quad y = \frac{1}{\sqrt{6}}, \quad z = 0$$

which when substituted in (11.3) gives

$$\mathbf{qpq}^{-1} = \begin{bmatrix} 1-(\frac{2}{6}) & 2(\frac{1}{6}) & 2(\frac{1}{3}) \\ 2(\frac{1}{6}) & 1-2(\frac{1}{6}) & 2(-\frac{1}{3}) \\ 2(-\frac{1}{3}) & 2(\frac{1}{3}) & 1-2(\frac{1}{6}+\frac{1}{6}) \end{bmatrix} \begin{bmatrix} x_u \\ y_u \\ z_u \end{bmatrix}$$

and

$$\mathbf{qpq}^{-1} = \begin{bmatrix} \frac{2}{3} & \frac{1}{3} & \frac{2}{3} \\ \frac{1}{3} & \frac{2}{3} & -\frac{2}{3} \\ -\frac{2}{3} & \frac{2}{3} & \frac{1}{3} \end{bmatrix} \begin{bmatrix} x_u \\ y_u \\ z_u \end{bmatrix}.$$

Substituting the coordinates (0, 1, 1) gives

$$\begin{bmatrix} 1 \\ 0 \\ 1 \end{bmatrix} = \begin{bmatrix} \frac{2}{3} & \frac{1}{3} & \frac{2}{3} \\ \frac{1}{3} & \frac{2}{3} & -\frac{2}{3} \\ -\frac{2}{3} & \frac{2}{3} & \frac{1}{3} \end{bmatrix} \begin{bmatrix} 0 \\ 1 \\ 1 \end{bmatrix} \tag{11.9}$$

which gives the point (1, 0, 1).

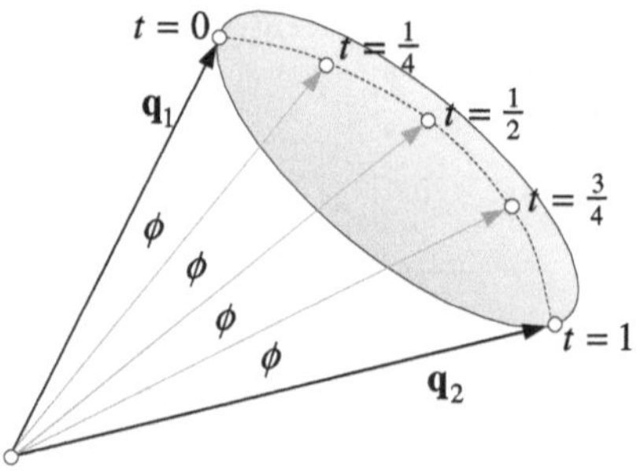

Fig. 11.13 Spherical interpolation between $\mathbf{q}_1$ and $\mathbf{q}_2$

One of the reasons for using this spherical interpolant is that it linearly interpolates the angle between the two quaternions, which creates a constant-angular velocity between the quaternions. However, one of the problems with visualising quaternions is that they reside in a four-dimensional space and create a hyper-sphere with a radius equal to the quaternion's magnitude. With our 3D brains, this is difficult to visualise. Nevertheless, we can convince ourselves into thinking we see what is going on with a simple sketch, as shown in Fig. 11.13, where we see part of the hyper-sphere and two quaternions $\mathbf{q}_1$ and $\mathbf{q}_2$. In this example, the angle ϕ is a constant angle between two values of the interpolant t. The spherical interpolant also ensures that the magnitude of the interpolated quaternion remains constant at unity and prevents any unwanted scaling.

Figure 11.14 provides another sketch to help visualise what is going on. For example, when $t = 0$, the interpolated quaternion is $\mathbf{q}_1$ which rotates the point $(0, 1, 1)$ to $(1, 1, 0)$, and when $t = 1$, the interpolated quaternion is $\mathbf{q}_2$ which rotates the point $(0, 1, 1)$ to $(0, -1, 1)$. When $t = 0.5$, the interpolated quaternion rotates the point $(0, 1, 1)$ to $(1, 0, 1)$ as computed above. Two other curves show what happens for $t = 0.25$ and $t = 0.75$.

A natural consequence of the interpolant is that the angle of rotation is 90° for $t = 0$ and $t = 1$, but for $t = 0.5$ the angle of rotation (eigenvalue) is approximately 70.5°. Corresponding angles arise for other values of t.

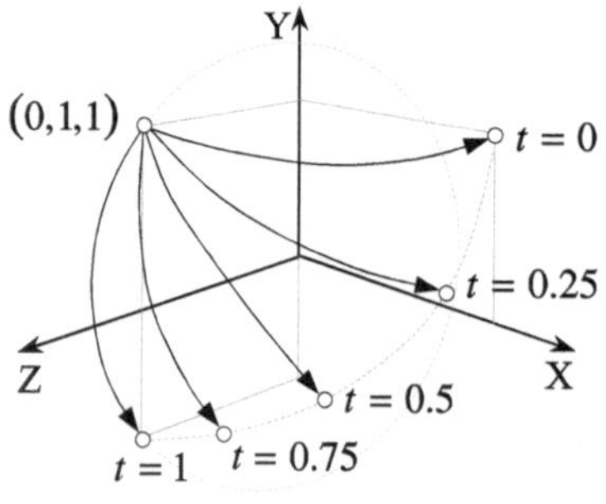

Fig. 11.14 Sketch showing the actions of the interpolated quaternions

11.9 Converting a Rotation Matrix to a Quaternion

The matrix transform equivalent to $\mathbf{qpq}^{-1}$ is

$$\mathbf{qpq}^{-1} = \begin{bmatrix} 2(s^2+x^2)-1 & 2(xy-sz) & 2(xz+sy) \\ 2(xy+sz) & 2(s^2+y^2)-1 & 2(yz-sx) \\ 2(xz-sy) & 2(yz+sx) & 2(s^2+z^2)-1 \end{bmatrix} \begin{bmatrix} x_u \\ y_u \\ z_u \end{bmatrix}$$
$$= \begin{bmatrix} a_{11} & a_{12} & a_{13} \\ a_{21} & a_{22} & a_{23} \\ a_{31} & a_{32} & a_{33} \end{bmatrix} \begin{bmatrix} x_u \\ y_u \\ z_u \end{bmatrix}.$$

Inspection of the matrix shows that by combining various elements we can isolate the terms of a quaternion s, x, y, z. For example, by adding the terms $a_{11}+a_{22}+a_{33}$ we obtain:

$$\begin{aligned} a_{11}+a_{22}+a_{33} &= \left(2\left(s^2+x^2\right)-1\right)+\left(2\left(s^2+y^2\right)-1\right)+\left(2\left(s^2+z^2\right)-1\right) \\ &= 6s^2+2\left(x^2+y^2+z^2\right)-3 \\ &= 4s^2-1 \end{aligned}$$

therefore,

$$s = \pm\frac{1}{2}\sqrt{1+a_{11}+a_{22}+a_{33}}.$$

To isolate x, y and z we use

$$\begin{aligned} x &= \frac{1}{4s}(a_{32}-a_{23}) \\ y &= \frac{1}{4s}(a_{13}-a_{31}) \\ z &= \frac{1}{4s}(a_{21}-a_{12}). \end{aligned}$$

We can confirm their accuracy using the matrix (11.9):

$$\begin{aligned} s &= \pm\frac{1}{2}\sqrt{1+\frac{2}{3}+\frac{2}{3}+\frac{1}{3}} = \frac{\sqrt{2}}{\sqrt{3}} \\ x &= \frac{\sqrt{3}}{4\sqrt{2}}\left(\frac{2}{3}+\frac{2}{3}\right) = \frac{1}{\sqrt{6}} \\ y &= \frac{\sqrt{3}}{4\sqrt{2}}\left(\frac{2}{3}+\frac{2}{3}\right) = \frac{1}{\sqrt{6}} \\ z &= \frac{\sqrt{3}}{4\sqrt{2}}\left(\frac{1}{3}-\frac{1}{3}\right) = 0 \end{aligned}$$

which agree with the original values.

Say, for example, the value of s had been close to zero, this could have made the values of x, y, z unreliable. Consequently, other combinations are available:

$$
\begin{aligned}
x &= \pm\frac{1}{2}\sqrt{1+a_{11}-a_{22}-a_{33}}\\
y &= \frac{1}{4x}(a_{12}+a_{21})\\
z &= \frac{1}{4x}(a_{13}+a_{31})\\
s &= \frac{1}{4x}(a_{32}-a_{23})
\end{aligned}
$$

$$
\begin{aligned}
y &= \pm\frac{1}{2}\sqrt{1-a_{11}+a_{22}-a_{33}}\\
x &= \frac{1}{4y}(a_{12}+a_{21})\\
z &= \frac{1}{4y}(a_{23}+a_{32})\\
s &= \frac{1}{4y}(a_{13}-a_{31})
\end{aligned}
$$

$$
\begin{aligned}
z &= \pm\frac{1}{2}\sqrt{1-a_{11}-a_{22}+a_{33}}\\
x &= \frac{1}{4z}(a_{13}+a_{31})\\
y &= \frac{1}{4z}(a_{23}+a_{32})\\
s &= \frac{1}{4z}(a_{21}-a_{12}).
\end{aligned}
$$

11.10 Summary

Quaternion algebra offers a simple and efficient way for computing rotations, but can also be evaluated in matrix form. We have also shown that it is possible to move between both forms of notation. It is left to the reader to code up some of these ideas and explore issues of accuracy and efficiency.

11.10.1 Summary of Quaternion Transforms

Given

$$
\begin{aligned}
\mathbf{q} &= s+\hat{\mathbf{v}} = \cos(\theta/2)+\sin(\theta/2)(x\mathbf{i}+y\mathbf{j}+z\mathbf{k})\\
\mathbf{p} &= 0+\mathbf{u}.
\end{aligned}
$$

Rotating a point about a vector

$$\mathbf{qpq}^{-1} = (1-\cos\theta)(\hat{\mathbf{v}}\cdot\mathbf{u})\hat{\mathbf{v}} + \cos\theta\mathbf{u} + \sin\theta\hat{\mathbf{v}}\times\mathbf{u}.$$

Rotating a frame about a vector

$$\mathbf{q}^{-1}\mathbf{pq} = (1-\cos\theta)(\hat{\mathbf{v}}\cdot\mathbf{u})\hat{\mathbf{v}} + \cos\theta\mathbf{u} - \sin\theta\hat{\mathbf{v}}\times\mathbf{u}.$$

Matrix for rotating a point about a vector

$$\mathbf{qpq}^{-1} = \begin{bmatrix} 1-2(y^2+z^2) & 2(xy-sz) & 2(xz+sy) \\ 2(xy+sz) & 1-2(x^2+z^2) & 2(yz-sx) \\ 2(xz-sy) & 2(yz+sx) & 1-2(x^2+y^2) \end{bmatrix} \begin{bmatrix} x_u \\ y_u \\ z_u \end{bmatrix}.$$

Matrix for rotating a frame about a vector

$$\mathbf{q}^{-1}\mathbf{pq} = \begin{bmatrix} 1-2(y^2+z^2) & 2(xy+sz) & 2(xz-sy) \\ 2(xy-sz) & 1-2(x^2+z^2) & 2(yz+sx) \\ 2(xz+sy) & 2(yz-sx) & 1-2(x^2+y^2) \end{bmatrix} \begin{bmatrix} x_u \\ y_u \\ z_u \end{bmatrix}.$$

Matrix for a quaternion product

$$\mathbf{q}_1\mathbf{q}_2 = L(\mathbf{q}_1)\mathbf{q}_2 = \begin{bmatrix} s_1 & -x_1 & -y_1 & -z_1 \\ x_1 & s_1 & -z_1 & y_1 \\ y_1 & z_1 & s_1 & -x_1 \\ z_1 & -y_1 & x_1 & s_1 \end{bmatrix} \begin{bmatrix} s_2 \\ x_2 \\ y_2 \\ z_2 \end{bmatrix}$$

$$\mathbf{q}_1\mathbf{q}_2 = R(\mathbf{q}_2)\mathbf{q}_1 = \begin{bmatrix} s_2 & -x_2 & -y_2 & -z_2 \\ x_2 & s_2 & z_2 & -y_2 \\ y_2 & -z_2 & s_2 & x_2 \\ z_2 & y_2 & -x_2 & s_2 \end{bmatrix} \begin{bmatrix} s_1 \\ x_1 \\ y_1 \\ z_1 \end{bmatrix}.$$

Interpolating two quaternions

$$\mathbf{q} = \frac{\sin(1-t)\theta}{\sin\theta}\mathbf{q}_1 + \frac{\sin t\theta}{\sin\theta}\mathbf{q}_2$$

where

$$\cos\theta = \frac{\mathbf{q}_1\cdot\mathbf{q}_2}{|\mathbf{q}_1||\mathbf{q}_2|}$$

$$\cos\theta = \frac{s_1s_2 + x_1x_2 + y_1y_2 + z_1z_2}{|\mathbf{q}_1||\mathbf{q}_2|}.$$

Quaternion from a rotation matrix

$$s = \pm\frac{1}{2}\sqrt{1+a_{11}+a_{22}+a_{33}}$$

$$x = \frac{1}{4s}(a_{32}-a_{23})$$

$$y = \frac{1}{4s}(a_{13}-a_{31})$$

$$z = \frac{1}{4s}(a_{21}-a_{12})$$

$$
\begin{aligned}
x &= \pm\frac{1}{2}\sqrt{1+a_{11}-a_{22}-a_{33}} \\
y &= \frac{1}{4x}(a_{12}+a_{21}) \\
z &= \frac{1}{4x}(a_{13}+a_{31}) \\
s &= \frac{1}{4x}(a_{32}-a_{23})
\end{aligned}
$$

$$
\begin{aligned}
y &= \pm\frac{1}{2}\sqrt{1-a_{11}+a_{22}-a_{33}} \\
x &= \frac{1}{4y}(a_{12}+a_{21}) \\
z &= \frac{1}{4y}(a_{23}+a_{32}) \\
s &= \frac{1}{4y}(a_{13}-a_{31})
\end{aligned}
$$

$$
\begin{aligned}
z &= \pm\frac{1}{2}\sqrt{1-a_{11}-a_{22}+a_{33}} \\
x &= \frac{1}{4z}(a_{13}+a_{31}) \\
y &= \frac{1}{4z}(a_{23}+a_{32}) \\
s &= \frac{1}{4z}(a_{21}-a_{12}).
\end{aligned}
$$

Chapter 12
Bivector Rotors

12.1 Introduction

In Chap. 6 we explored multivectors, and in Chap. 11 we saw how quaternions are used to rotate points and frames of reference about an arbitrary vector. In this chapter we will see how these two ideas merge into one to form bivector rotors. In order to show how such rotors operate, we begin with reflections and show how these can effect a rotation.

12.2 The Three Reflections Theorem

The three reflections theorem states that '*each isometry of the Euclidean plane is the composite of one, two, or three reflections.*' To begin with, an *isometry of the Euclidean plane* is a way of transforming the plane that preserves length. Such isometries include rotation, translation, reflection and glide reflections. The latter is a combination of a reflection in a line and a translation along that line. John Stillwell provides an elegant proof for this theorem in his book *Numbers and Geometry* [8].

The isometry we are particularly interested in is reflection, where the distance between two points is preserved in their reflection. Consider, for example, the 2D scenario shown in Fig. 12.1 where two lines M and N are imaginary mirrors separated by an angle θ. The real point P subtends an angle α to mirror M and creates a virtual image P_R which subtends an equal but opposite angle.

Although it is not physically possible, we can imagine that the virtual image P_R is reflected in the second mirror N. To begin with, P_R subtends an angle $\theta - \alpha$ to mirror N and creates another virtual image P' which subtends an equal and opposite angle. What is interesting about this configuration is that although the mirrors are separated by θ, the angle between P and P' is 2θ. In order to take advantage of this effect we need to know how vectors are reflected using multivectors, which is the subject of the next section.

J. Vince, *Rotation Transforms for Computer Graphics*,
DOI 10.1007/978-0-85729-154-7_12,

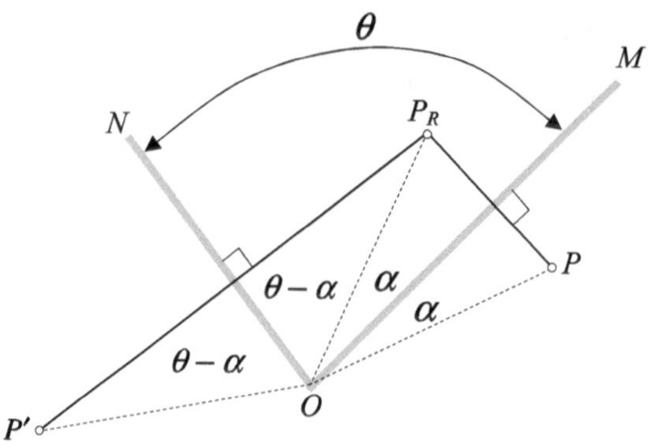

Fig. 12.1 Rotating a point by a double reflection

12.3 Reflecting a Vector

Figure 12.2 shows a mirror with a unit normal vector $\hat{\mathbf{n}}$ and a vector $\mathbf{a}$ with its reflection $\mathbf{a}'$. Vector $\mathbf{a}$ has a perpendicular component $\mathbf{a}_\perp$ and a parallel component $\mathbf{a}_\parallel$ with $\hat{\mathbf{n}}$, and our objective is to derive a definition of the reflection $\mathbf{a}'$ in terms of vector $\mathbf{a}$ and any other essential vectors.

From our knowledge of multivectors, we know that $\hat{\mathbf{n}}^2 = 1$ which permits us to write

$$\mathbf{a} = \hat{\mathbf{n}}^2\mathbf{a} = \hat{\mathbf{n}}(\hat{\mathbf{n}}\mathbf{a}).$$

This has created the geometric product $\hat{\mathbf{n}}\mathbf{a}$ which equals

$$\hat{\mathbf{n}}\mathbf{a} = \hat{\mathbf{n}} \cdot \mathbf{a} + \hat{\mathbf{n}} \wedge \mathbf{a} \tag{12.1}$$

therefore,

$$\mathbf{a} = \hat{\mathbf{n}}(\hat{\mathbf{n}} \cdot \mathbf{a} + \hat{\mathbf{n}} \wedge \mathbf{a}). \tag{12.2}$$

We can see that (12.2) has two parts: $\hat{\mathbf{n}}(\hat{\mathbf{n}} \cdot \mathbf{a})$ and $\hat{\mathbf{n}}(\hat{\mathbf{n}} \wedge \mathbf{a})$. The first part is another way of expressing $\mathbf{a}_\parallel$:

$$\mathbf{a}_\parallel = (\hat{\mathbf{n}} \cdot \mathbf{a})\hat{\mathbf{n}}$$

and as

$$\mathbf{a} = \mathbf{a}_\perp + \mathbf{a}_\parallel$$

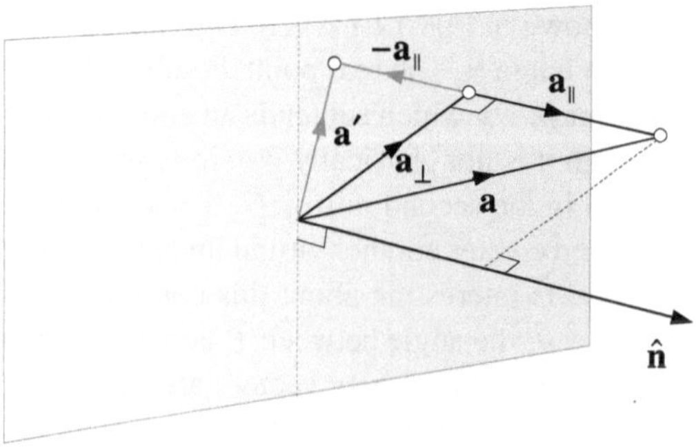

Fig. 12.2 Reflecting a vector in a mirror

the second part must be

$$\mathbf{a}_\perp = \hat{\mathbf{n}}(\hat{\mathbf{n}} \wedge \mathbf{a}).$$

From Fig. 12.2 we see that

$$\mathbf{a}' = \mathbf{a}_\perp - \mathbf{a}_\parallel \tag{12.3}$$

$$= \hat{\mathbf{n}}(\hat{\mathbf{n}} \wedge \mathbf{a}) - (\hat{\mathbf{n}} \cdot \mathbf{a})\hat{\mathbf{n}}. \tag{12.4}$$

Equation (12.4) contains the product of a vector $\hat{\mathbf{n}}$ and a bivector $\hat{\mathbf{n}} \wedge \hat{\mathbf{a}}$ which anti-commute:

$$\begin{aligned}\hat{\mathbf{n}}(\hat{\mathbf{n}} \wedge \mathbf{a}) &= \frac{\hat{\mathbf{n}}}{2}(\hat{\mathbf{n}}\mathbf{a} - \mathbf{a}\hat{\mathbf{n}}) \\ &= \frac{1}{2}(\mathbf{a} - \hat{\mathbf{n}}\mathbf{a}\hat{\mathbf{n}})\end{aligned}$$

whereas,

$$\begin{aligned}(\hat{\mathbf{n}} \wedge \mathbf{a})\hat{\mathbf{n}} &= \frac{1}{2}(\hat{\mathbf{n}}\mathbf{a} - \mathbf{a}\hat{\mathbf{n}})\hat{\mathbf{n}} \\ &= \frac{1}{2}(\hat{\mathbf{n}}\mathbf{a}\hat{\mathbf{n}} - \mathbf{a})\end{aligned}$$

therefore, we can write (12.4) as

$$\mathbf{a}' = -(\hat{\mathbf{n}} \cdot \mathbf{a})\hat{\mathbf{n}} - (\hat{\mathbf{n}} \wedge \mathbf{a})\hat{\mathbf{n}}$$

which simplifies to

$$\mathbf{a}' = -(\hat{\mathbf{n}} \cdot \mathbf{a} + \hat{\mathbf{n}} \wedge \mathbf{a})\hat{\mathbf{n}}. \tag{12.5}$$

By substituting (12.1) in (12.5) we have

$$\mathbf{a}' = -\hat{\mathbf{n}}\mathbf{a}\hat{\mathbf{n}} \tag{12.6}$$

which is rather elegant!

To illustrate (12.6), consider the scenario shown in Fig. 12.3 where we see a mirror placed on the zx-plane with normal vector $\mathbf{j}$ or $\mathbf{e}_2$. The vector to be reflected is

$$\mathbf{a} = \mathbf{i} + \mathbf{j} - \mathbf{k}$$

which can also be expressed as

$$\mathbf{a} = \mathbf{e}_1 + \mathbf{e}_2 - \mathbf{e}_3.$$

Using (12.6) we have

$$\mathbf{a}' = -\mathbf{e}_2(\mathbf{e}_1 + \mathbf{e}_2 - \mathbf{e}_3)\mathbf{e}_2$$

which, using the rules of multivectors simplifies to

$$\begin{aligned}\mathbf{a}' &= -\mathbf{e}_2\mathbf{e}_1\mathbf{e}_2 - \mathbf{e}_2\mathbf{e}_2\mathbf{e}_2 + \mathbf{e}_2\mathbf{e}_3\mathbf{e}_2 \\ &= \mathbf{e}_1 - \mathbf{e}_2 - \mathbf{e}_3 \\ &= \mathbf{i} - \mathbf{j} - \mathbf{k}\end{aligned}$$

and is confirmed by Fig. 12.3. Now let's see how these ideas can be generalised into 3D rotations.

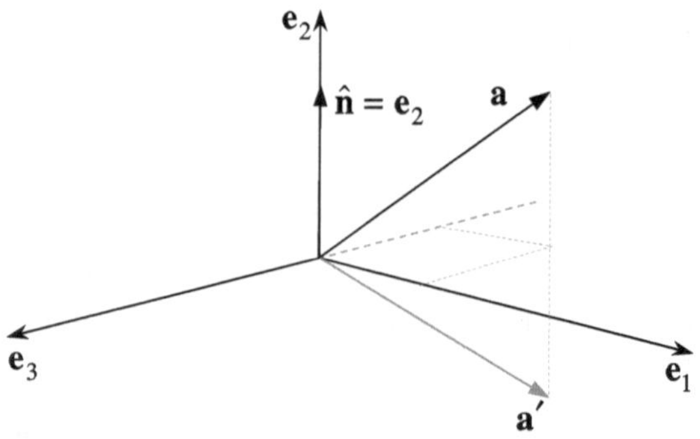

Fig. 12.3 Reflecting a vector in a mirror

12.4 3D Rotations

Figure 12.4 shows a plan view of two mirrors M and N with their respective unit normal vectors $\hat{\mathbf{m}}$ and $\hat{\mathbf{n}}$ separated by an angle θ. The plane containing $\hat{\mathbf{m}}$ and $\hat{\mathbf{n}}$ is readily defined by their wedge product $\hat{\mathbf{m}} \wedge \hat{\mathbf{n}}$. Using (12.6) we can compute vector **a**'s reflection by

$$\mathbf{b} = -\hat{\mathbf{m}}\mathbf{a}\hat{\mathbf{m}} \tag{12.7}$$

and **b**'s reflection by

$$\mathbf{a}' = -\hat{\mathbf{n}}\mathbf{b}\hat{\mathbf{n}}. \tag{12.8}$$

Substituting (12.7) in (12.8) we obtain

$$\mathbf{a}' = \hat{\mathbf{n}}\hat{\mathbf{m}}\mathbf{a}\hat{\mathbf{m}}\hat{\mathbf{n}}, \tag{12.9}$$

which is extremely compact. However, we must remember that **a** is rotated twice the angle separating the mirrors, i.e. 2θ. Within geometric algebra $\hat{\mathbf{n}}\hat{\mathbf{m}}$ is called a *rotor*, and is represented by **R**, which means that using the reverse operation † we can write (12.9) as

$$\mathbf{a}' = \mathbf{R}\mathbf{a}\mathbf{R}^\dagger \tag{12.10}$$

which reminds us of the way quaternions work.

To illustrate the action of (12.10) consider the 2D scenario shown in Fig. 12.5 with two mirrors M and N and their unit normal vectors $\hat{\mathbf{m}}$, $\hat{\mathbf{n}}$ and position vector **p**:

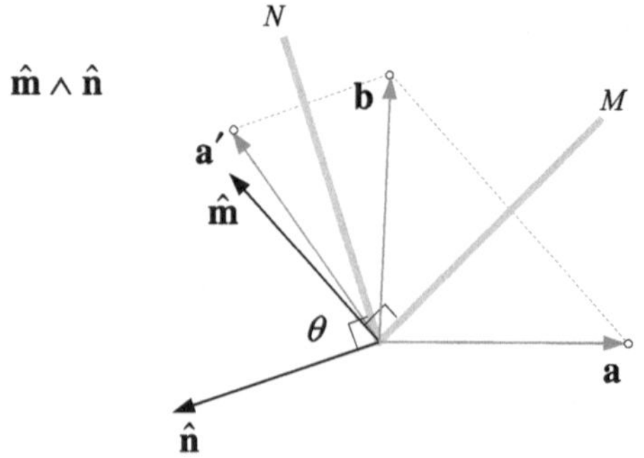

Fig. 12.4 Rotating a point by a double reflection

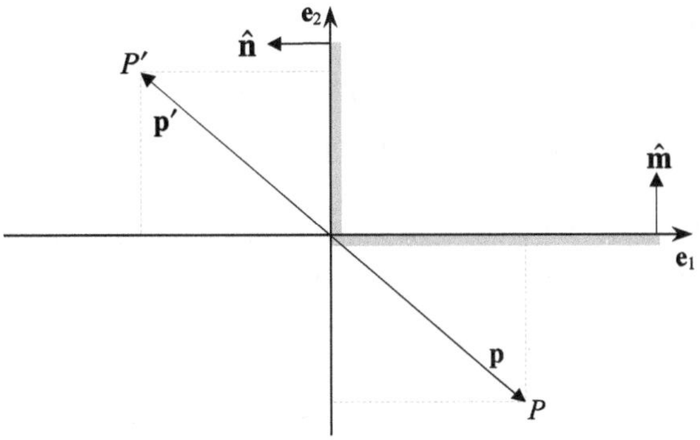

Fig. 12.5 Rotating a point by 180°

$$\hat{\mathbf{m}} = \mathbf{e}_2$$
$$\hat{\mathbf{n}} = -\mathbf{e}_1$$
$$\mathbf{p} = \mathbf{e}_1 - \mathbf{e}_2.$$

As the mirrors are separated by 90° the point P is rotated 180°:

$$\begin{aligned}\mathbf{p}' &= \hat{\mathbf{n}}\hat{\mathbf{m}}\mathbf{p}\hat{\mathbf{m}}\hat{\mathbf{n}} \\ &= -\mathbf{e}_1\mathbf{e}_2(\mathbf{e}_1 - \mathbf{e}_2)\mathbf{e}_2(-\mathbf{e}_1) \\ &= \mathbf{e}_{12121} - \mathbf{e}_{12221} \\ \mathbf{p}' &= -\mathbf{e}_1 + \mathbf{e}_2.\end{aligned}$$

Let's now define a rotor in terms of its bivector and the actual angle a vector is rotated as follows. The bivector defining the plane is $\hat{\mathbf{m}} \wedge \hat{\mathbf{n}}$ and θ is the rotor angle, which means that the bivector angle is $\theta/2$. Let

$$\mathbf{R}_\theta = \hat{\mathbf{n}}\hat{\mathbf{m}}$$
$$\mathbf{R}_\theta^\dagger = \hat{\mathbf{m}}\hat{\mathbf{n}}$$

where

$$\begin{aligned}\hat{\mathbf{n}}\hat{\mathbf{m}} &= \hat{\mathbf{n}} \cdot \hat{\mathbf{m}} - \hat{\mathbf{m}} \wedge \hat{\mathbf{n}} \\ \hat{\mathbf{m}}\hat{\mathbf{n}} &= \hat{\mathbf{n}} \cdot \hat{\mathbf{m}} + \hat{\mathbf{m}} \wedge \hat{\mathbf{n}} \\ \hat{\mathbf{n}} \cdot \hat{\mathbf{m}} &= \cos(\theta/2) \\ \hat{\mathbf{m}} \wedge \hat{\mathbf{n}} &= \sin(\theta/2)\hat{\mathbf{B}}.\end{aligned}$$

Therefore,

$$\mathbf{R}_\theta = \cos(\theta/2) - \sin(\theta/2)\hat{\mathbf{B}}$$
$$\mathbf{R}_\theta^\dagger = \cos(\theta/2) + \sin(\theta/2)\hat{\mathbf{B}}.$$

We now have an equation that rotates a vector $\mathbf{p}$ through an angle θ about an axis defined by $\hat{\mathbf{B}}$:

$$\mathbf{p}' = \mathbf{R}_\theta\mathbf{p}\mathbf{R}_\theta^\dagger$$

or

$$\mathbf{p}' = \big(\cos(\theta/2) - \sin(\theta/2)\hat{\mathbf{B}}\big)\mathbf{p}\big(\cos(\theta/2) + \sin(\theta/2)\hat{\mathbf{B}}\big). \tag{12.11}$$

Let's demonstrate how (12.11) works with two examples.

Example 1 Figure 12.6 shows a scenario where vector $\mathbf{p}$ is to be rotated 90° about $\mathbf{e}_2$ which is perpendicular to $\hat{\mathbf{B}}$, where

$$\theta = 90^\circ, \quad \mathbf{a} = \mathbf{e}_2, \quad \mathbf{p} = \mathbf{e}_1 + \mathbf{e}_2, \quad \hat{\mathbf{B}} = \mathbf{e}_{31}.$$

Therefore,

$$\begin{aligned}
\mathbf{p}' &= \big(\cos 45^\circ - \sin 45^\circ \mathbf{e}_{31}\big)(\mathbf{e}_1 + \mathbf{e}_2)\big(\cos 45^\circ + \sin 45^\circ \mathbf{e}_{31}\big)\\
&= \left(\frac{\sqrt{2}}{2} - \frac{\sqrt{2}}{2}\mathbf{e}_{31}\right)(\mathbf{e}_1 + \mathbf{e}_2)\left(\frac{\sqrt{2}}{2} + \frac{\sqrt{2}}{2}\mathbf{e}_{31}\right)\\
&= \left(\frac{\sqrt{2}}{2}\mathbf{e}_1 + \frac{\sqrt{2}}{2}\mathbf{e}_2 - \frac{\sqrt{2}}{2}\mathbf{e}_3 - \frac{\sqrt{2}}{2}\mathbf{e}_{312}\right)\left(\frac{\sqrt{2}}{2} + \frac{\sqrt{2}}{2}\mathbf{e}_{31}\right)\\
&= \frac{1}{2}(\mathbf{e}_1 - \mathbf{e}_3 + \mathbf{e}_2 + \mathbf{e}_{231} - \mathbf{e}_3 - \mathbf{e}_1 - \mathbf{e}_{312} - \mathbf{e}_{31231})\\
&= \mathbf{e}_2 - \mathbf{e}_3
\end{aligned}$$

which is correct.

Observe what happens when the bivector's sign is reversed to $-\mathbf{e}_{31}$:

$$\begin{aligned}
\mathbf{p}' &= \big(\cos 45^\circ + \sin 45^\circ \mathbf{e}_{31}\big)(\mathbf{e}_1 + \mathbf{e}_2)\big(\cos 45^\circ - \sin 45^\circ \mathbf{e}_{31}\big)\\
&= \frac{1}{2}(1 + \mathbf{e}_{31})(\mathbf{e}_1 + \mathbf{e}_2)(1 - \mathbf{e}_{31})\\
&= \frac{1}{2}(\mathbf{e}_1 + \mathbf{e}_2 + \mathbf{e}_3 + \mathbf{e}_{312})(1 - \mathbf{e}_{31})
\end{aligned}$$

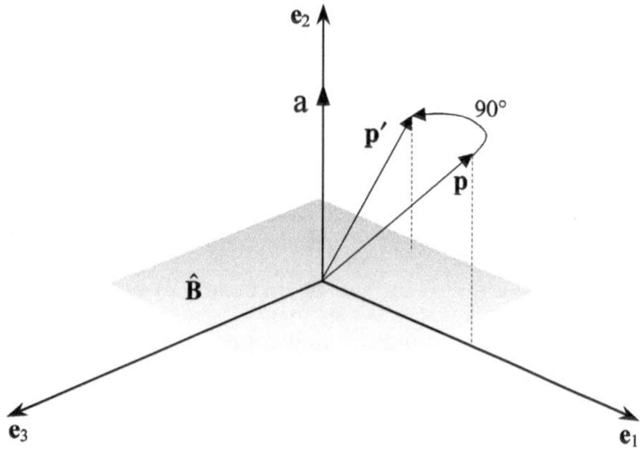

Fig. 12.6 Rotating a vector by 90°

Fig. 12.7 Rotating a vector by 120°

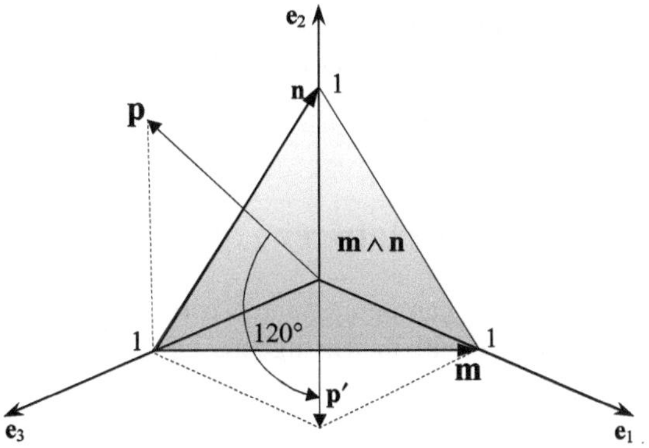

$$
\begin{aligned}
&= \frac{1}{2}(\mathbf{e}_1 + \mathbf{e}_3 + \mathbf{e}_2 + \mathbf{e}_{231} + \mathbf{e}_3 - \mathbf{e}_1 + \mathbf{e}_{312} - \mathbf{e}_{31231}) \\
\mathbf{p}' &= \mathbf{e}_2 + \mathbf{e}_3
\end{aligned}
$$

the rotation is clockwise about $\mathbf{e}_2$.

Example 2 Figure 12.7 shows a scenario where vector $\mathbf{p}$ is to be rotated 120° about the bivector $\mathbf{B}$, where

$$
\mathbf{m} = \mathbf{e}_1 - \mathbf{e}_3, \quad \mathbf{n} = \mathbf{e}_2 - \mathbf{e}_3, \quad \theta = 120^\circ, \quad \mathbf{p} = \mathbf{e}_2 + \mathbf{e}_3.
$$

First, we compute the bivector:

$$
\begin{aligned}
\mathbf{B} &= \mathbf{m} \wedge \mathbf{n} \\
&= (\mathbf{e}_1 - \mathbf{e}_3) \wedge (\mathbf{e}_2 - \mathbf{e}_3) \\
&= \mathbf{e}_{12} + \mathbf{e}_{23} + \mathbf{e}_{31}.
\end{aligned}
$$

Next, we normalise $\mathbf{B}$ to $\hat{\mathbf{B}}$:

$$
\hat{\mathbf{B}} = \frac{1}{\sqrt{3}}(\mathbf{e}_{12} + \mathbf{e}_{23} + \mathbf{e}_{31})
$$

and

$$
\begin{aligned}
\mathbf{p}' &= \big(\cos 60^\circ - \sin 60^\circ \hat{\mathbf{B}}\big)\mathbf{p}\big(\cos 60^\circ + \sin 60^\circ \hat{\mathbf{B}}\big) \\
&= \left(\frac{1}{2} - \frac{\sqrt{3}}{2}\frac{1}{\sqrt{3}}(\mathbf{e}_{12} + \mathbf{e}_{23} + \mathbf{e}_{31})\right)(\mathbf{e}_2 + \mathbf{e}_3)\left(\frac{1}{2} + \frac{\sqrt{3}}{2}\frac{1}{\sqrt{3}}(\mathbf{e}_{12} + \mathbf{e}_{23} + \mathbf{e}_{31})\right) \\
&= \left(\frac{1}{2} - \frac{\mathbf{e}_{12}}{2} - \frac{\mathbf{e}_{23}}{2} - \frac{\mathbf{e}_{31}}{2}\right)(\mathbf{e}_2 + \mathbf{e}_3)\left(\frac{1}{2} + \frac{\mathbf{e}_{12}}{2} + \frac{\mathbf{e}_{23}}{2} + \frac{\mathbf{e}_{31}}{2}\right) \\
&= \frac{1}{4}(\mathbf{e}_2 + \mathbf{e}_3 - \mathbf{e}_1 - \mathbf{e}_{123} + \mathbf{e}_3 - \mathbf{e}_2 - \mathbf{e}_{312} + \mathbf{e}_1)(1 + \mathbf{e}_{12} + \mathbf{e}_{23} + \mathbf{e}_{31}) \\
&= \frac{1}{2}(\mathbf{e}_3 - \mathbf{e}_{123})(1 + \mathbf{e}_{12} + \mathbf{e}_{23} + \mathbf{e}_{31}) \\
&= \frac{1}{2}(\mathbf{e}_3 - \mathbf{e}_2 + \mathbf{e}_1 + \mathbf{e}_3 + \mathbf{e}_1 + \mathbf{e}_2) \\
&= \mathbf{e}_1 + \mathbf{e}_3.
\end{aligned}
$$

These examples show that rotors behave just like quaternions. Rotors not only rotate vectors, but they can be used to rotate any multivector, irrespective of their dimension.

12.5 Rotors as Matrices

Although rotors can be computed using geometric algebra, there is a one-to-one correspondence with matrix algebra, which we will now demonstrate.

12.5.1 2D Rotor

To begin with we will show that a 2D rotor is nothing more that a 2×2 matrix in disguise for rotating a point 2θ about the origin.

Given

$$\begin{aligned}\hat{\mathbf{m}} &= m_1\mathbf{e}_1 + m_2\mathbf{e}_2\\ \hat{\mathbf{n}} &= n_1\mathbf{e}_1 + n_2\mathbf{e}_2\\ \mathbf{p} &= p_1\mathbf{e}_1 + p_2\mathbf{e}_2\end{aligned}$$

and θ is the angle between $\hat{\mathbf{m}}$ and $\hat{\mathbf{n}}$. Therefore, we can write

$$\begin{aligned}\hat{\mathbf{n}}\hat{\mathbf{m}} &= \hat{\mathbf{n}} \cdot \hat{\mathbf{m}} - \hat{\mathbf{m}} \wedge \hat{\mathbf{n}}\\ \hat{\mathbf{m}}\hat{\mathbf{n}} &= \hat{\mathbf{n}} \cdot \hat{\mathbf{m}} + \hat{\mathbf{m}} \wedge \hat{\mathbf{n}}\end{aligned}$$

where

$$\begin{aligned}\hat{\mathbf{n}} \cdot \hat{\mathbf{m}} &= \cos\theta\\ \hat{\mathbf{m}} \wedge \hat{\mathbf{n}} &= \sin\theta\mathbf{e}_{12}.\end{aligned}$$

Therefore, using the definition of a rotor

$$\begin{aligned}\mathbf{p}' &= \hat{\mathbf{n}}\hat{\mathbf{m}}\mathbf{p}\hat{\mathbf{m}}\hat{\mathbf{n}}\\ &= (\cos\theta - \sin\theta\mathbf{e}_{12})(p_1\mathbf{e}_1 + p_2\mathbf{e}_2)(\cos\theta + \sin\theta\mathbf{e}_{12})\\ &= (p_1\cos\theta\mathbf{e}_1 + p_2\cos\theta\mathbf{e}_2 + p_1\sin\theta\mathbf{e}_2 - p_2\sin\theta\mathbf{e}_1)(\cos\theta + \sin\theta\mathbf{e}_{12})\\ &= \big((p_1\cos\theta - p_2\sin\theta)\mathbf{e}_1 + (p_1\sin\theta + p_2\cos\theta)\mathbf{e}_2\big)(\cos\theta + \sin\theta\mathbf{e}_{12})\\ &= \big((\cos^2\theta - \sin^2\theta)p_1 - 2\cos\theta\sin\theta p_2\big)\mathbf{e}_1\\ &\quad + \big(2\cos\theta\sin\theta p_1 + (\cos^2\theta - \sin^2\theta)p_2\big)\mathbf{e}_2\\ &= (p_1\cos 2\theta - p_2\sin 2\theta)\mathbf{e}_1 + (p_1\sin 2\theta + p_2\cos 2\theta)\mathbf{e}_2\end{aligned}$$

or in matrix form

$$\begin{bmatrix} p_1' \\ p_2' \end{bmatrix} = \begin{bmatrix} \cos 2\theta & -\sin 2\theta \\ \sin 2\theta & \cos 2\theta \end{bmatrix} \begin{bmatrix} p_1 \\ p_2 \end{bmatrix}$$

which is the matrix for rotating a point 2θ about the origin. Now let's do the same for a 3D rotor.

12.5.2 3D Rotor

We begin with a unit bivector defining the plane $\hat{\mathbf{m}} \wedge \hat{\mathbf{n}}$, about which the rotation is effected, where

$$\begin{aligned}\hat{\mathbf{m}} &= m_1\mathbf{e}_1 + m_2\mathbf{e}_2 + m_3\mathbf{e}_3\\ \hat{\mathbf{n}} &= n_1\mathbf{e}_1 + n_2\mathbf{e}_2 + n_3\mathbf{e}_3\end{aligned}$$

and we deliberately define

$$\mathbf{R}_\theta = \hat{\mathbf{n}}\hat{\mathbf{m}}$$

where θ is half the angle between $\hat{\mathbf{m}}$ and $\hat{\mathbf{n}}$.

The rotor will take the form

$$\mathbf{R}_\theta = s - x\mathbf{e}_{23} - y\mathbf{e}_{31} - z\mathbf{e}_{12}$$

which permits us to define

$$\mathbf{R}_\theta^\dagger = \hat{\mathbf{m}}\hat{\mathbf{n}}$$

which is

$$\mathbf{R}_\theta^\dagger = s + x\mathbf{e}_{23} + y\mathbf{e}_{31} + z\mathbf{e}_{12}.$$

Therefore, given an arbitrary vector

$$\mathbf{v} = v_1\mathbf{e}_1 + v_2\mathbf{e}_2 + v_3\mathbf{e}_3$$

the rotated vector is given by

$$\mathbf{v}' = \mathbf{R}_\theta\mathbf{v}\mathbf{R}_\theta^\dagger.$$

To keep the algebra simple it is best to compute the individual components of $\mathbf{v}'$ using $\mathbf{R}_\theta v_1\mathbf{e}_1\mathbf{R}_\theta^\dagger$, $\mathbf{R}_\theta v_2\mathbf{e}_2\mathbf{R}_\theta^\dagger$ and $\mathbf{R}_\theta v_3\mathbf{e}_3\mathbf{R}_\theta^\dagger$:

$$\begin{aligned}\mathbf{R}_\theta v_1\mathbf{e}_1\mathbf{R}_\theta^\dagger &= (s - x\mathbf{e}_{23} - y\mathbf{e}_{31} - z\mathbf{e}_{12})v_1\mathbf{e}_1(s + x\mathbf{e}_{23} + y\mathbf{e}_{31} + z\mathbf{e}_{12})\\ &= v_1(s\mathbf{e}_1 - x\mathbf{e}_{123} - y\mathbf{e}_3 + z\mathbf{e}_2)(s + x\mathbf{e}_{23} + y\mathbf{e}_{31} + z\mathbf{e}_{12})\\ &= v_1\big((s^2 + x^2 - y^2 - z^2)\mathbf{e}_1 + 2(xy + sz)\mathbf{e}_2 + 2(xz - sy)\mathbf{e}_3\big)\end{aligned}$$

but

$$s^2 + x^2 = 1 - y^2 - z^2$$

therefore,

$$\mathbf{R}_\theta v_1\mathbf{e}_1\mathbf{R}_\theta^\dagger = v_1\big(\big(1 - 2(y^2 + z^2)\big)\mathbf{e}_1 + 2(xy + sz)\mathbf{e}_2 + 2(xz - sy)\mathbf{e}_3\big).$$

Next,

$$\begin{aligned}\mathbf{R}_\theta v_2\mathbf{e}_2\mathbf{R}_\theta^\dagger &= (s - x\mathbf{e}_{23} - y\mathbf{e}_{31} - z\mathbf{e}_{12})v_2\mathbf{e}_2(s + x\mathbf{e}_{23} + y\mathbf{e}_{31} + z\mathbf{e}_{12})\\ &= v_2(s\mathbf{e}_2 + x\mathbf{e}_3 - y\mathbf{e}_{123} + z\mathbf{e}_1)(s + x\mathbf{e}_{23} + y\mathbf{e}_{31} + z\mathbf{e}_{12})\\ &= v_2\big(2(xy - sz)\mathbf{e}_1 + \big(s^2 - x^2 + y^2 - z^2\big)\mathbf{e}_2 + 2(yz + sx)\mathbf{e}_3\big).\end{aligned}$$

Substituting

$$s^2 + y^2 = 1 - x^2 - z^2$$

we have

$$\mathbf{R}_\theta v_2\mathbf{e}_2\mathbf{R}_\theta^\dagger = v_2\big(2(xy - sz)\mathbf{e}_1 + \big(1 - 2\big(x^2 + z^2\big)\big)\mathbf{e}_2 + 2(yz + sx)\mathbf{e}_3\big).$$

Next,

$$\begin{aligned}\mathbf{R}_\theta v_3\mathbf{e}_3\mathbf{R}_\theta^\dagger &= (s - x\mathbf{e}_{23} - y\mathbf{e}_{31} - z\mathbf{e}_{12})v_3\mathbf{e}_3(s + x\mathbf{e}_{23} + y\mathbf{e}_{31} + z\mathbf{e}_{12})\\ &= v_3(s\mathbf{e}_3 - x\mathbf{e}_2 + y\mathbf{e}_1 - z\mathbf{e}_{123})(s + x\mathbf{e}_{23} + y\mathbf{e}_{31} + z\mathbf{e}_{12})\\ &= v_3\big(2(xz + sy)\mathbf{e}_1 + 2(yz - sx)\mathbf{e}_2 + \big(s^2 - x^2 - y^2 + z^2\big)\mathbf{e}_3\big).\end{aligned}$$

Substituting

$$s^2 + z^2 = 1 - x^2 - y^2$$

we have

$$\mathbf{R}_\theta v_3\mathbf{e}_3\mathbf{R}_\theta^\dagger = v_3\big(2(xz - sy)\mathbf{e}_1 + 2(yz - sx)\mathbf{e}_2 + \big(1 - 2\big(x^2 + y^2\big)\big)\mathbf{e}_3\big).$$

Therefore,

$$\mathbf{R}_\theta\mathbf{v}\mathbf{R}_\theta^\dagger = \mathbf{R}v_1\mathbf{e}_1\mathbf{R}^\dagger + \mathbf{R}v_2\mathbf{e}_2\mathbf{R}^\dagger + \mathbf{R}v_3\mathbf{e}_3\mathbf{R}^\dagger$$

or as a matrix

$$\begin{bmatrix} v_1' \\ v_2' \\ v_3' \end{bmatrix} = \begin{bmatrix} 1 - 2(y^2 + z^2) & 2(xy - sz) & 2(xz + sy) \\ 2(xy + sz) & 1 - 2(x^2 + z^2) & 2(yz - sx) \\ 2(xz - sy) & 2(yz + sx) & 1 - 2(x^2 + y^2) \end{bmatrix} \begin{bmatrix} v_1 \\ v_2 \\ v_3 \end{bmatrix}$$

which is the same matrix representing the quaternion triple $\mathbf{q}\mathbf{p}\mathbf{q}^{-1}$.

The reader should not be put off by the above algebraic proof. It has been included to demonstrate that bivector rotors behave just like quaternions and are represented by identical matrices.

You may wish to investigate the matrix for the reverse rotor triple $\mathbf{R}_\theta^\dagger\mathbf{p}\mathbf{R}_\theta$, which you will discover is

$$\begin{bmatrix} v_1' \\ v_2' \\ v_3' \end{bmatrix} = \begin{bmatrix} 1 - 2(y^2 + z^2) & 2(xy + sz) & 2(xz - sy) \\ 2(xy - sz) & 1 - 2(x^2 + z^2) & 2(yz + sx) \\ 2(xz + sy) & 2(yz - sx) & 1 - 2(x^2 + y^2) \end{bmatrix} \begin{bmatrix} v_1 \\ v_2 \\ v_3 \end{bmatrix}$$

and is the transpose of the above matrix for $\mathbf{R}_\theta\mathbf{v}\mathbf{R}_\theta^\dagger$. Thus the matrices confirm that

$\mathbf{R}_\theta \mathbf{v} \mathbf{R}_\theta^\dagger$ rotates a vector anticlockwise by θ

$\mathbf{R}_\theta^\dagger \mathbf{v} \mathbf{R}_\theta$ rotates a vector clockwise by θ.

Furthermore, maintaining our convention about rotating points and frames:

$\mathbf{R}_\theta \mathbf{v} \mathbf{R}_\theta^\dagger$ rotates a frame clockwise by θ

$\mathbf{R}_\theta^\dagger \mathbf{v} \mathbf{R}_\theta$ rotates a frame anticlockwise by θ.

12.5.3 Extracting a Rotor

Say we are presented with

$$\hat{\mathbf{b}} = \mathbf{R}_\theta \hat{\mathbf{a}} \mathbf{R}_\theta^\dagger$$

where we know $\hat{\mathbf{a}}$ and $\hat{\mathbf{b}}$ and have to discover $\mathbf{R}_\theta$. Here is one way we can undertake the task, which is cunning, rather than obvious!

Figure 12.8 shows vectors $\hat{\mathbf{a}}$ and $\hat{\mathbf{b}}$ and a third vector $\hat{\mathbf{n}}$, mid-way between the two vectors. Vector $\hat{\mathbf{n}}$ bisects the angle θ separating $\hat{\mathbf{a}}$ and $\hat{\mathbf{b}}$, therefore, the product $\hat{\mathbf{b}}\hat{\mathbf{n}}$ must be a rotor capable of rotating any vector in the plane $\hat{\mathbf{n}} \wedge \hat{\mathbf{b}}$ by θ, which permits us to write

$$\hat{\mathbf{b}} = \hat{\mathbf{b}}\hat{\mathbf{n}}\hat{\mathbf{a}}\hat{\mathbf{n}}\hat{\mathbf{b}}$$

or

$$\hat{\mathbf{b}} = \mathbf{R}_\theta \hat{\mathbf{a}} \mathbf{R}_\theta^\dagger$$

where

$$\mathbf{R}_\theta = \hat{\mathbf{b}}\hat{\mathbf{n}} \quad (12.12)$$

$$\mathbf{R}_\theta^\dagger = \hat{\mathbf{n}}\hat{\mathbf{b}}. \quad (12.13)$$

Next, to eliminate $\hat{\mathbf{n}}$ we compute

$$\hat{\mathbf{n}} = \frac{\hat{\mathbf{a}} + \hat{\mathbf{b}}}{|\hat{\mathbf{a}} + \hat{\mathbf{b}}|}$$

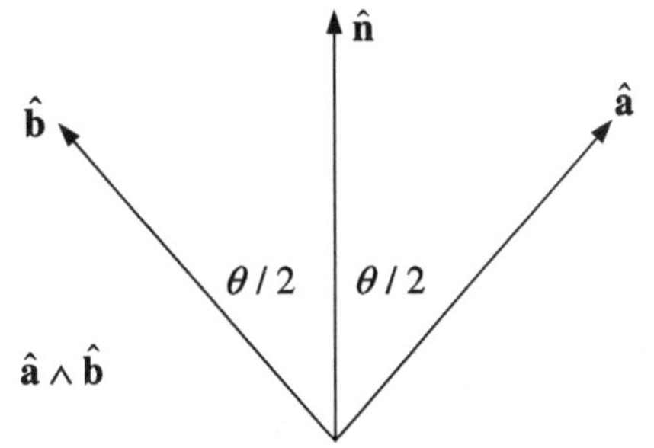

Fig. 12.8 Vector $\hat{\mathbf{n}}$ bisects θ

and substitute it in (12.12):

$$\begin{aligned}\mathbf{R}_\theta &= \hat{\mathbf{b}}\hat{\mathbf{n}}\\ &= \hat{\mathbf{b}}\left(\frac{\hat{\mathbf{a}}+\hat{\mathbf{b}}}{|\hat{\mathbf{a}}+\hat{\mathbf{b}}|}\right)\\ &= \frac{1+\hat{\mathbf{b}}\hat{\mathbf{a}}}{|\hat{\mathbf{a}}+\hat{\mathbf{b}}|}.\end{aligned}$$

Similarly,

$$\begin{aligned}\mathbf{R}^\dagger_\theta &= \hat{\mathbf{n}}\hat{\mathbf{b}}\\ &= \left(\frac{\hat{\mathbf{a}}+\hat{\mathbf{b}}}{|\hat{\mathbf{a}}+\hat{\mathbf{b}}|}\right)\hat{\mathbf{b}}\\ &= \frac{1+\hat{\mathbf{a}}\hat{\mathbf{b}}}{|\hat{\mathbf{a}}+\hat{\mathbf{b}}|}.\end{aligned}$$

It is possible to show that

$$|\hat{\mathbf{a}}+\hat{\mathbf{b}}| = \sqrt{2(1+\hat{\mathbf{a}}\cdot\hat{\mathbf{b}})}$$

which permits us to propose an alternative solution

$$\mathbf{R}_\theta = \frac{1+\hat{\mathbf{b}}\hat{\mathbf{a}}}{\sqrt{2(1+\hat{\mathbf{a}}\cdot\hat{\mathbf{b}})}} \tag{12.14}$$

$$\mathbf{R}^\dagger_\theta = \frac{1+\hat{\mathbf{a}}\hat{\mathbf{b}}}{\sqrt{2(1+\hat{\mathbf{a}}\cdot\hat{\mathbf{b}})}}. \tag{12.15}$$

Now let's put these definitions to the test.

Figure 12.9 shows vector $\hat{\mathbf{a}}$ aligned with the $\mathbf{e}_1$ axis and $\hat{\mathbf{b}}$ aligned with the $\mathbf{e}_2$ axis. Therefore, the rotor is acting in the $\mathbf{e}_{12}$ plane with an angle of 45° to effect a rotation of 90°. Using our knowledge of rotors, it is obvious that

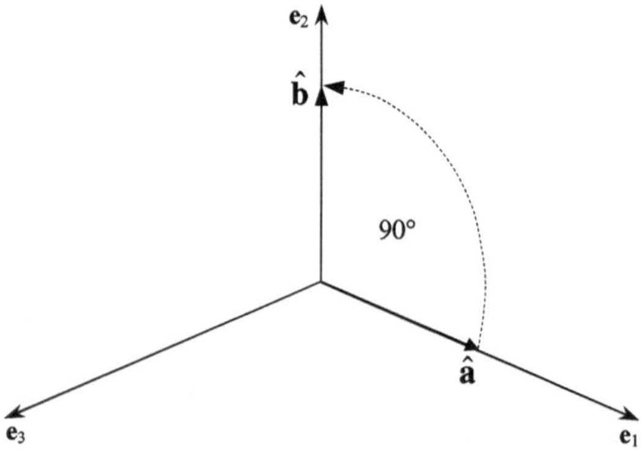

Fig. 12.9 Vector $\hat{\mathbf{a}}$ rotates to $\hat{\mathbf{b}}$

$$\begin{aligned}\mathbf{R}_{90^\circ} &= \cos 45^\circ - \sin 45^\circ \mathbf{e}_{12}\\ &= \frac{1-\mathbf{e}_{12}}{\sqrt{2}}\\ \mathbf{R}^\dagger_{90^\circ} &= \cos 45^\circ + \sin 45^\circ \mathbf{e}_{12}\\ &= \frac{1+\mathbf{e}_{12}}{\sqrt{2}}.\end{aligned}$$

So let's confirm these using (12.14) and (12.15):

$$\begin{aligned}\mathbf{R}_{90^\circ} &= \frac{1+\mathbf{e}_2\mathbf{e}_1}{\sqrt{2(1+\mathbf{e}_1\cdot\mathbf{e}_2)}}\\ &= \frac{1-\mathbf{e}_{12}}{\sqrt{2}}\\ \mathbf{R}^\dagger_{90^\circ} &= \frac{1+\mathbf{e}_1\mathbf{e}_2}{\sqrt{2(1+\mathbf{e}_1\cdot\mathbf{e}_2)}}\\ &= \frac{1+\mathbf{e}_{12}}{\sqrt{2}}\end{aligned}$$

which confirm our predictions.

In a previous example above, we used

$$\begin{aligned}\mathbf{R}_{120^\circ} &= \cos 60^\circ - \sin 60^\circ \hat{\mathbf{B}}\\ \mathbf{R}^\dagger_{120^\circ} &= \cos 60^\circ + \sin 60^\circ \hat{\mathbf{B}}\\ \hat{\mathbf{B}} &= \frac{1}{\sqrt{3}}(\mathbf{e}_{12}+\mathbf{e}_{23}+\mathbf{e}_{31})\end{aligned}$$

to rotate $\mathbf{e}_2+\mathbf{e}_3$ to $\mathbf{e}_1+\mathbf{e}_3$.

Let's use (12.14) and (12.15) to invert the process, but remember that we are dealing with unit vectors, which means that we have to normalise $\mathbf{a}$ and $\mathbf{b}$:

$$\begin{aligned}\hat{\mathbf{a}} &= \frac{1}{\sqrt{2}}(\mathbf{e}_2+\mathbf{e}_3)\\ \hat{\mathbf{b}} &= \frac{1}{\sqrt{2}}(\mathbf{e}_1+\mathbf{e}_3).\end{aligned}$$

Furthermore, although the bivector $\hat{\mathbf{B}}$ formed the plane of rotation in the previous example, this time, the plane of rotation is $\hat{\mathbf{a}}\wedge\hat{\mathbf{b}}$. Therefore,

$$\begin{aligned}\mathbf{R}_{120^\circ} &= \frac{1+\hat{\mathbf{b}}\hat{\mathbf{a}}}{\sqrt{2(1+\hat{\mathbf{a}}\cdot\hat{\mathbf{b}})}}\\ &= \frac{1+\frac{1}{\sqrt{2}}(\mathbf{e}_1+\mathbf{e}_3)\frac{1}{\sqrt{2}}(\mathbf{e}_2+\mathbf{e}_3)}{\sqrt{2(1+\frac{1}{\sqrt{2}}(\mathbf{e}_2+\mathbf{e}_3)\cdot\frac{1}{\sqrt{2}}(\mathbf{e}_1+\mathbf{e}_3))}}\end{aligned}$$

$$= \frac{1 + \frac{1}{2}(\mathbf{e}_{12} - \mathbf{e}_{31} - \mathbf{e}_{23} + 1)}{\sqrt{2(1 + \frac{1}{2})}}$$

$$= \frac{1 + \frac{1}{2}\mathbf{e}_{12} - \frac{1}{2}\mathbf{e}_{23} - \frac{1}{2}\mathbf{e}_{31} + \frac{1}{2}}{\sqrt{3}}$$

$$= \frac{\sqrt{3}}{2} + \frac{\sqrt{3}}{6}\mathbf{e}_{12} - \frac{\sqrt{3}}{6}\mathbf{e}_{23} - \frac{\sqrt{3}}{6}\mathbf{e}_{31}$$

which makes

$$\mathbf{R}^{\dagger}_{120^\circ} = \frac{\sqrt{3}}{2} - \frac{\sqrt{3}}{6}\mathbf{e}_{12} + \frac{\sqrt{3}}{6}\mathbf{e}_{23} + \frac{\sqrt{3}}{6}\mathbf{e}_{31}.$$

But does it work? Well, let's find out by forming the product

$$\begin{aligned}
\mathbf{R}_{120^\circ}\frac{1}{\sqrt{2}}(\mathbf{e}_2 + \mathbf{e}_3)\mathbf{R}^{\dagger}_{120^\circ} &= \frac{1}{12\sqrt{2}}(3 + \mathbf{e}_{12} + \mathbf{e}_{32} + \mathbf{e}_{13}) \\
&\quad \times (\mathbf{e}_2 + \mathbf{e}_3)(3 - \mathbf{e}_{12} - \mathbf{e}_{32} - \mathbf{e}_{13}) \\
&= \frac{1}{6\sqrt{2}}(\mathbf{e}_1 + \mathbf{e}_2 + 2\mathbf{e}_3)(3 - \mathbf{e}_{12} - \mathbf{e}_{32} - \mathbf{e}_{13}) \\
&= \frac{1}{6\sqrt{2}}(6\mathbf{e}_1 + 6\mathbf{e}_3) \\
&= \frac{1}{\sqrt{2}}(\mathbf{e}_1 + \mathbf{e}_3)
\end{aligned}$$

which is correct.

Finally, let's employ the rotor matrix. But remember that its definition of $\mathbf{R}_\theta$ has a negative bivector term, which means that we have to switch the bivector terms in $\mathbf{R}_{120^\circ}$ or use the bivector terms from $\mathbf{R}^{\dagger}_{120^\circ}$:

$$\mathbf{R}^{\dagger}_{120^\circ} = \frac{\sqrt{3}}{2} - \frac{\sqrt{3}}{6}\mathbf{e}_{12} + \frac{\sqrt{3}}{6}\mathbf{e}_{23} + \frac{\sqrt{3}}{6}\mathbf{e}_{31}$$

where

$$s = \frac{\sqrt{3}}{2}, \quad x = \frac{\sqrt{3}}{6}, \quad y = \frac{\sqrt{3}}{6}, \quad z = -\frac{\sqrt{3}}{6}$$

$$\begin{aligned}
\mathbf{R}_{120^\circ}\mathbf{v}\mathbf{R}^{\dagger}_{120^\circ} &= \begin{bmatrix} 1-2(y^2+z^2) & 2(xy-sz) & 2(xz+sy) \\ 2(xy+sz) & 1-2(x^2+z^2) & 2(yz-sx) \\ 2(xz-sy) & 2(yz+sx) & 1-2(x^2+y^2) \end{bmatrix} \begin{bmatrix} v_1 \\ v_2 \\ v_3 \end{bmatrix} \\
&= \begin{bmatrix} 1-2(\frac{1}{12}+\frac{1}{12}) & 2(\frac{1}{12}+\frac{3}{12}) & 2(-\frac{1}{12}+\frac{3}{12}) \\ 2(\frac{1}{12}-\frac{3}{12}) & 1-2(\frac{1}{12}+\frac{1}{12}) & 2(-\frac{1}{12}-\frac{3}{12}) \\ 2(-\frac{1}{12}-\frac{3}{12}) & 2(-\frac{1}{12}+\frac{3}{12}) & 1-2(\frac{1}{12}+\frac{1}{12}) \end{bmatrix} \begin{bmatrix} v_1 \\ v_2 \\ v_3 \end{bmatrix} \\
&= \begin{bmatrix} \frac{2}{3} & \frac{2}{3} & \frac{1}{3} \\ -\frac{1}{3} & \frac{2}{3} & -\frac{2}{3} \\ -\frac{2}{3} & \frac{1}{3} & \frac{2}{3} \end{bmatrix} \begin{bmatrix} v_1 \\ v_2 \\ v_3 \end{bmatrix}.
\end{aligned}$$

Substituting $\mathbf{a} = [\mathbf{e}_2 + \mathbf{e}_3]$ we obtain

$$\begin{bmatrix} 1 \\ 0 \\ 1 \end{bmatrix} = \begin{bmatrix} \frac{2}{3} & \frac{2}{3} & \frac{1}{3} \\ -\frac{1}{3} & \frac{2}{3} & -\frac{2}{3} \\ -\frac{2}{3} & \frac{1}{3} & \frac{2}{3} \end{bmatrix} \begin{bmatrix} 0 \\ 1 \\ 1 \end{bmatrix}$$

which is correct.

You may also like to verify that the determinant of the matrix is 1.

12.6 Summary

It is very interesting to see the close relationship between quaternions and geometric algebra. It demonstrates that although it is possible to describe the low-level arithmetic that actually does the work behind the scenes, such as a matrix, it is also possible to invent objects such as quaternions or bivectors, trivectors, etc., that provide a conceptual high-level framework that allow mathematicians to work more productively and creatively. In the end, Hamilton, Grassman and Clifford have provided us with some extraordinary mathematical inventions that have found their way into computer graphics, and I hope that this chapter has shown you another way of handling rotations.

12.6.1 Summary of Bivector Transforms

Reflecting a vector in a plane

$$\mathbf{v}' = -\hat{\mathbf{n}}\mathbf{v}\hat{\mathbf{n}}.$$

Rotating a vector using rotors

$$\mathbf{v}' = \mathbf{R}_\theta \mathbf{v} \mathbf{R}_\theta^\dagger$$

where

$$\mathbf{R}_\theta = \cos(\theta/2) - \sin(\theta/2)\hat{\mathbf{B}}$$
$$\mathbf{R}_\theta^\dagger = \cos(\theta/2) + \sin(\theta/2)\hat{\mathbf{B}}.$$

Rotor as a matrix

$$\mathbf{R}_\theta \mathbf{v} \mathbf{R}_\theta^\dagger = \begin{bmatrix} 1-2(y^2+z^2) & 2(xy-sz) & 2(xz+sy) \\ 2(xy+sz) & 1-2(x^2+z^2) & 2(yz-sx) \\ 2(xz-sy) & 2(yz+sx) & 1-2(x^2+y^2) \end{bmatrix} \begin{bmatrix} v_1 \\ v_2 \\ v_3 \end{bmatrix}$$

$$\mathbf{R}_\theta^\dagger \mathbf{v} \mathbf{R}_\theta = \begin{bmatrix} 1-2(y^2+z^2) & 2(xy+sz) & 2(xz-sy) \\ 2(xy-sz) & 1-2(x^2+z^2) & 2(yz+sx) \\ 2(xz+sy) & 2(yz-sx) & 1-2(x^2+y^2) \end{bmatrix} \begin{bmatrix} v_1 \\ v_2 \\ v_3 \end{bmatrix}$$

where

$$\begin{aligned}\mathbf{R}_\theta &= s - x\mathbf{e}_{23} - y\mathbf{e}_{31} - z\mathbf{e}_{12}\\ \mathbf{R}_\theta^\dagger &= s + x\mathbf{e}_{23} + y\mathbf{e}_{31} + z\mathbf{e}_{12}.\end{aligned}$$

Extracting a rotor
If

$$\hat{\mathbf{b}} = \mathbf{R}_\theta \hat{\mathbf{a}} \mathbf{R}_\theta^\dagger$$

then

$$\begin{aligned}\mathbf{R}_\theta &= \frac{1+\hat{\mathbf{b}}\hat{\mathbf{a}}}{|\hat{\mathbf{a}}+\hat{\mathbf{b}}|} = \frac{1+\hat{\mathbf{b}}\hat{\mathbf{a}}}{\sqrt{2(1+\hat{\mathbf{a}}\cdot\hat{\mathbf{b}})}}\\ \mathbf{R}_\theta^\dagger &= \frac{1+\hat{\mathbf{a}}\hat{\mathbf{b}}}{|\hat{\mathbf{a}}+\hat{\mathbf{b}}|} = \frac{1+\hat{\mathbf{a}}\hat{\mathbf{b}}}{\sqrt{2(1+\hat{\mathbf{a}}\cdot\hat{\mathbf{b}})}}.\end{aligned}$$

Chapter 13
Conclusion

The aim of this book was to take the reader through the important ideas and mathematical techniques associated with rotation transforms. I mentioned that I would not be too pedantic about mathematical terminology and would not swamp the reader with high-level concepts and axioms that pervade the real world of mathematics. My prime objective was to make the reader confident and comfortable with complex numbers, vectors, matrices, quaternions and bivector rotors. I knew that this was a challenge, but as they all share rotation as a common thread, hopefully, this has not been too onerous for the reader.

The worked examples will provide the reader with real problems to explore. As far as I know, they all produce correct results. But that was not always the case, as it is so easy to switch a sign during an algebraic expansion that creates a false result. However, repeated examination eventually leads one to the mistake, and the correct answer emerges so naturally.

The real challenge for the reader is the next level. There are some excellent books, technical papers and websites that introduce more advanced topics such as the B-spline interpolation of quaternions, the kinematics of moving frames, exponential rotors and conformal geometry. Hopefully, the contents of this book has prepared the reader for such journeys.

What I have tried to show throughout the previous dozen chapters is that rotations are about sines and cosines, which are ratios associated with a line sweeping the unit circle. These, in turn, can be expressed in various identities, especially half-angle identities.

Imaginary quantities also seem to play an important role in rotations, and it is just as well that they exist otherwise life would be extremely difficult! We have seen that complex numbers, quaternions and bivector rotors all include imaginary quantities, and at the end of the day, they just seem to be different ways of controlling sines and cosines. I am certain that you now appreciate that quaternions are just one of four possible algebras that require an n-square identity, and that they are closely related to Clifford algebra. Which one is best for computer graphics? I don't know. But I am certain that if you attempt to implement these ideas, you will discover the answer.

J. Vince, *Rotation Transforms for Computer Graphics*,
DOI 10.1007/978-0-85729-154-7_13, © Springer-Verlag London Limited 2011

Chapter 15
Conclusion

Appendix A
Composite Point Rotation Sequences

A.1 Euler Rotations

In Chap. 9 we considered composite Euler rotations comprising individual rotations about the x, y and z-axes such as $\mathbf{R}_{\gamma,x}\mathbf{R}_{\beta,y}\mathbf{R}_{\alpha,z}$ and $\mathbf{R}_{\gamma,z}\mathbf{R}_{\beta,y}\mathbf{R}_{\alpha,x}$. However, there is nothing preventing us from creating other combinations such as $\mathbf{R}_{\gamma,x}\mathbf{R}_{\beta,y}\mathbf{R}_{\alpha,x}$ or $\mathbf{R}_{\gamma,z}\mathbf{R}_{\beta,y}\mathbf{R}_{\alpha,z}$ that do not include two consecutive rotations about the same axis. In all, there are twelve possible combinations:

$$
\begin{array}{llll}
\mathbf{R}_{\gamma,x}\mathbf{R}_{\beta,y}\mathbf{R}_{\alpha,x}, & \mathbf{R}_{\gamma,x}\mathbf{R}_{\beta,y}\mathbf{R}_{\alpha,z}, & \mathbf{R}_{\gamma,x}\mathbf{R}_{\beta,z}\mathbf{R}_{\alpha,x}, & \mathbf{R}_{\gamma,x}\mathbf{R}_{\beta,z}\mathbf{R}_{\alpha,y} \\
\mathbf{R}_{\gamma,y}\mathbf{R}_{\beta,x}\mathbf{R}_{\alpha,y}, & \mathbf{R}_{\gamma,y}\mathbf{R}_{\beta,x}\mathbf{R}_{\alpha,z}, & \mathbf{R}_{\gamma,y}\mathbf{R}_{\beta,z}\mathbf{R}_{\alpha,x}, & \mathbf{R}_{\gamma,y}\mathbf{R}_{\beta,z}\mathbf{R}_{\alpha,y} \\
\mathbf{R}_{\gamma,z}\mathbf{R}_{\beta,x}\mathbf{R}_{\alpha,y}, & \mathbf{R}_{\gamma,z}\mathbf{R}_{\beta,x}\mathbf{R}_{\alpha,z}, & \mathbf{R}_{\gamma,z}\mathbf{R}_{\beta,y}\mathbf{R}_{\alpha,x}, & \mathbf{R}_{\gamma,z}\mathbf{R}_{\beta,y}\mathbf{R}_{\alpha,z}
\end{array}
$$

which we now cover in detail.

For each combination there are three Euler rotation matrices, the resulting composite matrix, a matrix where the three angles equal 90°, the coordinates of the rotated unit cube, the axis and angle of rotation and a figure illustrating the stages of rotation. To compute the axis of rotation $[v_1 \quad v_2 \quad v_3]^{\mathrm{T}}$ we use

$$
\begin{aligned}
v_1 &= (a_{22}-1)(a_{33}-1)-a_{23}a_{32} \\
v_2 &= (a_{33}-1)(a_{11}-1)-a_{31}a_{13} \\
v_3 &= (a_{11}-1)(a_{22}-1)-a_{12}a_{21}
\end{aligned}
$$

where

$$
\mathbf{R} = \begin{bmatrix} a_{11} & a_{12} & a_{13} \\ a_{21} & a_{22} & a_{23} \\ a_{31} & a_{32} & a_{33} \end{bmatrix}
$$

and for the angle of rotation δ we use

$$
\cos\delta = \frac{1}{2}\big(\mathrm{Tr}(\mathbf{R})-1\big).
$$

We begin by defining the three principal Euler rotations:

J. Vince, *Rotation Transforms for Computer Graphics*,
DOI 10.1007/978-0-85729-154-7, © Springer-Verlag London Limited 2011

$$\text{rotate } \alpha \text{ about the } x\text{-axis} \quad \mathbf{R}_{\alpha,x} = \begin{bmatrix} 1 & 0 & 0 \\ 0 & c_\alpha & -s_\alpha \\ 0 & s_\alpha & c_\alpha \end{bmatrix}$$

$$\text{rotate } \beta \text{ about the } y\text{-axis} \quad \mathbf{R}_{\beta,y} = \begin{bmatrix} c_\beta & 0 & s_\beta \\ 0 & 1 & 0 \\ -s_\beta & 0 & c_\beta \end{bmatrix}$$

$$\text{rotate } \gamma \text{ about the } z\text{-axis} \quad \mathbf{R}_{\gamma,z} = \begin{bmatrix} c_\gamma & -s_\gamma & 0 \\ s_\gamma & c_\gamma & 0 \\ 0 & 0 & 1 \end{bmatrix}$$

where $c_\alpha = \cos\alpha$ and $s_\alpha = \sin\alpha$, etc.

Remember that the right-most transform is applied first and the left-most transform last. In terms of angles, the sequence is always α, β, γ.

For each composite transform you can verify that when $\alpha = \beta = \gamma = 0$ the result is the identity transform **I**.

We now examine the twelve combinations in turn.

A.2 $\mathbf{R}_{\gamma,x}\mathbf{R}_{\beta,y}\mathbf{R}_{\alpha,x}$

$$\mathbf{R}_{\gamma,x}\mathbf{R}_{\beta,y}\mathbf{R}_{\alpha,x} = \begin{bmatrix} 1 & 0 & 0 \\ 0 & c_\gamma & -s_\gamma \\ 0 & s_\gamma & c_\gamma \end{bmatrix} \begin{bmatrix} c_\beta & 0 & s_\beta \\ 0 & 1 & 0 \\ -s_\beta & 0 & c_\beta \end{bmatrix} \begin{bmatrix} 1 & 0 & 0 \\ 0 & c_\alpha & -s_\alpha \\ 0 & s_\alpha & c_\alpha \end{bmatrix}$$

$$= \begin{bmatrix} c_\beta & s_\beta s_\alpha & s_\beta c_\alpha \\ s_\gamma s_\beta & (c_\gamma c_\alpha - s_\gamma c_\beta s_\alpha) & (-c_\gamma s_\alpha - s_\gamma c_\beta c_\alpha) \\ -c_\gamma s_\beta & (s_\gamma c_\alpha + c_\gamma c_\beta s_\alpha) & (-s_\gamma s_\alpha + c_\gamma c_\beta c_\alpha) \end{bmatrix}$$

$$\mathbf{R}_{90^\circ,x}\mathbf{R}_{90^\circ,y}\mathbf{R}_{90^\circ,x} = \begin{bmatrix} 0 & 1 & 0 \\ 1 & 0 & 0 \\ 0 & 0 & -1 \end{bmatrix}$$

$$\begin{bmatrix} 0 & 1 & 0 \\ 1 & 0 & 0 \\ 0 & 0 & -1 \end{bmatrix} \begin{bmatrix} 0 & 0 & 0 & 0 & 1 & 1 & 1 & 1 \\ 0 & 0 & 1 & 1 & 0 & 0 & 1 & 1 \\ 0 & 1 & 0 & 1 & 0 & 1 & 0 & 1 \end{bmatrix}$$

$$= \begin{bmatrix} 0 & 0 & 1 & 1 & 0 & 0 & 1 & 1 \\ 0 & 0 & 0 & 0 & 1 & 1 & 1 & 1 \\ 0 & -1 & 0 & -1 & 0 & -1 & 0 & -1 \end{bmatrix}.$$

This rotation sequence is illustrated in Fig. A.1 (a)–(d), where the axis of rotation is $[2 \quad 2 \quad 0]^{\mathrm{T}}$ and the angle of rotation 180°.

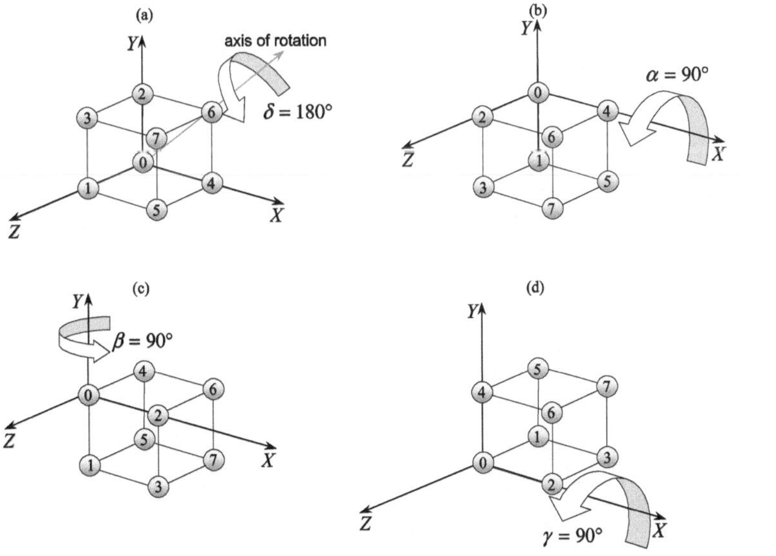

Fig. A.1 Four views of the unit cube before and during the three rotations $\mathbf{R}_{90^\circ,x}\mathbf{R}_{90^\circ,y}\mathbf{R}_{90^\circ,x}$

A.3 $\mathbf{R}_{\gamma,x}\mathbf{R}_{\beta,y}\mathbf{R}_{\alpha,z}$

$$\mathbf{R}_{\gamma,x}\mathbf{R}_{\beta,y}\mathbf{R}_{\alpha,z} = \begin{bmatrix} 1 & 0 & 0 \\ 0 & c_\gamma & -s_\gamma \\ 0 & s_\gamma & c_\gamma \end{bmatrix} \begin{bmatrix} c_\beta & 0 & s_\beta \\ 0 & 1 & 0 \\ -s_\beta & 0 & c_\beta \end{bmatrix} \begin{bmatrix} c_\alpha & -s_\alpha & 0 \\ s_\alpha & c_\alpha & 0 \\ 0 & 0 & 1 \end{bmatrix}$$

$$= \begin{bmatrix} c_\beta c_\alpha & -c_\beta s_\alpha & s_\beta \\ (c_\gamma s_\alpha + s_\gamma s_\beta c_\alpha) & (c_\gamma c_\alpha - s_\gamma s_\beta s_\alpha) & -s_\gamma c_\beta \\ (s_\gamma s_\alpha - c_\gamma s_\beta c_\alpha) & (s_\gamma c_\alpha + c_\gamma s_\beta s_\alpha) & c_\gamma c_\beta \end{bmatrix}$$

$$\mathbf{R}_{90°,x}\mathbf{R}_{90°,y}\mathbf{R}_{90°,z} = \begin{bmatrix} 0 & 0 & 1 \\ 0 & -1 & 0 \\ 1 & 0 & 0 \end{bmatrix}$$

$$\begin{bmatrix} 0 & 0 & 1 \\ 0 & -1 & 0 \\ 1 & 0 & 0 \end{bmatrix} \begin{bmatrix} 0 & 0 & 0 & 0 & 1 & 1 & 1 & 1 \\ 0 & 0 & 1 & 1 & 0 & 0 & 1 & 1 \\ 0 & 1 & 0 & 1 & 0 & 1 & 0 & 1 \end{bmatrix}$$

$$= \begin{bmatrix} 0 & 1 & 0 & 1 & 0 & 1 & 0 & 1 \\ 0 & 0 & -1 & -1 & 0 & 0 & -1 & -1 \\ 0 & 0 & 0 & 0 & 1 & 1 & 1 & 1 \end{bmatrix}.$$

This rotation sequence is illustrated in Fig. A.2 (a)–(d), where the axis of rotation is $[2 \quad 0 \quad 2]^{\mathrm{T}}$ and the angle of rotation 180°.

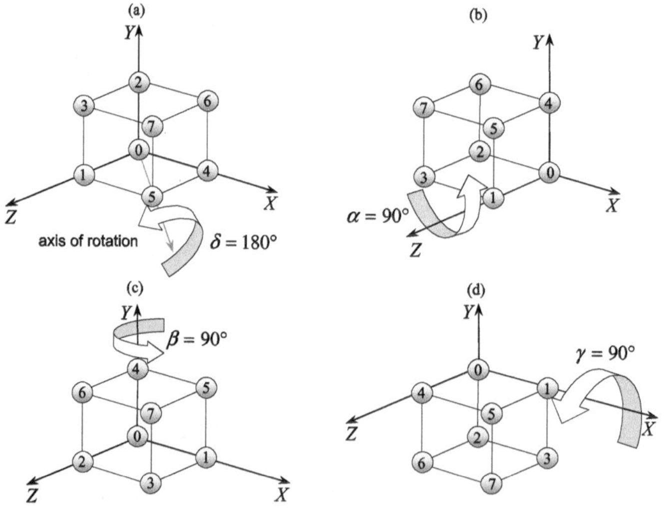

Fig. A.2 Four views of the unit cube before and during the three rotations $\mathbf{R}_{90°,x}\mathbf{R}_{90°,y}\mathbf{R}_{90°,z}$

A.4 $\mathbf{R}_{\gamma,x}\mathbf{R}_{\beta,z}\mathbf{R}_{\alpha,x}$

$$\mathbf{R}_{\gamma,x}\mathbf{R}_{\beta,z}\mathbf{R}_{\alpha,x} = \begin{bmatrix} 1 & 0 & 0 \\ 0 & c_\gamma & -s_\gamma \\ 0 & s_\gamma & c_\gamma \end{bmatrix} \begin{bmatrix} c_\beta & -s_\beta & 0 \\ s_\beta & c_\beta & 0 \\ 0 & 0 & 1 \end{bmatrix} \begin{bmatrix} 1 & 0 & 0 \\ 0 & c_\alpha & -s_\alpha \\ 0 & s_\alpha & c_\alpha \end{bmatrix}$$

$$= \begin{bmatrix} c_\beta & -s_\beta c_\alpha & s_\beta s_\alpha \\ c_\gamma s_\beta & (-s_\gamma s_\alpha + c_\gamma c_\beta c_\alpha) & (-s_\gamma c_\alpha - c_\gamma c_\beta s_\alpha) \\ s_\gamma s_\beta & (c_\gamma s_\alpha + s_\gamma c_\beta c_\alpha) & (c_\gamma c_\alpha - s_\gamma c_\beta s_\alpha) \end{bmatrix}$$

$$\mathbf{R}_{90^\circ,x}\mathbf{R}_{90^\circ,z}\mathbf{R}_{90^\circ,x} = \begin{bmatrix} 0 & 0 & 1 \\ 0 & -1 & 0 \\ 1 & 0 & 0 \end{bmatrix}$$

$$\begin{bmatrix} 0 & 0 & 1 \\ 0 & -1 & 0 \\ 1 & 0 & 0 \end{bmatrix} \begin{bmatrix} 0 & 0 & 0 & 0 & 1 & 1 & 1 & 1 \\ 0 & 0 & 1 & 1 & 0 & 0 & 1 & 1 \\ 0 & 1 & 0 & 1 & 0 & 1 & 0 & 1 \end{bmatrix}$$

$$= \begin{bmatrix} 0 & 1 & 0 & 1 & 0 & 1 & 0 & 1 \\ 0 & 0 & -1 & -1 & 0 & 0 & -1 & -1 \\ 0 & 0 & 0 & 0 & 1 & 1 & 1 & 1 \end{bmatrix}.$$

This rotation sequence is illustrated in Fig. A.3 (a)–(d), where the axis of rotation is $[2\quad 0\quad 2]^{\mathrm{T}}$ and the angle of rotation 180°.

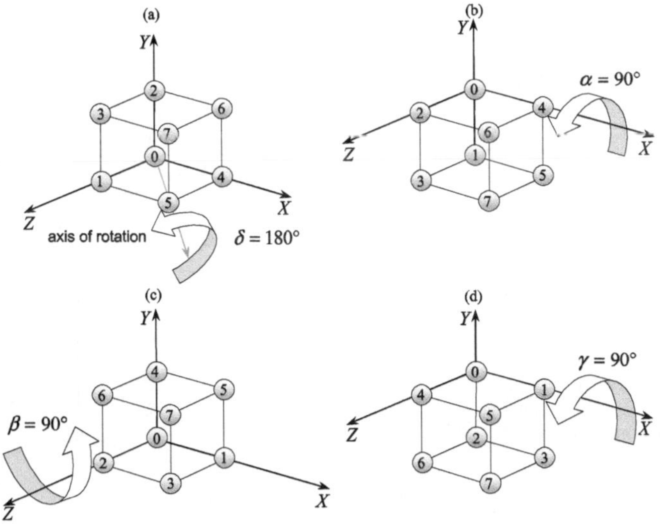

Fig. A.3 Four views of the unit cube before and during the three rotations $\mathbf{R}_{90^\circ,x}\mathbf{R}_{90^\circ,z}\mathbf{R}_{90^\circ,x}$

A.5 $\mathbf{R}_{\gamma,x}\mathbf{R}_{\beta,z}\mathbf{R}_{\alpha,y}$

$$\mathbf{R}_{\gamma,x}\mathbf{R}_{\beta,z}\mathbf{R}_{\alpha,y} = \begin{bmatrix} 1 & 0 & 0 \\ 0 & c_\gamma & -s_\gamma \\ 0 & s_\gamma & c_\gamma \end{bmatrix} \begin{bmatrix} c_\beta & -s_\beta & 0 \\ s_\beta & c_\beta & 0 \\ 0 & 0 & 1 \end{bmatrix} \begin{bmatrix} c_\alpha & 0 & s_\alpha \\ 0 & 1 & 0 \\ -s_\alpha & 0 & c_\alpha \end{bmatrix}$$

$$= \begin{bmatrix} c_\beta c_\alpha & -s_\beta & c_\beta s_\alpha \\ (s_\gamma s_\alpha + c_\gamma s_\beta c_\alpha) & c_\gamma c_\beta & (-s_\gamma c_\alpha + c_\gamma s_\beta s_\alpha) \\ (-c_\gamma s_\alpha + s_\gamma s_\beta c_\alpha) & s_\gamma c_\beta & (c_\gamma c_\alpha + s_\gamma s_\beta s_\alpha) \end{bmatrix}$$

$$\mathbf{R}_{90^\circ,x}\mathbf{R}_{90^\circ,z}\mathbf{R}_{90^\circ,y} = \begin{bmatrix} 0 & -1 & 0 \\ 1 & 0 & 0 \\ 0 & 0 & 1 \end{bmatrix}$$

$$\begin{bmatrix} 0 & -1 & 0 \\ 1 & 0 & 0 \\ 0 & 0 & 1 \end{bmatrix} \begin{bmatrix} 0 & 0 & 0 & 0 & 1 & 1 & 1 & 1 \\ 0 & 0 & 1 & 1 & 0 & 0 & 1 & 1 \\ 0 & 1 & 0 & 1 & 0 & 1 & 0 & 1 \end{bmatrix}$$

$$= \begin{bmatrix} 0 & 0 & -1 & -1 & 0 & 0 & -1 & -1 \\ 0 & 0 & 0 & 0 & 1 & 1 & 1 & 1 \\ 0 & 1 & 0 & 1 & 0 & 1 & 0 & 1 \end{bmatrix}.$$

This rotation sequence is illustrated in Fig. A.4 (a)–(d), where the axis of rotation is $[0 \quad 0 \quad 2]^{\mathrm{T}}$ and the angle of rotation 90°.

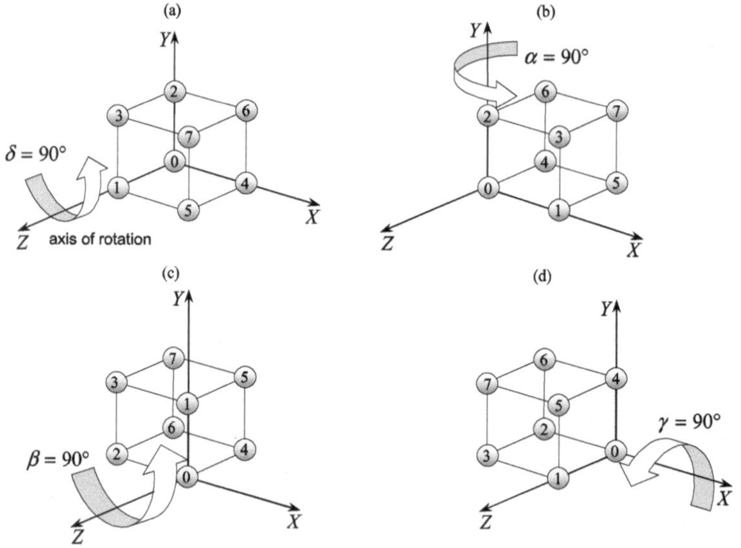

Fig. A.4 Four views of the unit cube before and during the three rotations $\mathbf{R}_{90^\circ,x}\mathbf{R}_{90^\circ,z}\mathbf{R}_{90^\circ,y}$

A.6 $\mathbf{R}_{\gamma,y}\mathbf{R}_{\beta,x}\mathbf{R}_{\alpha,y}$

$$\mathbf{R}_{\gamma,y}\mathbf{R}_{\beta,x}\mathbf{R}_{\alpha,y} = \begin{bmatrix} c_\gamma & 0 & s_\gamma \\ 0 & 1 & 0 \\ -s_\gamma & 0 & c_\gamma \end{bmatrix} \begin{bmatrix} 1 & 0 & 0 \\ 0 & c_\beta & -s_\beta \\ 0 & s_\beta & c_\beta \end{bmatrix} \begin{bmatrix} c_\alpha & 0 & s_\alpha \\ 0 & 1 & 0 \\ -s_\alpha & 0 & c_\alpha \end{bmatrix}$$

$$= \begin{bmatrix} (c_\gamma c_\alpha - s_\gamma c_\beta s_\alpha) & s_\gamma s_\beta & (c_\gamma s_\alpha + s_\gamma c_\beta c_\alpha) \\ s_\beta s_\alpha & c_\beta & -s_\beta c_\alpha \\ (-s_\gamma c_\alpha - c_\gamma c_\beta s_\alpha) & c_\gamma s_\beta & (-s_\gamma s_\alpha + c_\gamma c_\beta c_\alpha) \end{bmatrix}$$

$$\mathbf{R}_{90^\circ,y}\mathbf{R}_{90^\circ,x}\mathbf{R}_{90^\circ,y} = \begin{bmatrix} 0 & 1 & 0 \\ 1 & 0 & 0 \\ 0 & 0 & -1 \end{bmatrix}$$

$$\begin{bmatrix} 0 & 1 & 0 \\ 1 & 0 & 0 \\ 0 & 0 & -1 \end{bmatrix} \begin{bmatrix} 0 & 0 & 0 & 0 & 1 & 1 & 1 & 1 \\ 0 & 0 & 1 & 1 & 0 & 0 & 1 & 1 \\ 0 & 1 & 0 & 1 & 0 & 1 & 0 & 1 \end{bmatrix}$$

$$= \begin{bmatrix} 0 & 0 & 1 & 1 & 0 & 0 & 1 & 1 \\ 0 & 0 & 0 & 0 & 1 & 1 & 1 & 1 \\ 0 & -1 & 0 & -1 & 0 & -1 & 0 & -1 \end{bmatrix}.$$

This rotation sequence is illustrated in Fig. A.5 (a)–(d), where the axis of rotation is $[2 \quad 2 \quad 0]^{\mathrm{T}}$ and the angle of rotation 180°.

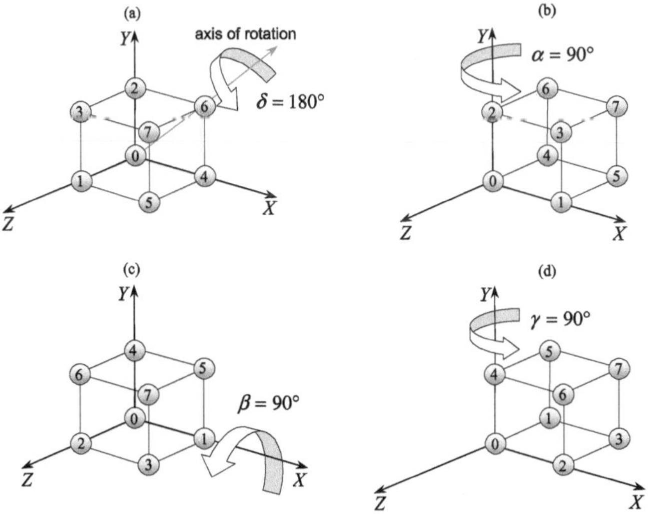

Fig. A.5 Four views of the unit cube before and during the three rotations $\mathbf{R}_{90^\circ,y}\mathbf{R}_{90^\circ,x}\mathbf{R}_{90^\circ,y}$

A.7 $\mathbf{R}_{\gamma,y}\mathbf{R}_{\beta,x}\mathbf{R}_{\alpha,z}$

$$\mathbf{R}_{\gamma,y}\mathbf{R}_{\beta,x}\mathbf{R}_{\alpha,z} = \begin{bmatrix} c_\gamma & 0 & s_\gamma \\ 0 & 1 & 0 \\ -s_\gamma & 0 & c_\gamma \end{bmatrix} \begin{bmatrix} 1 & 0 & 0 \\ 0 & c_\beta & -s_\beta \\ 0 & s_\beta & c_\beta \end{bmatrix} \begin{bmatrix} c_\alpha & -s_\alpha & 0 \\ s_\alpha & c_\alpha & 0 \\ 0 & 0 & 1 \end{bmatrix}$$

$$= \begin{bmatrix} (c_\gamma c_\alpha + s_\gamma s_\beta s_\alpha) & (-c_\gamma s_\alpha + s_\gamma s_\beta c_\alpha) & s_\gamma c_\beta \\ c_\beta s_\alpha & c_\beta c_\alpha & -s_\beta \\ (-s_\gamma c_\alpha + c_\gamma s_\beta s_\alpha) & (s_\gamma s_\alpha + c_\gamma s_\beta c_\alpha) & c_\gamma c_\beta \end{bmatrix}$$

$$\mathbf{R}_{90^\circ,y}\mathbf{R}_{90^\circ,x}\mathbf{R}_{90^\circ,z} = \begin{bmatrix} 1 & 0 & 0 \\ 0 & 0 & -1 \\ 0 & 1 & 0 \end{bmatrix}$$

$$\begin{bmatrix} 1 & 0 & 0 \\ 0 & 0 & -1 \\ 0 & 1 & 0 \end{bmatrix} \begin{bmatrix} 0 & 0 & 0 & 0 & 1 & 1 & 1 & 1 \\ 0 & 0 & 1 & 1 & 0 & 0 & 1 & 1 \\ 0 & 1 & 0 & 1 & 0 & 1 & 0 & 1 \end{bmatrix}$$

$$= \begin{bmatrix} 0 & 0 & 0 & 0 & 1 & 1 & 1 & 1 \\ 0 & -1 & 0 & -1 & 0 & -1 & 0 & -1 \\ 0 & 0 & 1 & 1 & 0 & 0 & 1 & 1 \end{bmatrix}.$$

This rotation sequence is illustrated in Fig. A.6 (a)–(d), where the axis of rotation is $[2 \quad 0 \quad 0]^{\mathrm{T}}$ and the angle of rotation 90°.

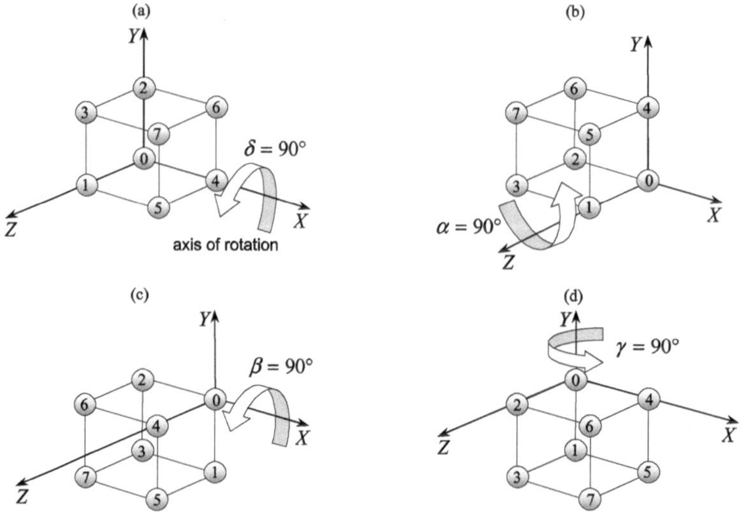

Fig. A.6 Four views of the unit cube before and during the three rotations $\mathbf{R}_{90^\circ,y}\mathbf{R}_{90^\circ,x}\mathbf{R}_{90^\circ,z}$

A.8 $\mathbf{R}_{\gamma,y}\mathbf{R}_{\beta,z}\mathbf{R}_{\alpha,x}$

$$\mathbf{R}_{\gamma,y}\mathbf{R}_{\beta,z}\mathbf{R}_{\alpha,x} = \begin{bmatrix} c_\gamma & 0 & s_\gamma \\ 0 & 1 & 0 \\ -s_\gamma & 0 & c_\gamma \end{bmatrix} \begin{bmatrix} c_\beta & -s_\beta & 0 \\ s_\beta & c_\beta & 0 \\ 0 & 0 & 1 \end{bmatrix} \begin{bmatrix} 1 & 0 & 0 \\ 0 & c_\alpha & -s_\alpha \\ 0 & s_\alpha & c_\alpha \end{bmatrix}$$

$$= \begin{bmatrix} c_\gamma c_\beta & (s_\gamma s_\alpha - c_\gamma s_\beta c_\alpha) & (s_\gamma c_\alpha + c_\gamma s_\beta s_\alpha) \\ s_\beta & c_\beta c_\alpha & -c_\beta s_\alpha \\ -s_\gamma c_\beta & (c_\gamma s_\alpha + s_\gamma s_\beta c_\alpha) & (c_\gamma c_\alpha - s_\gamma s_\beta s_\alpha) \end{bmatrix}$$

$$\mathbf{R}_{90^\circ,y}\mathbf{R}_{90^\circ,z}\mathbf{R}_{90^\circ,x} = \begin{bmatrix} 0 & 1 & 0 \\ 1 & 0 & 0 \\ 0 & 0 & -1 \end{bmatrix}$$

$$\begin{bmatrix} 0 & 1 & 0 \\ 1 & 0 & 0 \\ 0 & 0 & -1 \end{bmatrix} \begin{bmatrix} 0 & 0 & 0 & 0 & 1 & 1 & 1 & 1 \\ 0 & 0 & 1 & 1 & 0 & 0 & 1 & 1 \\ 0 & 1 & 0 & 1 & 0 & 1 & 0 & 1 \end{bmatrix}$$

$$= \begin{bmatrix} 0 & 0 & 1 & 1 & 0 & 0 & 1 & 1 \\ 0 & 0 & 0 & 0 & 1 & 1 & 1 & 1 \\ 0 & -1 & 0 & -1 & 0 & -1 & 0 & -1 \end{bmatrix}.$$

This rotation sequence is illustrated in Fig. A.7 (a)–(d), where the axis of rotation is $[2 \quad 2 \quad 0]^T$ and the angle of rotation 180°.

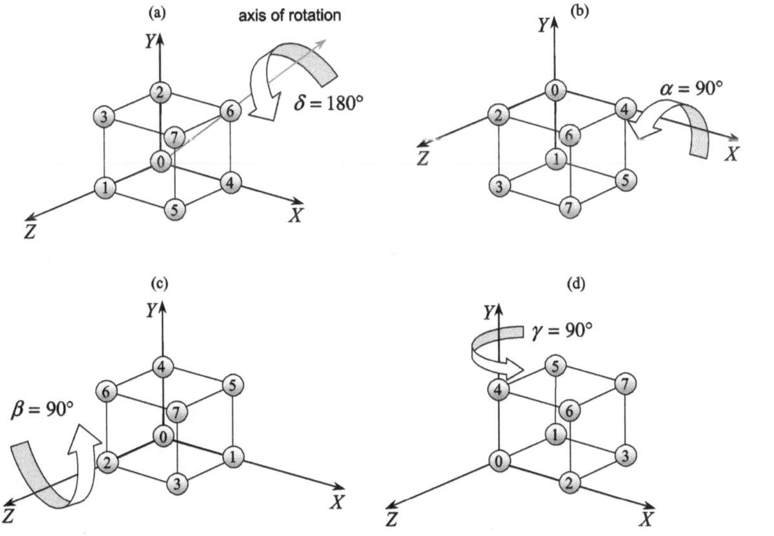

Fig. A.7 Four views of the unit cube before and during the three rotations $\mathbf{R}_{90^\circ,y}\mathbf{R}_{90^\circ,z}\mathbf{R}_{90^\circ,x}$

A.9 $\mathbf{R}_{\gamma,y}\mathbf{R}_{\beta,z}\mathbf{R}_{\alpha,y}$

$$\mathbf{R}_{\gamma,y}\mathbf{R}_{\beta,z}\mathbf{R}_{\alpha,y} = \begin{bmatrix} c_\gamma & 0 & s_\gamma \\ 0 & 1 & 0 \\ -s_\gamma & 0 & c_\gamma \end{bmatrix} \begin{bmatrix} c_\beta & -s_\beta & 0 \\ s_\beta & c_\beta & 0 \\ 0 & 0 & 1 \end{bmatrix} \begin{bmatrix} c_\alpha & 0 & s_\alpha \\ 0 & 1 & 0 \\ -s_\alpha & 0 & c_\alpha \end{bmatrix}$$

$$= \begin{bmatrix} (-s_\gamma s_\alpha + c_\gamma c_\beta c_\alpha) & -c_\gamma s_\beta & (s_\gamma c_\alpha + c_\gamma c_\beta s_\alpha) \\ s_\beta c_\alpha & c_\beta & s_\beta s_\alpha \\ (-c_\gamma s_\alpha - s_\gamma c_\beta c_\alpha) & s_\gamma s_\beta & (c_\gamma c_\alpha - s_\gamma c_\beta s_\alpha) \end{bmatrix}$$

$$\mathbf{R}_{90^\circ,y}\mathbf{R}_{90^\circ,z}\mathbf{R}_{90^\circ,y} = \begin{bmatrix} -1 & 0 & 0 \\ 0 & 0 & 1 \\ 0 & 1 & 0 \end{bmatrix}$$

$$\begin{bmatrix} -1 & 0 & 0 \\ 0 & 0 & 1 \\ 0 & 1 & 0 \end{bmatrix} \begin{bmatrix} 0 & 0 & 0 & 0 & 1 & 1 & 1 & 1 \\ 0 & 0 & 1 & 1 & 0 & 0 & 1 & 1 \\ 0 & 1 & 0 & 1 & 0 & 1 & 0 & 1 \end{bmatrix}$$

$$= \begin{bmatrix} 0 & 0 & 0 & 0 & -1 & -1 & -1 & -1 \\ 0 & 1 & 0 & 1 & 0 & 1 & 0 & 1 \\ 0 & 0 & 1 & 1 & 0 & 0 & 1 & 1 \end{bmatrix}.$$

This rotation sequence is illustrated in Fig. A.8 (a)–(d), where the axis of rotation is $[0 \quad 2 \quad 2]^{\mathrm{T}}$ and the angle of rotation 180°.

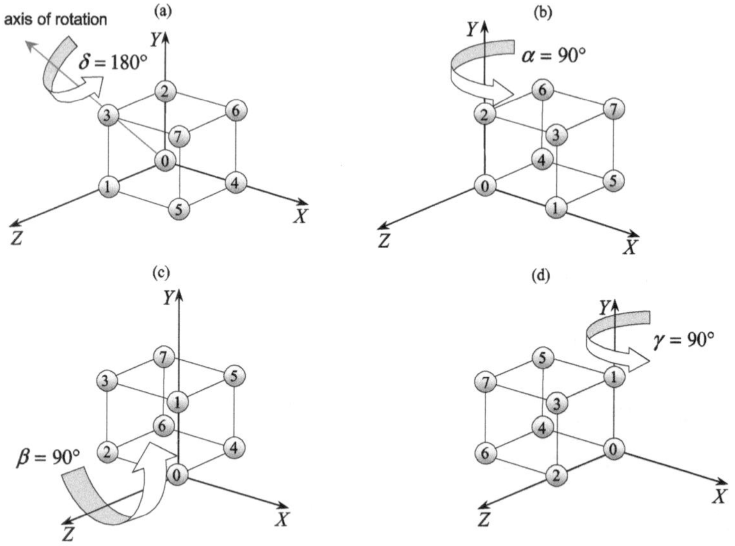

Fig. A.8 Four views of the unit cube before and during the three rotations $\mathbf{R}_{90^\circ,y}\mathbf{R}_{90^\circ,z}\mathbf{R}_{90^\circ,y}$

A.10 $\mathbf{R}_{\gamma,z}\mathbf{R}_{\beta,x}\mathbf{R}_{\alpha,y}$

$$\mathbf{R}_{\gamma,z}\mathbf{R}_{\beta,x}\mathbf{R}_{\alpha,y} = \begin{bmatrix} c_\gamma & -s_\gamma & 0 \\ s_\gamma & c_\gamma & 0 \\ 0 & 0 & 1 \end{bmatrix}\begin{bmatrix} 1 & 0 & 0 \\ 0 & c_\beta & -s_\beta \\ 0 & s_\beta & c_\beta \end{bmatrix}\begin{bmatrix} c_\alpha & 0 & s_\alpha \\ 0 & 1 & 0 \\ -s_\alpha & 0 & c_\alpha \end{bmatrix}$$

$$= \begin{bmatrix} (c_\gamma c_\alpha - s_\gamma s_\beta s_\alpha) & -s_\gamma c_\beta & (c_\gamma s_\alpha + s_\gamma s_\beta c_\alpha) \\ (s_\gamma c_\alpha + c_\gamma s_\beta s_\alpha) & c_\gamma c_\beta & (s_\gamma s_\alpha - c_\gamma s_\beta c_\alpha) \\ -c_\beta s_\alpha & s_\beta & c_\beta c_\alpha \end{bmatrix}$$

$$\mathbf{R}_{90^\circ,z}\mathbf{R}_{90^\circ,x}\mathbf{R}_{90^\circ,y} = \begin{bmatrix} -1 & 0 & 0 \\ 0 & 0 & 1 \\ 0 & 1 & 0 \end{bmatrix}$$

$$\begin{bmatrix} -1 & 0 & 0 \\ 0 & 0 & 1 \\ 0 & 1 & 0 \end{bmatrix}\begin{bmatrix} 0 & 0 & 0 & 0 & 1 & 1 & 1 & 1 \\ 0 & 0 & 1 & 1 & 0 & 0 & 1 & 1 \\ 0 & 1 & 0 & 1 & 0 & 1 & 0 & 1 \end{bmatrix}$$

$$= \begin{bmatrix} 0 & 0 & 0 & 0 & -1 & -1 & -1 & -1 \\ 0 & 1 & 0 & 1 & 0 & 1 & 0 & 1 \\ 0 & 0 & 1 & 1 & 0 & 0 & 1 & 1 \end{bmatrix}.$$

This rotation sequence is illustrated in Fig. A.9 (a)–(d), where the axis of rotation is $[0 \quad 2 \quad 2]^{\mathrm{T}}$ and the angle of rotation 180°.

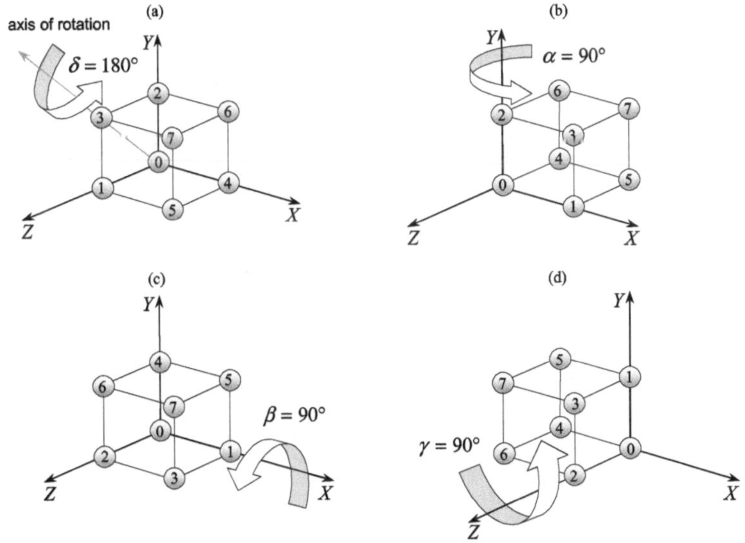

Fig. A.9 Four views of the unit cube before and during the three rotations $\mathbf{R}_{90^\circ,z}\mathbf{R}_{90^\circ,x}\mathbf{R}_{90^\circ,y}$

A.11 $\mathbf{R}_{\gamma,z}\mathbf{R}_{\beta,x}\mathbf{R}_{\alpha,z}$

$$\mathbf{R}_{\gamma,z}\mathbf{R}_{\beta,x}\mathbf{R}_{\alpha,z} = \begin{bmatrix} c_\gamma & -s_\gamma & 0 \\ s_\gamma & c_\gamma & 0 \\ 0 & 0 & 1 \end{bmatrix}\begin{bmatrix} 1 & 0 & 0 \\ 0 & c_\beta & -s_\beta \\ 0 & s_\beta & c_\beta \end{bmatrix}\begin{bmatrix} c_\alpha & -s_\alpha & 0 \\ s_\alpha & c_\alpha & 0 \\ 0 & 0 & 1 \end{bmatrix}$$

$$= \begin{bmatrix} (c_\gamma c_\alpha - s_\gamma c_\beta s_\alpha) & (-c_\gamma s_\alpha - s_\gamma c_\beta c_\alpha) & s_\gamma s_\beta \\ (s_\gamma c_\alpha + c_\gamma c_\beta s_\alpha) & (-s_\gamma s_\alpha + c_\gamma c_\beta c_\alpha) & -c_\gamma s_\beta \\ s_\beta s_\alpha & s_\beta c_\alpha & c_\beta \end{bmatrix}$$

$$\mathbf{R}_{90^\circ,z}\mathbf{R}_{90^\circ,x}\mathbf{R}_{90^\circ,z} = \begin{bmatrix} 0 & 0 & 1 \\ 0 & -1 & 0 \\ 1 & 0 & 0 \end{bmatrix}$$

$$\begin{bmatrix} 0 & 0 & 1 \\ 0 & -1 & 0 \\ 1 & 0 & 0 \end{bmatrix}\begin{bmatrix} 0 & 0 & 0 & 0 & 1 & 1 & 1 & 1 \\ 0 & 0 & 1 & 1 & 0 & 0 & 1 & 1 \\ 0 & 1 & 0 & 1 & 0 & 1 & 0 & 1 \end{bmatrix}$$

$$= \begin{bmatrix} 0 & 1 & 0 & 1 & 0 & 1 & 0 & 1 \\ 0 & 0 & -1 & -1 & 0 & 0 & -1 & -1 \\ 0 & 0 & 0 & 0 & 1 & 1 & 1 & 1 \end{bmatrix}.$$

This rotation sequence is illustrated in Fig. A.10 (a)–(d), where the axis of rotation is $[2 \quad 0 \quad 2]^{\mathrm{T}}$ and the angle of rotation 180°.

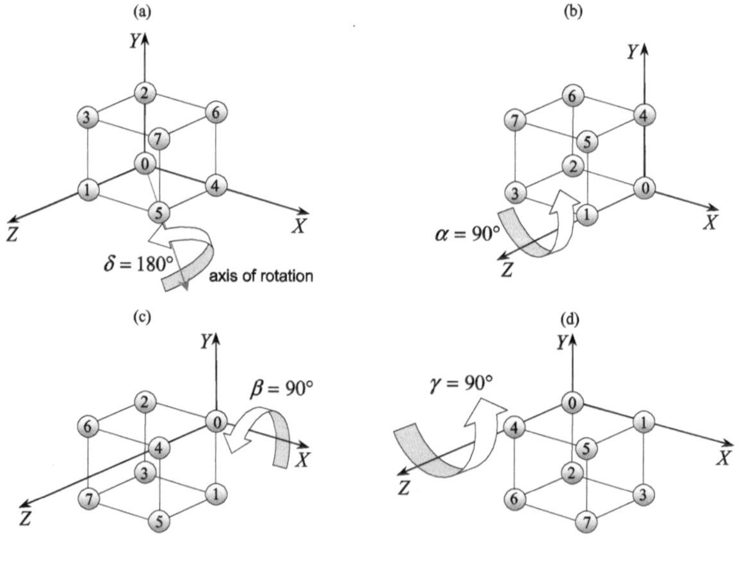

Fig. A.10 Four views of the unit cube before and during the three rotations $\mathbf{R}_{90^\circ,z}\mathbf{R}_{90^\circ,x}\mathbf{R}_{90^\circ,z}$

A.12 $\mathbf{R}_{\gamma,z}\mathbf{R}_{\beta,y}\mathbf{R}_{\alpha,x}$

$$\mathbf{R}_{\gamma,z}\mathbf{R}_{\beta,y}\mathbf{R}_{\alpha,x} = \begin{bmatrix} c_\gamma & -s_\gamma & 0 \\ s_\gamma & c_\gamma & 0 \\ 0 & 0 & 1 \end{bmatrix} \begin{bmatrix} c_\beta & 0 & s_\beta \\ 0 & 1 & 0 \\ -s_\beta & 0 & c_\beta \end{bmatrix} \begin{bmatrix} 1 & 0 & 0 \\ 0 & c_\alpha & -s_\alpha \\ 0 & s_\alpha & c_\alpha \end{bmatrix}$$

$$= \begin{bmatrix} c_\gamma c_\beta & (-s_\gamma c_\alpha + c_\gamma s_\beta s_\alpha) & (s_\gamma s_\alpha + c_\gamma s_\beta c_\alpha) \\ s_\gamma c_\beta & (c_\gamma c_\alpha + s_\gamma s_\beta s_\alpha) & (-c_\gamma s_\alpha + s_\gamma s_\beta c_\alpha) \\ -s_\beta & c_\beta s_\alpha & c_\beta c_\alpha \end{bmatrix}$$

$$\mathbf{R}_{90^\circ,z}\mathbf{R}_{90^\circ,y}\mathbf{R}_{90^\circ,x} = \begin{bmatrix} 0 & 0 & 1 \\ 0 & 1 & 0 \\ -1 & 0 & 0 \end{bmatrix}$$

$$\begin{bmatrix} 0 & 0 & 1 \\ 0 & 1 & 0 \\ -1 & 0 & 0 \end{bmatrix} \begin{bmatrix} 0 & 0 & 0 & 0 & 1 & 1 & 1 & 1 \\ 0 & 0 & 1 & 1 & 0 & 0 & 1 & 1 \\ 0 & 1 & 0 & 1 & 0 & 1 & 0 & 1 \end{bmatrix}$$

$$= \begin{bmatrix} 0 & 1 & 0 & 1 & 0 & 1 & 0 & 1 \\ 0 & 0 & 1 & 1 & 0 & 0 & 1 & 1 \\ 0 & 0 & 0 & 0 & -1 & -1 & -1 & -1 \end{bmatrix}.$$

This rotation sequence is illustrated in Fig. A.11 (a)–(d), where the axis of rotation is $[0 \quad 2 \quad 0]^{\mathrm{T}}$ and the angle of rotation 90°.

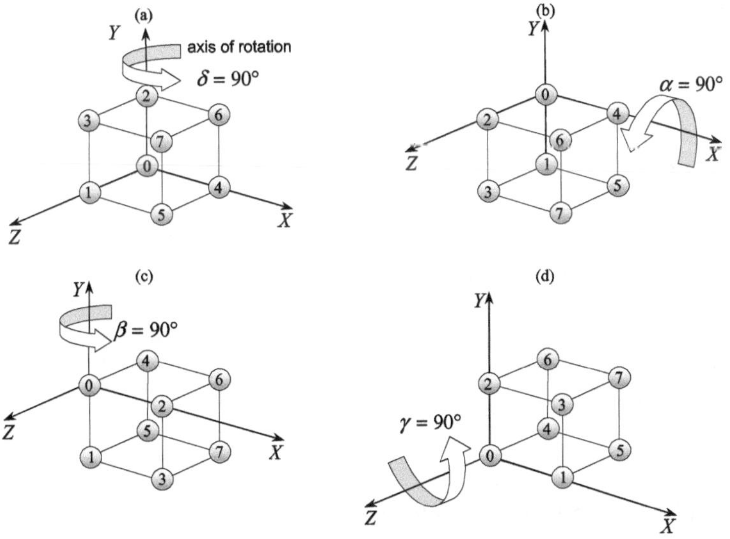

Fig. A.11 Four views of the unit cube before and during the three rotations $\mathbf{R}_{90^\circ,z}\mathbf{R}_{90^\circ,y}\mathbf{R}_{90^\circ,x}$

A.13 $\mathbf{R}_{\gamma,z}\mathbf{R}_{\beta,y}\mathbf{R}_{\alpha,z}$

$$\mathbf{R}_{\gamma,z}\mathbf{R}_{\beta,y}\mathbf{R}_{\alpha,z} = \begin{bmatrix} c_\gamma & -s_\gamma & 0 \\ s_\gamma & c_\gamma & 0 \\ 0 & 0 & 1 \end{bmatrix}\begin{bmatrix} c_\beta & 0 & s_\beta \\ 0 & 1 & 0 \\ -s_\beta & 0 & c_\beta \end{bmatrix}\begin{bmatrix} c_\alpha & -s_\alpha & 0 \\ s_\alpha & c_\alpha & 0 \\ 0 & 0 & 1 \end{bmatrix}$$

$$= \begin{bmatrix} (-s_\gamma s_\alpha + c_\gamma c_\beta c_\alpha) & (-s_\gamma c_\alpha - c_\gamma c_\beta s_\alpha) & c_\gamma s_\beta \\ (c_\gamma s_\alpha + s_\gamma c_\beta c_\alpha) & (c_\gamma c_\alpha - s_\gamma c_\beta s_\alpha) & s_\gamma s_\beta \\ -s_\beta c_\alpha & s_\beta s_\alpha & c_\beta \end{bmatrix}$$

$$\mathbf{R}_{90^\circ,z}\mathbf{R}_{90^\circ,y}\mathbf{R}_{90^\circ,z} = \begin{bmatrix} -1 & 0 & 0 \\ 0 & 0 & 1 \\ 0 & 1 & 0 \end{bmatrix}$$

$$\begin{bmatrix} -1 & 0 & 0 \\ 0 & 0 & 1 \\ 0 & 1 & 0 \end{bmatrix}\begin{bmatrix} 0 & 0 & 0 & 0 & 1 & 1 & 1 & 1 \\ 0 & 0 & 1 & 1 & 0 & 0 & 1 & 1 \\ 0 & 1 & 0 & 1 & 0 & 1 & 0 & 1 \end{bmatrix}$$

$$= \begin{bmatrix} 0 & 0 & 0 & 0 & -1 & -1 & -1 & -1 \\ 0 & 1 & 0 & 1 & 0 & 1 & 0 & 1 \\ 0 & 0 & 1 & 1 & 0 & 0 & 1 & 1 \end{bmatrix}.$$

This rotation sequence is illustrated in Fig. A.12 (a)–(d), where the axis of rotation is $[0 \quad 2 \quad 2]^{\mathrm{T}}$ and the angle of rotation 180°.

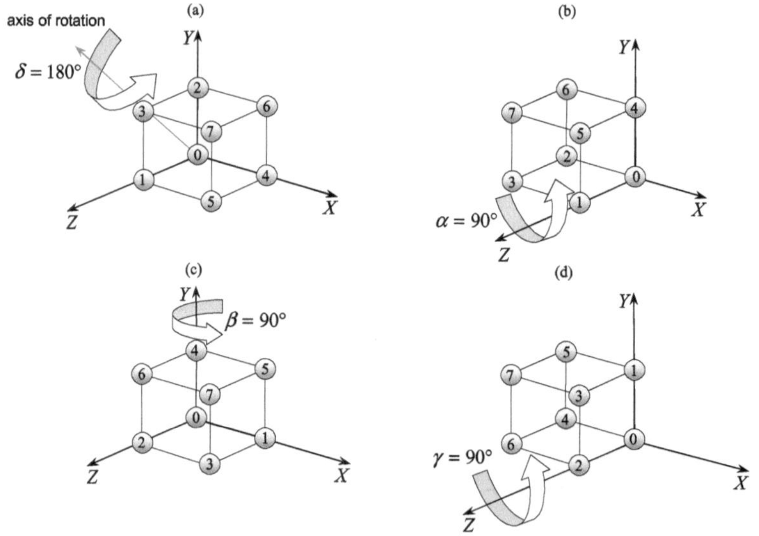

Fig. A.12 Four views of the unit cube before and during the three rotations $\mathbf{R}_{90^\circ,z}\mathbf{R}_{90^\circ,y}\mathbf{R}_{90^\circ,z}$

Appendix B
Composite Frame Rotation Sequences

B.1 Euler Rotations

This appendix lists the twelve combinations of creating a composite frame rotation sequence from $\mathbf{R}_{\alpha,x}^{-1}$, $\mathbf{R}_{\beta,y}^{-1}$ and $\mathbf{R}_{\gamma,z}^{-1}$, which are

$$\begin{array}{llll}
\mathbf{R}_{\gamma,x}^{-1}\mathbf{R}_{\beta,y}^{-1}\mathbf{R}_{\alpha,x}^{-1}, & \mathbf{R}_{\gamma,x}^{-1}\mathbf{R}_{\beta,y}^{-1}\mathbf{R}_{\alpha,z}^{-1}, & \mathbf{R}_{\gamma,x}^{-1}\mathbf{R}_{\beta,z}^{-1}\mathbf{R}_{\alpha,x}^{-1}, & \mathbf{R}_{\gamma,x}^{-1}\mathbf{R}_{\beta,z}^{-1}\mathbf{R}_{\alpha,y}^{-1} \\
\mathbf{R}_{\gamma,y}^{-1}\mathbf{R}_{\beta,x}^{-1}\mathbf{R}_{\alpha,y}^{-1}, & \mathbf{R}_{\gamma,y}^{-1}\mathbf{R}_{\beta,x}^{-1}\mathbf{R}_{\alpha,z}^{-1}, & \mathbf{R}_{\gamma,y}^{-1}\mathbf{R}_{\beta,z}^{-1}\mathbf{R}_{\alpha,x}^{-1}, & \mathbf{R}_{\gamma,y}^{-1}\mathbf{R}_{\beta,z}^{-1}\mathbf{R}_{\alpha,y}^{-1} \\
\mathbf{R}_{\gamma,z}^{-1}\mathbf{R}_{\beta,x}^{-1}\mathbf{R}_{\alpha,y}^{-1}, & \mathbf{R}_{\gamma,z}^{-1}\mathbf{R}_{\beta,x}^{-1}\mathbf{R}_{\alpha,z}^{-1}, & \mathbf{R}_{\gamma,z}^{-1}\mathbf{R}_{\beta,y}^{-1}\mathbf{R}_{\alpha,x}^{-1}, & \mathbf{R}_{\gamma,z}^{-1}\mathbf{R}_{\beta,y}^{-1}\mathbf{R}_{\alpha,z}^{-1}.
\end{array}$$

For each combination there are three Euler frame rotation matrices, the resulting composite matrix, a matrix where the three angles equal 90°, the coordinates of the unit cube in the rotated frame, the axis and angle of rotation and a figure illustrating the stages of rotation. To compute the axis of rotation $[v_1 \quad v_2 \quad v_3]^{\mathrm{T}}$ we use

$$\begin{aligned}
v_1 &= (a_{22}-1)(a_{33}-1) - a_{23}a_{32} \\
v_2 &= (a_{33}-1)(a_{11}-1) - a_{31}a_{13} \\
v_3 &= (a_{11}-1)(a_{22}-1) - a_{12}a_{21}
\end{aligned}$$

where

$$\mathbf{R} = \begin{bmatrix} a_{11} & a_{12} & a_{13} \\ a_{21} & a_{22} & a_{23} \\ a_{31} & a_{32} & a_{33} \end{bmatrix}$$

and for the angle of rotation δ we use

$$\cos\delta = \frac{1}{2}\big(\mathrm{Tr}(\mathbf{R}) - 1\big).$$

We begin by defining the three principal inverse Euler frame rotations:

$$\text{rotate the frame } \alpha \text{ about the } x\text{-axis} \quad \mathbf{R}_{\alpha,x}^{-1} = \begin{bmatrix} 1 & 0 & 0 \\ 0 & c_\alpha & s_\alpha \\ 0 & -s_\alpha & c_\alpha \end{bmatrix}$$

J. Vince, *Rotation Transforms for Computer Graphics*,
DOI 10.1007/978-0-85729-154-7, © Springer-Verlag London Limited 2011

rotate the frame β about the y-axis $\mathbf{R}_{\beta,y}^{-1} = \begin{bmatrix} c_\beta & 0 & -s_\beta \\ 0 & 1 & 0 \\ s_\beta & 0 & c_\beta \end{bmatrix}$

rotate the frame γ about the z-axis $\mathbf{R}_{\gamma,z}^{-1} = \begin{bmatrix} c_\gamma & s_\gamma & 0 \\ -s_\gamma & c_\gamma & 0 \\ 0 & 0 & 1 \end{bmatrix}$

where $c_\alpha = \cos\alpha$ and $s_\alpha = \sin\alpha$, etc.

Remember that the right-most transform is applied first and the left-most transform last. In terms of angles, the sequence is always α, β, γ.

For each composite transform you can verify that when $\alpha = \beta = \gamma = 0$ the result is the identity transform $\mathbf{I}$.

We now examine the twelve combinations in turn.

B.2 $\mathbf{R}^{-1}_{\gamma,x}\mathbf{R}^{-1}_{\beta,y}\mathbf{R}^{-1}_{\alpha,x}$

$$\mathbf{R}^{-1}_{\gamma,x}\mathbf{R}^{-1}_{\beta,y}\mathbf{R}^{-1}_{\alpha,x} = \begin{bmatrix} 1 & 0 & 0 \\ 0 & c_\gamma & s_\gamma \\ 0 & -s_\gamma & c_\gamma \end{bmatrix} \begin{bmatrix} c_\beta & 0 & -s_\beta \\ 0 & 1 & 0 \\ s_\beta & 0 & c_\beta \end{bmatrix} \begin{bmatrix} 1 & 0 & 0 \\ 0 & c_\alpha & s_\alpha \\ 0 & -s_\alpha & c_\alpha \end{bmatrix}$$

$$= \begin{bmatrix} c_\beta & s_\beta s_\alpha & -s_\beta c_\alpha \\ s_\gamma s_\beta & (c_\gamma c_\alpha - s_\gamma c_\beta s_\alpha) & (c_\gamma s_\alpha + s_\gamma c_\beta c_\alpha) \\ c_\gamma s_\beta & (-s_\gamma c_\alpha - c_\gamma c_\beta s_\alpha) & (-s_\gamma s_\alpha + c_\gamma c_\beta c_\alpha) \end{bmatrix}$$

$$\mathbf{R}^{-1}_{90^\circ,x}\mathbf{R}^{-1}_{90^\circ,y}\mathbf{R}^{-1}_{90^\circ,x} = \begin{bmatrix} 0 & 1 & 0 \\ 1 & 0 & 0 \\ 0 & 0 & -1 \end{bmatrix}$$

$$\begin{bmatrix} 0 & 1 & 0 \\ 1 & 0 & 0 \\ 0 & 0 & -1 \end{bmatrix} \begin{bmatrix} 0 & 0 & 0 & 0 & 1 & 1 & 1 & 1 \\ 0 & 0 & 1 & 1 & 0 & 0 & 1 & 1 \\ 0 & 1 & 0 & 1 & 0 & 1 & 0 & 1 \end{bmatrix}$$

$$= \begin{bmatrix} 0 & 0 & 1 & 1 & 0 & 0 & 1 & 1 \\ 0 & 0 & 0 & 0 & 1 & 1 & 1 & 1 \\ 0 & -1 & 0 & -1 & 0 & -1 & 0 & -1 \end{bmatrix}.$$

This rotation sequence is illustrated in Fig. B.1 (a)–(d), where the axis of rotation is $[2 \quad 2 \quad 0]^T$ and the angle of rotation 180°.

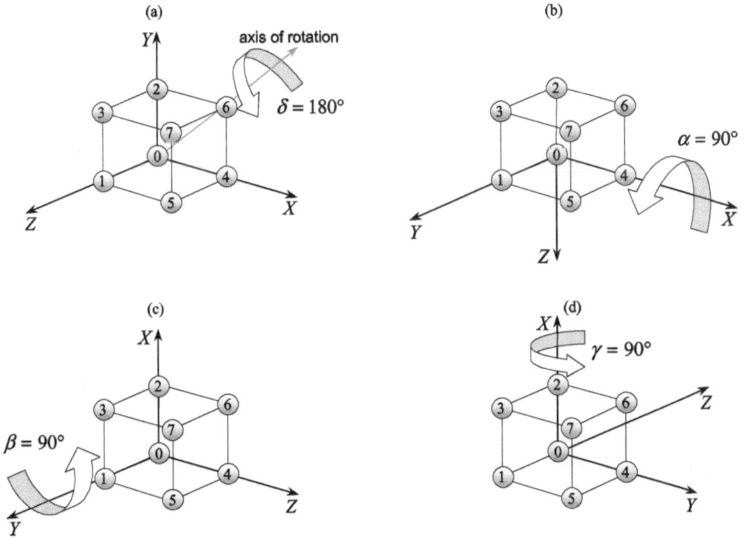

Fig. B.1 Four views of the frame before and during the three rotations $\mathbf{R}^{-1}_{90^\circ,x}\mathbf{R}^{-1}_{90^\circ,y}\mathbf{R}^{-1}_{90^\circ,x}$

B.3 $\mathbf{R}_{\gamma,x}^{-1}\mathbf{R}_{\beta,y}^{-1}\mathbf{R}_{\alpha,z}^{-1}$

$$\mathbf{R}_{\gamma,x}^{-1}\mathbf{R}_{\beta,y}^{-1}\mathbf{R}_{\alpha,z}^{-1} = \begin{bmatrix} 1 & 0 & 0 \\ 0 & c_\gamma & s_\gamma \\ 0 & -s_\gamma & c_\gamma \end{bmatrix} \begin{bmatrix} c_\beta & 0 & -s_\beta \\ 0 & 1 & 0 \\ s_\beta & 0 & c_\beta \end{bmatrix} \begin{bmatrix} c_\alpha & s_\alpha & 0 \\ -s_\alpha & c_\alpha & 0 \\ 0 & 0 & 1 \end{bmatrix}$$

$$= \begin{bmatrix} c_\beta c_\alpha & c_\beta s_\alpha & -s_\beta \\ (-c_\gamma s_\alpha + s_\gamma s_\beta c_\alpha) & (c_\gamma c_\alpha + s_\gamma s_\beta s_\alpha) & +s_\gamma c_\beta \\ (s_\gamma s_\alpha + c_\gamma s_\beta c_\alpha) & (-s_\gamma c_\alpha + c_\gamma s_\beta s_\alpha) & c_\gamma c_\beta \end{bmatrix}$$

$$\mathbf{R}_{90^\circ,x}^{-1}\mathbf{R}_{90^\circ,y}^{-1}\mathbf{R}_{90^\circ,z}^{-1} = \begin{bmatrix} 0 & 0 & -1 \\ 0 & 1 & 0 \\ 1 & 0 & 0 \end{bmatrix}$$

$$\begin{bmatrix} 0 & 0 & -1 \\ 0 & 1 & 0 \\ 1 & 0 & 0 \end{bmatrix} \begin{bmatrix} 0 & 0 & 0 & 0 & 1 & 1 & 1 & 1 \\ 0 & 0 & 1 & 1 & 0 & 0 & 1 & 1 \\ 0 & 1 & 0 & 1 & 0 & 1 & 0 & 1 \end{bmatrix}$$

$$= \begin{bmatrix} 0 & -1 & 0 & -1 & 0 & -1 & 0 & -1 \\ 0 & 0 & 1 & 1 & 0 & 0 & 1 & 1 \\ 0 & 0 & 0 & 0 & 1 & 1 & 1 & 1 \end{bmatrix}.$$

This rotation sequence is illustrated in Fig. B.2 (a)–(d), where the axis of rotation is $[0 \quad 2 \quad 0]^{\mathrm{T}}$ and the angle of rotation 90°.

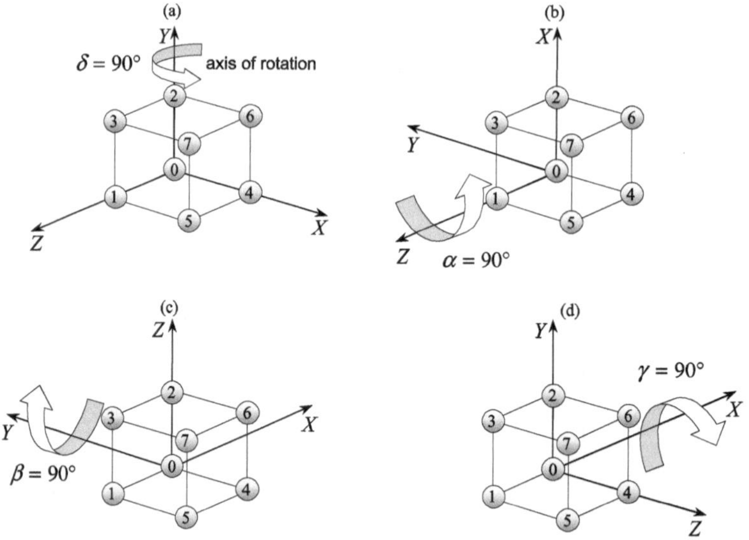

Fig. B.2 Four views of the frame before and during the three rotations $\mathbf{R}_{90^\circ,x}^{-1}\mathbf{R}_{90^\circ,y}^{-1}\mathbf{R}_{90^\circ,z}^{-1}$

B.4 $\mathbf{R}_{\gamma,x}^{-1}\mathbf{R}_{\beta,z}^{-1}\mathbf{R}_{\alpha,x}^{-1}$

$$\mathbf{R}_{\gamma,x}^{-1}\mathbf{R}_{\beta,z}^{-1}\mathbf{R}_{\alpha,x}^{-1} = \begin{bmatrix} 1 & 0 & 0 \\ 0 & c_\gamma & s_\gamma \\ 0 & -s_\gamma & c_\gamma \end{bmatrix} \begin{bmatrix} c_\beta & s_\beta & 0 \\ -s_\beta & c_\beta & 0 \\ 0 & 0 & 1 \end{bmatrix} \begin{bmatrix} 1 & 0 & 0 \\ 0 & c_\alpha & s_\alpha \\ 0 & -s_\alpha & c_\alpha \end{bmatrix}$$

$$= \begin{bmatrix} c_\beta & s_\beta c_\alpha & s_\beta s_\alpha \\ -c_\gamma s_\beta & (-s_\gamma s_\alpha + c_\gamma c_\beta c_\alpha) & (-s_\gamma c_\alpha + c_\gamma c_\beta s_\alpha) \\ s_\gamma s_\beta & (-c_\gamma s_\alpha - s_\gamma c_\beta c_\alpha) & (c_\gamma c_\alpha - s_\gamma c_\beta s_\alpha) \end{bmatrix}$$

$$\mathbf{R}_{90^\circ,x}^{-1}\mathbf{R}_{90^\circ,z}^{-1}\mathbf{R}_{90^\circ,x}^{-1} = \begin{bmatrix} 0 & 0 & 1 \\ 0 & -1 & 0 \\ 1 & 0 & 0 \end{bmatrix}$$

$$\begin{bmatrix} 0 & 0 & 1 \\ 0 & -1 & 0 \\ 1 & 0 & 0 \end{bmatrix} \begin{bmatrix} 0 & 0 & 0 & 0 & 1 & 1 & 1 & 1 \\ 0 & 0 & 1 & 1 & 0 & 0 & 1 & 1 \\ 0 & 1 & 0 & 1 & 0 & 1 & 0 & 1 \end{bmatrix}$$

$$= \begin{bmatrix} 0 & 1 & 0 & 1 & 0 & 1 & 0 & 1 \\ 0 & 0 & -1 & -1 & 0 & 0 & -1 & -1 \\ 0 & 0 & 0 & 0 & 1 & 1 & 1 & 1 \end{bmatrix}.$$

This rotation sequence is illustrated in Fig. B.3 (a)–(d), where the axis of rotation is $[2 \quad 0 \quad 2]^T$ and the angle of rotation 180°.

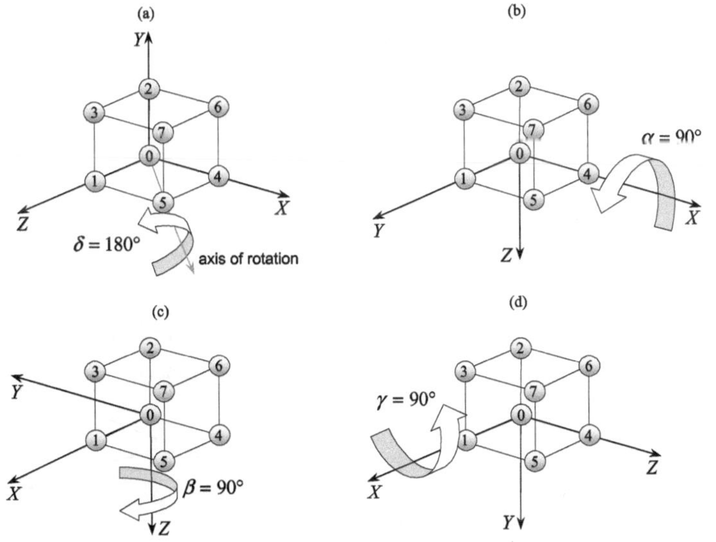

Fig. B.3 Four views of the frame before and during the three rotations $\mathbf{R}_{90^\circ,x}^{-1}\mathbf{R}_{90^\circ,z}^{-1}\mathbf{R}_{90^\circ,x}^{-1}$

B.5 $\mathbf{R}_{\gamma,x}^{-1}\mathbf{R}_{\beta,z}^{-1}\mathbf{R}_{\alpha,y}^{-1}$

$$\mathbf{R}_{\gamma,x}^{-1}\mathbf{R}_{\beta,z}^{-1}\mathbf{R}_{\alpha,y}^{-1} = \begin{bmatrix} 1 & 0 & 0 \\ 0 & c_\gamma & s_\gamma \\ 0 & -s_\gamma & c_\gamma \end{bmatrix} \begin{bmatrix} c_\beta & s_\beta & 0 \\ -s_\beta & c_\beta & 0 \\ 0 & 0 & 1 \end{bmatrix} \begin{bmatrix} c_\alpha & 0 & -s_\alpha \\ 0 & 1 & 0 \\ s_\alpha & 0 & c_\alpha \end{bmatrix}$$

$$= \begin{bmatrix} c_\beta c_\alpha & s_\beta & -c_\beta s_\alpha \\ (s_\gamma s_\alpha - c_\gamma s_\beta c_\alpha) & c_\gamma c_\beta & (s_\gamma c_\alpha + c_\gamma s_\beta s_\alpha) \\ (c_\gamma s_\alpha + s_\gamma s_\beta c_\alpha) & -s_\gamma c_\beta & (c_\gamma c_\alpha - s_\gamma s_\beta s_\alpha) \end{bmatrix}$$

$$\mathbf{R}_{90^\circ,x}^{-1}\mathbf{R}_{90^\circ,z}^{-1}\mathbf{R}_{90^\circ,y}^{-1} = \begin{bmatrix} 0 & 1 & 0 \\ 1 & 0 & 0 \\ 0 & 0 & -1 \end{bmatrix}$$

$$\begin{bmatrix} 0 & 1 & 0 \\ 1 & 0 & 0 \\ 0 & 0 & -1 \end{bmatrix} \begin{bmatrix} 0 & 0 & 0 & 0 & 1 & 1 & 1 & 1 \\ 0 & 0 & 1 & 1 & 0 & 0 & 1 & 1 \\ 0 & 1 & 0 & 1 & 0 & 1 & 0 & 1 \end{bmatrix}$$

$$= \begin{bmatrix} 0 & 0 & 1 & 1 & 0 & 0 & 1 & 1 \\ 0 & 0 & 0 & 0 & 1 & 1 & 1 & 1 \\ 0 & -1 & 0 & -1 & 0 & -1 & 0 & -1 \end{bmatrix}.$$

This rotation sequence is illustrated in Fig. B.4 (a)–(d), where the axis of rotation is $[2 \quad 2 \quad 0]^T$ and the angle of rotation 180°.

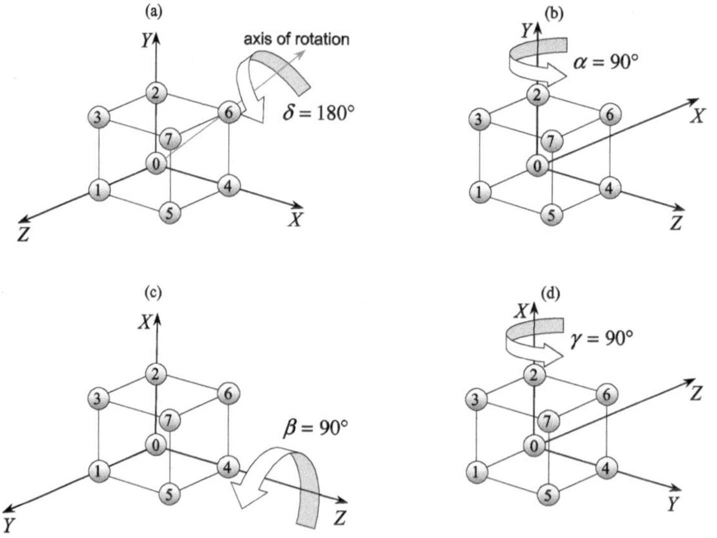

Fig. B.4 Four views of the frame before and during the three rotations $\mathbf{R}_{90^\circ,x}^{-1}\mathbf{R}_{90^\circ,z}^{-1}\mathbf{R}_{90^\circ,y}^{-1}$

B.6 $\mathbf{R}_{\gamma,y}^{-1}\mathbf{R}_{\beta,x}^{-1}\mathbf{R}_{\alpha,y}^{-1}$

$$\mathbf{R}_{\gamma,y}^{-1}\mathbf{R}_{\beta,x}^{-1}\mathbf{R}_{\alpha,y}^{-1} = \begin{bmatrix} c_\gamma & 0 & -s_\gamma \\ 0 & 1 & 0 \\ s_\gamma & 0 & c_\gamma \end{bmatrix}\begin{bmatrix} 1 & 0 & 0 \\ 0 & c_\beta & s_\beta \\ 0 & -s_\beta & c_\beta \end{bmatrix}\begin{bmatrix} c_\alpha & 0 & -s_\alpha \\ 0 & 1 & 0 \\ s_\alpha & 0 & c_\alpha \end{bmatrix}$$

$$= \begin{bmatrix} (c_\gamma c_\alpha - s_\gamma c_\beta s_\alpha) & s_\gamma s_\beta & (-c_\gamma s_\alpha - s_\gamma c_\beta c_\alpha) \\ s_\beta s_\alpha & c_\beta & s_\beta c_\alpha \\ (s_\gamma c_\alpha + c_\gamma c_\beta s_\alpha) & -c_\gamma s_\beta & (-s_\gamma s_\alpha + c_\gamma c_\beta c_\alpha) \end{bmatrix}$$

$$\mathbf{R}_{90^\circ,y}^{-1}\mathbf{R}_{90^\circ,x}^{-1}\mathbf{R}_{90^\circ,y}^{-1} = \begin{bmatrix} 0 & 1 & 0 \\ 1 & 0 & 0 \\ 0 & 0 & -1 \end{bmatrix}$$

$$\begin{bmatrix} 0 & 1 & 0 \\ 1 & 0 & 0 \\ 0 & 0 & -1 \end{bmatrix}\begin{bmatrix} 0 & 0 & 0 & 0 & 1 & 1 & 1 & 1 \\ 0 & 0 & 1 & 1 & 0 & 0 & 1 & 1 \\ 0 & 1 & 0 & 1 & 0 & 1 & 0 & 1 \end{bmatrix}$$

$$= \begin{bmatrix} 0 & 0 & 1 & 1 & 0 & 0 & 1 & 1 \\ 0 & 0 & 0 & 0 & 1 & 1 & 1 & 1 \\ 0 & -1 & 0 & -1 & 0 & -1 & 0 & -1 \end{bmatrix}.$$

This rotation sequence is illustrated in Fig. B.5 (a)–(d), where the axis of rotation is $[2 \quad 2 \quad 0]^{\mathrm{T}}$ and the angle of rotation 180°.

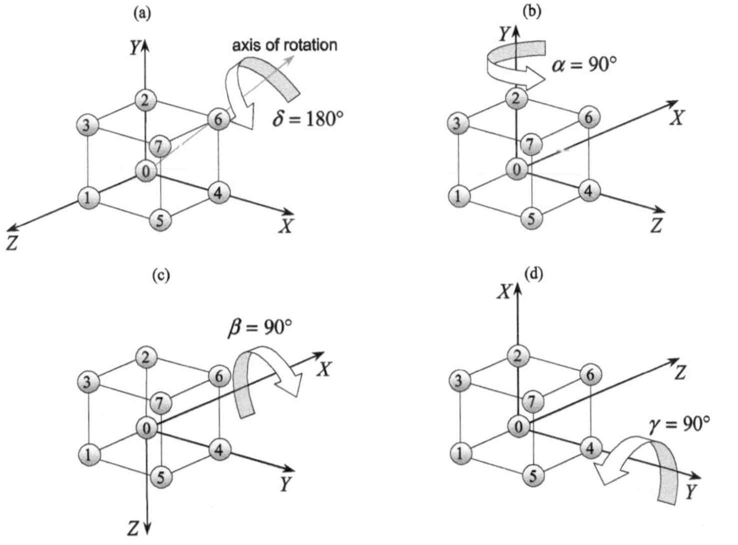

Fig. B.5 Four views of the frame before and during the three rotations $\mathbf{R}_{90^\circ,y}^{-1}\mathbf{R}_{90^\circ,x}^{-1}\mathbf{R}_{90^\circ,y}^{-1}$

B.7 $\mathbf{R}^{-1}_{\gamma,y}\mathbf{R}^{-1}_{\beta,x}\mathbf{R}^{-1}_{\alpha,z}$

$$\mathbf{R}^{-1}_{\gamma,y}\mathbf{R}^{-1}_{\beta,x}\mathbf{R}^{-1}_{\alpha,z} = \begin{bmatrix} c_\gamma & 0 & -s_\gamma \\ 0 & 1 & 0 \\ s_\gamma & 0 & c_\gamma \end{bmatrix} \begin{bmatrix} 1 & 0 & 0 \\ 0 & c_\beta & s_\beta \\ 0 & -s_\beta & c_\beta \end{bmatrix} \begin{bmatrix} c_\alpha & s_\alpha & 0 \\ -s_\alpha & c_\alpha & 0 \\ 0 & 0 & 1 \end{bmatrix}$$

$$= \begin{bmatrix} (c_\gamma c_\alpha - s_\gamma s_\beta s_\alpha) & (c_\gamma s_\alpha + s_\gamma s_\beta c_\alpha) & -s_\gamma c_\beta \\ -c_\beta s_\alpha & c_\beta c_\alpha & s_\beta \\ (s_\gamma c_\alpha + c_\gamma s_\beta s_\alpha) & (s_\gamma s_\alpha - c_\gamma s_\beta c_\alpha) & c_\gamma c_\beta \end{bmatrix}$$

$$\mathbf{R}^{-1}_{90^\circ,y}\mathbf{R}^{-1}_{90^\circ,x}\mathbf{R}^{-1}_{90^\circ,z} = \begin{bmatrix} -1 & 0 & 0 \\ 0 & 0 & 1 \\ 0 & 1 & 0 \end{bmatrix}$$

$$\begin{bmatrix} -1 & 0 & 0 \\ 0 & 0 & 1 \\ 0 & 1 & 0 \end{bmatrix} \begin{bmatrix} 0 & 0 & 0 & 0 & 1 & 1 & 1 & 1 \\ 0 & 0 & 1 & 1 & 0 & 0 & 1 & 1 \\ 0 & 1 & 0 & 1 & 0 & 1 & 0 & 1 \end{bmatrix}$$

$$= \begin{bmatrix} 0 & 0 & 0 & 0 & -1 & -1 & -1 & -1 \\ 0 & 1 & 0 & 1 & 0 & 1 & 0 & 1 \\ 0 & 0 & 1 & 1 & 0 & 0 & 1 & 1 \end{bmatrix}.$$

This rotation sequence is illustrated in Fig. B.6 (a)–(d), where the axis of rotation is $[0 \quad 2 \quad 2]^{\mathrm{T}}$ and the angle of rotation 180°.

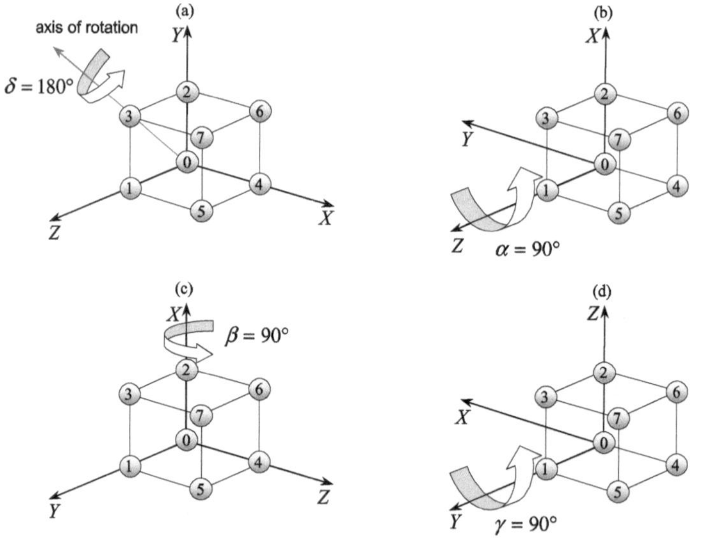

Fig. B.6 Four views of the frame before and during the three rotations $\mathbf{R}^{-1}_{90^\circ,y}\mathbf{R}^{-1}_{90^\circ,x}\mathbf{R}^{-1}_{90^\circ,z}$

B.8 $\mathbf{R}_{\gamma,y}^{-1}\mathbf{R}_{\beta,z}^{-1}\mathbf{R}_{\alpha,x}^{-1}$

$$\mathbf{R}_{\gamma,y}^{-1}\mathbf{R}_{\beta,z}^{-1}\mathbf{R}_{\alpha,x}^{-1} = \begin{bmatrix} c_\gamma & 0 & -s_\gamma \\ 0 & 1 & 0 \\ s_\gamma & 0 & c_\gamma \end{bmatrix} \begin{bmatrix} c_\beta & s_\beta & 0 \\ -s_\beta & c_\beta & 0 \\ 0 & 0 & 1 \end{bmatrix} \begin{bmatrix} 1 & 0 & 0 \\ 0 & c_\alpha & s_\alpha \\ 0 & -s_\alpha & c_\alpha \end{bmatrix}$$

$$= \begin{bmatrix} c_\gamma c_\beta & (s_\gamma s_\alpha + c_\gamma s_\beta c_\alpha) & (-s_\gamma c_\alpha + c_\gamma s_\beta s_\alpha) \\ -s_\beta & c_\beta c_\alpha & c_\beta s_\alpha \\ s_\gamma c_\beta & (-c_\gamma s_\alpha + s_\gamma s_\beta c_\alpha) & (c_\gamma c_\alpha + s_\gamma s_\beta s_\alpha) \end{bmatrix}$$

$$\mathbf{R}_{90^\circ,y}^{-1}\mathbf{R}_{90^\circ,z}^{-1}\mathbf{R}_{90^\circ,x}^{-1} = \begin{bmatrix} 0 & 1 & 0 \\ -1 & 0 & 0 \\ 0 & 0 & 1 \end{bmatrix}$$

$$\begin{bmatrix} 0 & 1 & 0 \\ -1 & 0 & 0 \\ 0 & 0 & 1 \end{bmatrix} \begin{bmatrix} 0 & 0 & 0 & 0 & 1 & 1 & 1 & 1 \\ 0 & 0 & 1 & 1 & 0 & 0 & 1 & 1 \\ 0 & 1 & 0 & 1 & 0 & 1 & 0 & 1 \end{bmatrix}$$

$$= \begin{bmatrix} 0 & 0 & 1 & 1 & 0 & 0 & 1 & 1 \\ 0 & 0 & 0 & 0 & -1 & -1 & -1 & -1 \\ 0 & 1 & 0 & 1 & 0 & 1 & 0 & 1 \end{bmatrix}.$$

This rotation sequence is illustrated in Fig. B.7 (a)–(d), where the axis of rotation is $[2 \quad 2 \quad 0]^{\mathrm{T}}$ and the angle of rotation 180°.

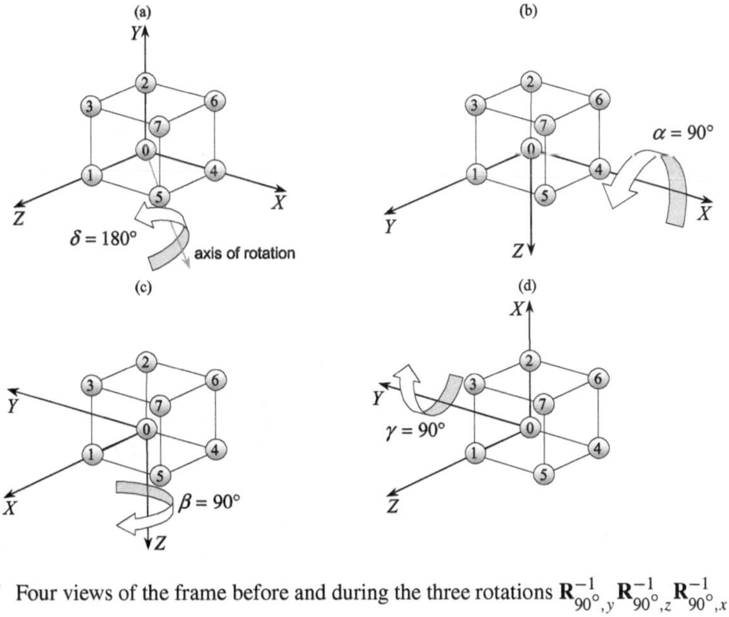

Fig. B.7 Four views of the frame before and during the three rotations $\mathbf{R}_{90^\circ,y}^{-1}\mathbf{R}_{90^\circ,z}^{-1}\mathbf{R}_{90^\circ,x}^{-1}$

B.9 $\mathbf{R}_{\gamma,y}^{-1}\mathbf{R}_{\beta,z}^{-1}\mathbf{R}_{\alpha,y}^{-1}$

$$\mathbf{R}_{\gamma,y}^{-1}\mathbf{R}_{\beta,z}^{-1}\mathbf{R}_{\alpha,y}^{-1} = \begin{bmatrix} c_\gamma & 0 & -s_\gamma \\ 0 & 1 & 0 \\ s_\gamma & 0 & c_\gamma \end{bmatrix} \begin{bmatrix} c_\beta & s_\beta & 0 \\ -s_\beta & c_\beta & 0 \\ 0 & 0 & 1 \end{bmatrix} \begin{bmatrix} c_\alpha & 0 & -s_\alpha \\ 0 & 1 & 0 \\ s_\alpha & 0 & c_\alpha \end{bmatrix}$$

$$= \begin{bmatrix} (-s_\gamma s_\alpha + c_\gamma c_\beta c_\alpha) & c_\gamma s_\beta & (-s_\gamma c_\alpha - c_\gamma c_\beta s_\alpha) \\ -s_\beta c_\alpha & c_\beta & s_\beta s_\alpha \\ (c_\gamma s_\alpha + s_\gamma c_\beta c_\alpha) & s_\gamma s_\beta & (c_\gamma c_\alpha - s_\gamma c_\beta s_\alpha) \end{bmatrix}$$

$$\mathbf{R}_{90^\circ,y}^{-1}\mathbf{R}_{90^\circ,z}^{-1}\mathbf{R}_{90^\circ,y}^{-1} = \begin{bmatrix} -1 & 0 & 0 \\ 0 & 0 & 1 \\ 0 & 1 & 0 \end{bmatrix}$$

$$\begin{bmatrix} -1 & 0 & 0 \\ 0 & 0 & 1 \\ 0 & 1 & 0 \end{bmatrix} \begin{bmatrix} 0 & 0 & 0 & 0 & 1 & 1 & 1 & 1 \\ 0 & 0 & 1 & 1 & 0 & 0 & 1 & 1 \\ 0 & 1 & 0 & 1 & 0 & 1 & 0 & 1 \end{bmatrix}$$

$$= \begin{bmatrix} 0 & 0 & 0 & 0 & -1 & -1 & -1 & -1 \\ 0 & 1 & 0 & 1 & 0 & 1 & 0 & 1 \\ 0 & 0 & 1 & 1 & 0 & 0 & 1 & 1 \end{bmatrix}.$$

This rotation sequence is illustrated in Fig. B.8 (a)–(d), where the axis of rotation is $[0 \quad 2 \quad 2]^{\mathrm{T}}$ and the angle of rotation 180°.

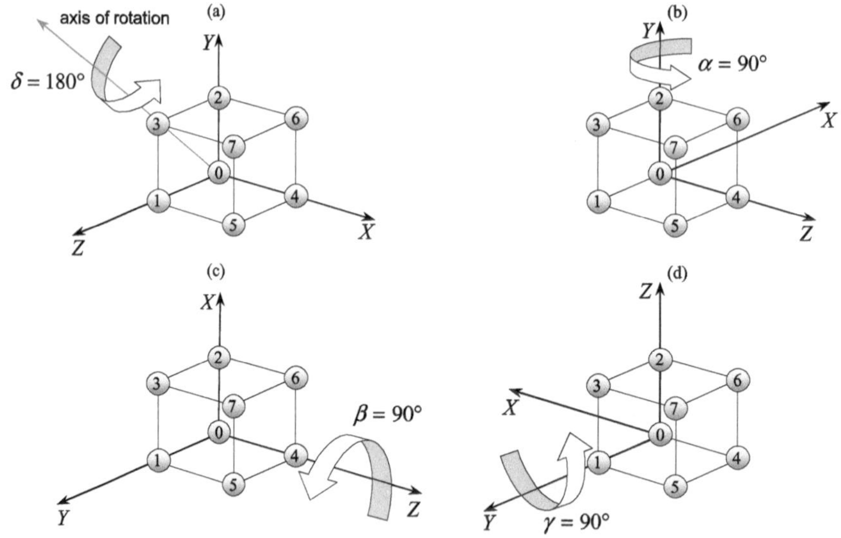

Fig. B.8 Four views of the frame before and during the three rotations $\mathbf{R}_{90^\circ,y}^{-1}\mathbf{R}_{90^\circ,z}^{-1}\mathbf{R}_{90^\circ,y}^{-1}$

B.10 $\mathbf{R}_{\gamma,z}^{-1}\mathbf{R}_{\beta,x}^{-1}\mathbf{R}_{\alpha,y}^{-1}$

$$\mathbf{R}_{\gamma,z}^{-1}\mathbf{R}_{\beta,x}^{-1}\mathbf{R}_{\alpha,y}^{-1} = \begin{bmatrix} c_\gamma & s_\gamma & 0 \\ -s_\gamma & c_\gamma & 0 \\ 0 & 0 & 1 \end{bmatrix} \begin{bmatrix} 1 & 0 & 0 \\ 0 & c_\beta & s_\beta \\ 0 & -s_\beta & c_\beta \end{bmatrix} \begin{bmatrix} c_\alpha & 0 & -s_\alpha \\ 0 & 1 & 0 \\ s_\alpha & 0 & c_\alpha \end{bmatrix}$$

$$= \begin{bmatrix} (c_\gamma c_\alpha + s_\gamma s_\beta s_\alpha) & s_\gamma c_\beta & (-c_\gamma s_\alpha + s_\gamma s_\beta c_\alpha) \\ (-s_\gamma c_\alpha + c_\gamma s_\beta s_\alpha) & c_\gamma c_\beta & (s_\gamma s_\alpha + c_\gamma s_\beta c_\alpha) \\ c_\beta s_\alpha & -s_\beta & c_\beta c_\alpha \end{bmatrix}$$

$$\mathbf{R}_{90^\circ,z}^{-1}\mathbf{R}_{90^\circ,x}^{-1}\mathbf{R}_{90^\circ,y}^{-1} = \begin{bmatrix} 1 & 0 & 0 \\ 0 & 0 & 1 \\ 0 & -1 & 0 \end{bmatrix}$$

$$\begin{bmatrix} 1 & 0 & 0 \\ 0 & 0 & 1 \\ 0 & -1 & 0 \end{bmatrix} \begin{bmatrix} 0 & 0 & 0 & 0 & 1 & 1 & 1 & 1 \\ 0 & 0 & 1 & 1 & 0 & 0 & 1 & 1 \\ 0 & 1 & 0 & 1 & 0 & 1 & 0 & 1 \end{bmatrix}$$

$$= \begin{bmatrix} 0 & 0 & 0 & 0 & 1 & 1 & 1 & 1 \\ 0 & 1 & 0 & 1 & 0 & 1 & 0 & 1 \\ 0 & 0 & -1 & -1 & 0 & 0 & -1 & -1 \end{bmatrix}.$$

This rotation sequence is illustrated in Fig. B.9 (a)–(d), where the axis of rotation is $[2 \quad 0 \quad 0]^{\mathrm{T}}$ and the angle of rotation 90°.

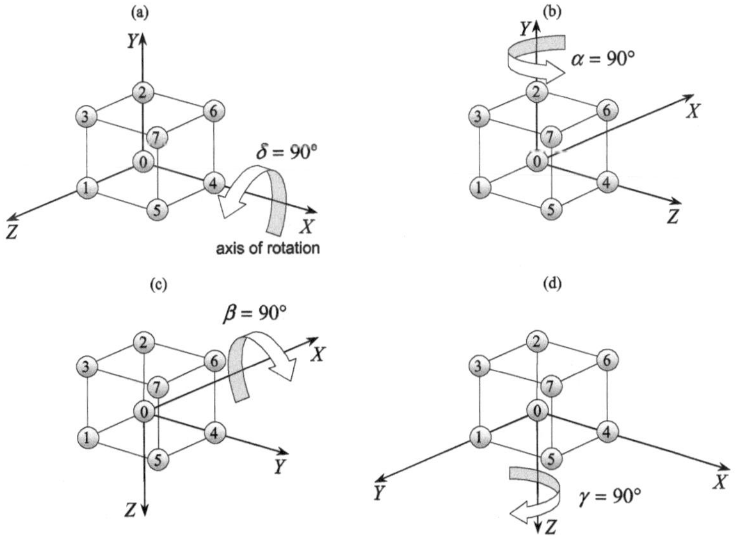

Fig. B.9 Four views of the frame before and during the three rotations $\mathbf{R}_{90^\circ,z}^{-1}\mathbf{R}_{90^\circ,x}^{-1}\mathbf{R}_{90^\circ,y}^{-1}$

B.11 $\mathbf{R}^{-1}_{\gamma,z}\mathbf{R}^{-1}_{\beta,x}\mathbf{R}^{-1}_{\alpha,z}$

$$\mathbf{R}^{-1}_{\gamma,z}\mathbf{R}^{-1}_{\beta,x}\mathbf{R}^{-1}_{\alpha,z} = \begin{bmatrix} c_\gamma & s_\gamma & 0 \\ -s_\gamma & c_\gamma & 0 \\ 0 & 0 & 1 \end{bmatrix}\begin{bmatrix} 1 & 0 & 0 \\ 0 & c_\beta & s_\beta \\ 0 & -s_\beta & c_\beta \end{bmatrix}\begin{bmatrix} c_\alpha & s_\alpha & 0 \\ -s_\alpha & c_\alpha & 0 \\ 0 & 0 & 1 \end{bmatrix}$$

$$= \begin{bmatrix} (c_\gamma c_\alpha - s_\gamma c_\beta s_\alpha) & (c_\gamma s_\alpha + s_\gamma c_\beta c_\alpha) & s_\gamma s_\beta \\ (-s_\gamma c_\alpha - c_\gamma c_\beta s_\alpha) & (-s_\gamma s_\alpha + c_\gamma c_\beta c_\alpha) & +c_\gamma s_\beta \\ s_\beta s_\alpha & -s_\beta c_\alpha & c_\beta \end{bmatrix}$$

$$\mathbf{R}^{-1}_{90^\circ,z}\mathbf{R}^{-1}_{90^\circ,x}\mathbf{R}^{-1}_{90^\circ,z} = \begin{bmatrix} 0 & 0 & 1 \\ 0 & -1 & 0 \\ 1 & 0 & 0 \end{bmatrix}$$

$$\begin{bmatrix} 0 & 0 & 1 \\ 0 & -1 & 0 \\ 1 & 0 & 0 \end{bmatrix}\begin{bmatrix} 0 & 0 & 0 & 0 & 1 & 1 & 1 & 1 \\ 0 & 0 & 1 & 1 & 0 & 0 & 1 & 1 \\ 0 & 1 & 0 & 1 & 0 & 1 & 0 & 1 \end{bmatrix}$$

$$= \begin{bmatrix} 0 & 1 & 0 & 1 & 0 & 1 & 0 & 1 \\ 0 & 0 & -1 & -1 & 0 & 0 & -1 & -1 \\ 0 & 0 & 0 & 0 & 1 & 1 & 1 & 1 \end{bmatrix}.$$

This rotation sequence is illustrated in Fig. B.10 (a)–(d), where the axis of rotation is $[2 \quad 0 \quad 2]^T$ and the angle of rotation 180°.

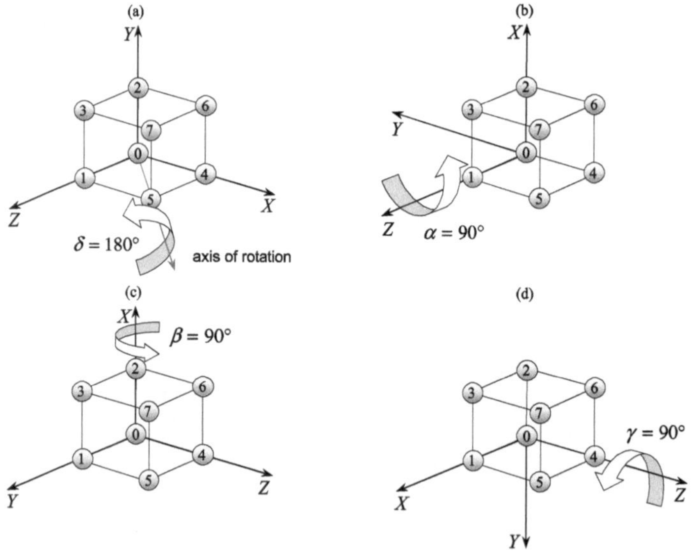

Fig. B.10 Four views of the frame before and during the three rotations $\mathbf{R}^{-1}_{90^\circ,z}\mathbf{R}^{-1}_{90^\circ,x}\mathbf{R}^{-1}_{90^\circ,z}$

B.12 $\mathbf{R}_{\gamma,z}^{-1}\mathbf{R}_{\beta,y}^{-1}\mathbf{R}_{\alpha,x}^{-1}$

$$\mathbf{R}_{\gamma,z}^{-1}\mathbf{R}_{\beta,y}^{-1}\mathbf{R}_{\alpha,x}^{-1} = \begin{bmatrix} c_\gamma & s_\gamma & 0 \\ -s_\gamma & c_\gamma & 0 \\ 0 & 0 & 1 \end{bmatrix}\begin{bmatrix} c_\beta & 0 & -s_\beta \\ 0 & 1 & 0 \\ s_\beta & 0 & c_\beta \end{bmatrix}\begin{bmatrix} 1 & 0 & 0 \\ 0 & c_\alpha & s_\alpha \\ 0 & -s_\alpha & c_\alpha \end{bmatrix}$$

$$= \begin{bmatrix} c_\gamma c_\beta & (s_\gamma c_\alpha + c_\gamma s_\beta s_\alpha) & (s_\gamma s_\alpha - c_\gamma s_\beta c_\alpha) \\ -s_\gamma c_\beta & (c_\gamma c_\alpha - s_\gamma s_\beta s_\alpha) & (c_\gamma s_\alpha + s_\gamma s_\beta c_\alpha) \\ s_\beta & -c_\beta s_\alpha & c_\beta c_\alpha \end{bmatrix}$$

$$\mathbf{R}_{90^\circ,z}^{-1}\mathbf{R}_{90^\circ,y}^{-1}\mathbf{R}_{90^\circ,x}^{-1} = \begin{bmatrix} 0 & 0 & 1 \\ 0 & -1 & 0 \\ 1 & 0 & 0 \end{bmatrix}$$

$$\begin{bmatrix} 0 & 0 & 1 \\ 0 & -1 & 0 \\ 1 & 0 & 0 \end{bmatrix}\begin{bmatrix} 0 & 0 & 0 & 0 & 1 & 1 & 1 & 1 \\ 0 & 0 & 1 & 1 & 0 & 0 & 1 & 1 \\ 0 & 1 & 0 & 1 & 0 & 1 & 0 & 1 \end{bmatrix}$$

$$= \begin{bmatrix} 0 & 1 & 0 & 1 & 0 & 1 & 0 & 1 \\ 0 & 0 & -1 & -1 & 0 & 0 & -1 & -1 \\ 0 & 0 & 0 & 0 & 1 & 1 & 1 & 1 \end{bmatrix}.$$

This rotation sequence is illustrated in Fig. B.11 (a)–(d), where the axis of rotation is $[2 \quad 0 \quad 2]^{\mathrm{T}}$ and the angle of rotation 180°.

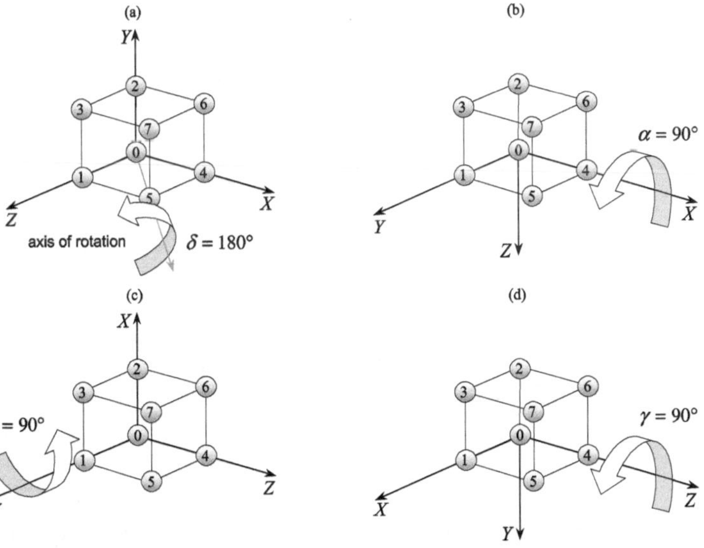

Fig. B.11 Four views of the frame before and during the three rotations $\mathbf{R}_{90^\circ,z}^{-1}\mathbf{R}_{90^\circ,y}^{-1}\mathbf{R}_{90^\circ,x}^{-1}$

B.13 $\mathbf{R}_{\gamma,z}^{-1}\mathbf{R}_{\beta,y}^{-1}\mathbf{R}_{\alpha,z}^{-1}$

$$\mathbf{R}_{\gamma,z}^{-1}\mathbf{R}_{\beta,y}^{-1}\mathbf{R}_{\alpha,z}^{-1} = \begin{bmatrix} c_\gamma & s_\gamma & 0 \\ -s_\gamma & c_\gamma & 0 \\ 0 & 0 & 1 \end{bmatrix} \begin{bmatrix} c_\beta & 0 & -s_\beta \\ 0 & 1 & 0 \\ s_\beta & 0 & c_\beta \end{bmatrix} \begin{bmatrix} c_\alpha & s_\alpha & 0 \\ -s_\alpha & c_\alpha & 0 \\ 0 & 0 & 1 \end{bmatrix}$$

$$= \begin{bmatrix} (-s_\gamma s_\alpha + c_\gamma c_\beta c_\alpha) & (s_\gamma c_\alpha + c_\gamma c_\beta s_\alpha) & -c_\gamma s_\beta \\ (-c_\gamma s_\alpha - s_\gamma c_\beta c_\alpha) & (c_\gamma c_\alpha - s_\gamma c_\beta s_\alpha) & s_\gamma s_\beta \\ s_\beta c_\alpha & s_\beta s_\alpha & c_\beta \end{bmatrix}$$

$$\mathbf{R}_{90^\circ,z}^{-1}\mathbf{R}_{90^\circ,y}^{-1}\mathbf{R}_{90^\circ,z}^{-1} = \begin{bmatrix} -1 & 0 & 0 \\ 0 & 0 & 1 \\ 0 & 1 & 0 \end{bmatrix}$$

$$\begin{bmatrix} -1 & 0 & 0 \\ 0 & 0 & 1 \\ 0 & 1 & 0 \end{bmatrix} \begin{bmatrix} 0 & 0 & 0 & 0 & 1 & 1 & 1 & 1 \\ 0 & 0 & 1 & 1 & 0 & 0 & 1 & 1 \\ 0 & 1 & 0 & 1 & 0 & 1 & 0 & 1 \end{bmatrix}$$

$$= \begin{bmatrix} 0 & 0 & 0 & 0 & -1 & -1 & -1 & -1 \\ 0 & 1 & 0 & 1 & 0 & 1 & 0 & 1 \\ 0 & 0 & 1 & 1 & 0 & 0 & 1 & 1 \end{bmatrix}.$$

This rotation sequence is illustrated in Fig. B.12 (a)–(d), where the axis of rotation is $[0 \quad 2 \quad 2]^T$ and the angle of rotation 180°.

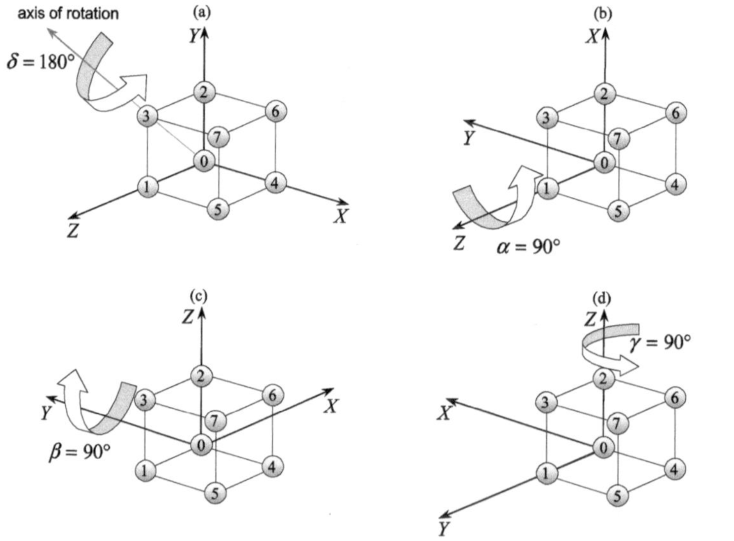

Fig. B.12 Four views of the frame before and during the three rotations $\mathbf{R}_{90^\circ,z}^{-1}\mathbf{R}_{90^\circ,y}^{-1}\mathbf{R}_{90^\circ,z}^{-1}$

Appendix C
The Four n-Square Algebras

C.1 Introduction

The magnitude of a real quantity is its positive value. However, when dealing with objects such as complex numbers, quaternions and octonions, their magnitude is expressed using the Pythagorean formula which takes the square root of the sums of the terms squared.

For example, the magnitude of a complex number $z_1 = a + bi$ is

$$|z_1| = \sqrt{a^2 + b^2}$$

and the magnitude of a quaternion $\mathbf{q} = s + x\mathbf{i} + y\mathbf{j} + z\mathbf{k}$ is

$$|\mathbf{q}| = \sqrt{s^2 + x^2 + y^2 + z^2}$$

and something similar for an octonion, which has 8 terms.

In their book *On Quaternions and Octonions* [9], John Conway and Derek Smith use the Euclidean norm N to represent the sums of the squares, although other authors define the Euclidean norm as $\sqrt{N}$. However, for the purpose of this description I will employ Conway and Smith's definition. Thus

$$N(a + bi) = a^2 + b^2$$

and

$$N(\mathbf{q}) = s^2 + x^2 + y^2 + z^2.$$

We know from the algebra of complex numbers that

$$|z_1||z_2| = |z_1 z_2|$$

or

$$N(z_1)N(z_2) = N(z_1 z_2)$$

and from the algebra of quaternions that

$$|\mathbf{q}_1||\mathbf{q}_2| = |\mathbf{q}_1\mathbf{q}_2|$$

J. Vince, *Rotation Transforms for Computer Graphics*,
DOI 10.1007/978-0-85729-154-7, © Springer-Verlag London Limited 2011

or

$$N(\mathbf{q}_1)N(\mathbf{q}_2) = N(\mathbf{q}_1\mathbf{q}_2).$$

The algebra of octonions also includes this *composition law.*

In the algebra of reals $\mathbb{R}$, we have

$$x_1^2 y_1^2 = (x_1 y_1)^2.$$

In the algebra of complex numbers $\mathbb{C}$, we have

$$\begin{aligned} z_1 &= x_1 + x_2 i \\ z_2 &= y_1 + y_2 i \\ (x_1^2 + x_2^2)(y_1^2 + y_2^2) &= (x_1 y_1 - x_2 y_2)^2 + (x_1 y_2 + x_2 y_1)^2. \end{aligned}$$

In the algebra of quaternions $\mathbb{H}$, we have

$$\begin{aligned} \mathbf{q}_1 &= x_1 + x_2\mathbf{i} + x_3\mathbf{j} + x_4\mathbf{k} \\ \mathbf{q}_2 &= y_1 + y_2\mathbf{i} + y_3\mathbf{j} + y_4\mathbf{k} \end{aligned}$$

$$\begin{aligned} (x_1^2 + x_2^2 + x_3^2 + x_4^2)(y_1^2 + y_2^2 + y_3^2 + y_4^2) = {} & (x_1 y_1 - x_2 y_2 - x_3 y_3 - x_4 y_4)^2 \\ & + (x_1 y_2 + x_2 y_1 + x_3 y_4 - x_4 y_3)^2 \\ & + (x_1 y_3 - x_2 y_4 + x_3 y_1 + x_4 y_2)^2 \\ & + (x_1 y_4 + x_2 y_3 - x_3 y_2 + x_4 y_1)^2. \end{aligned}$$

And in the algebra of octonions $\mathbb{O}$, we have something similar, but with many more terms.

The above algebras are called composition algebras because of their inherent composition law, and Adolf Hurwitz proved that such algebras can only exist in 1, 2, 4 and 8 dimensions.

References

1. Crowe, M.J.: A History of Vector Analysis. Dover, New York (1994)
2. Hamilton, W.R.: Lectures on Quaternions. Hodges and Smith, Dublin (1853)
3. Gibbs, J.W.: Elements of Vector Analysis. Tuttle, Moorehouse & Taylor, New Haven (1884)
4. Vince, J.A.: Mathematics for Computer Graphics. Springer, London (2010)
5. Vince, J.A.: Geometric Algebra for Computer Graphics. Springer, London (2008)
6. Vince, J.A.: Geometric Algebra: An Algebraic System for Computer Games and Animation. Springer, London (2009)
7. Altmann, S.: Hamilton, Rodrigues and the quaternion scandal. Math. Mag. **62**(5), 291–308 (1989)
8. Stillwell, J.: Numbers and Geometry. Springer, New York (1998)
9. Conway, J., Smith, D.: On Quaternions and Octonions. AK Peters, Natick (2003)

J. Vince, *Rotation Transforms for Computer Graphics*,
DOI 10.1007/978-0-85729-154-7,

Index

J. Vince, *Rotation Transforms for Computer Graphics*,
DOI 10.1007/978-0-85729-154-7, © Springer-Verlag London Limited 2011